Springer Complexity

Springer Complexity is an interdisciplinary program publishing the best research and academic-level teaching on both fundamental and applied aspects of complex systems — cutting across all traditional disciplines of the natural and life sciences, engineering, economics, medicine, neuroscience, social and computer science.

Complex Systems are systems that comprise many interacting parts with the ability to generate a new quality of macroscopic collective behavior the manifestations of which are the spontaneous formation of distinctive temporal, spatial or functional structures. Models of such systems can be successfully mapped onto quite diverse "real-life" situations like the climate, the coherent emission of light from lasers, chemical reaction-diffusion systems, biological cellular networks, the dynamics of stock markets and of the internet, earthquake statistics and prediction, freeway traffic, the human brain, or the formation of opinions in social systems, to name just some of the popular applications.

Although their scope and methodologies overlap somewhat, one can distinguish the following main concepts and tools: self-organization, nonlinear dynamics, synergetics, turbulence, dynamical systems, catastrophes, instabilities, stochastic processes, chaos, graphs and networks, cellular automata, adaptive systems, genetic algorithms and computational intelligence.

The three major book publication platforms of the Springer Complexity program are the monograph series "Understanding Complex Systems" focusing on the various applications of complexity, the "Springer Series in Synergetics", which is devoted to the quantitative theoretical and methodological foundations, and the "SpringerBriefs in Complexity" which are concise and topical working reports, case-studies, surveys, essays and lecture notes of relevance to the field.

In addition to the books in these two core series, the program also incorporates individual titles ranging from textbooks to major reference works.

Editorial and Programme Advisory Board

Understanding Complex Systems

Founding Editor: J.A. Scott Kelso

Future scientific and technological developments in many fields will necessarily depend upon coming to grips with complex systems. Such systems are complex in both their composition – typically many different kinds of components interacting simultaneously and nonlinearly with each other and their environments on multiple levels – and in the rich diversity of behavior of which they are capable.

The Springer Series in Understanding Complex Systems series (UCS) promotes new strategies and paradigms for understanding and realizing applications of complex systems research in a wide variety of fields and endeavors. UCS is explicitly transdisciplinary. It has three main goals: First, to elaborate the concepts, methods and tools of complex systems at all levels of description and in all scientific fields, especially newly emerging areas within the life, social, behavioral, economic, neuro- and cognitive sciences (and derivatives thereof); second, to encourage novel applications of these ideas in various fields of engineering and computation such as robotics, nano-technology and informatics; third, to provide a single forum within which commonalities and differences in the workings of complex systems may be discerned, hence leading to deeper insight and understanding.

UCS will publish monographs, lecture notes and selected edited contributions aimed at communicating new findings to a large multidisciplinary audience.

For further volumes:
http://www.springer.com/series/5394

Dirk Helbing
Editor

Social Self-Organization

Agent-Based Simulations and Experiments to Study Emergent Social Behavior

Springer

Editor
Dirk Helbing
ETH Zurich, CLU E1
Chair of Sociology, in particular of Modeling and Simulation
Clausiusstrasse 50
8092 Zurich
Switzerland

ISSN 1860-0832 ISSN 1860-0840 (electronic)
ISBN 978-3-642-24003-4 ISBN 978-3-642-24004-1 (eBook)
DOI 10.1007/978-3-642-24004-1
Springer Heidelberg New York Dordrecht London

Library of Congress Control Number: 2012936485

Printed on acid-free paper

Springer is part of Springer Science+Business Media (www.springer.com)

Preface

Social systems are the most complex systems we know. They are even more complex than physical or biological systems. Their complexity results not only from multiple interactions of individuals, but also from the complexity of cognitive systems. I, therefore, agree with August Comte that sociology is the queen of science, the ultimate scientific challenge. I furthermore believe that this field will soon be one of the most dynamic scientific areas, not only for its interesting fundamental questions, but also because of the many practical problems humanity is facing in the twenty-first century.

The crucial question is, how one can make substantial progress in a field as complicated and multi-faceted as the social sciences. There are certainly different possibilities, as Chap. 1 discusses. Nevertheless, it seems that many characteristic features of complex social systems can be understood from simple models of social interactions, and I am convinced that a number of challenging scientific puzzles can be solved, using concepts from complexity theory, including self-organization, coevolution, and emergence. Agent-based computational models and behavioral experiments can reveal the mechanisms underlying such phenomena, and the role that different factors play for them.

Complex systems often display a counter-intuitive behavior. For example, as will be shown in this book, the same kinds of social interactions can lead to opposite conclusions, when interactions occur with neighbors, friends, colleagues, or business partners rather than with average interaction partners (see Chaps. 7 and 8). Therefore, a simple nonlinear model may explain phenomena, which even complicated linear models may fail to reproduce. Nonlinear models are expected to shed new light on such social phenomena. They may even lead to a paradigm shift in the way we interpret society.

While part of this book is a compilation of recently published papers, some chapters present variants of previous work or new materials, for example, on the technique and future of agent-based computer simulation and on coordination games in networks (a subject that is interesting to understand the competitive spreading of innovations). The chapters do not need to be read in sequential order, but the organization of this book has a clear logic:

- Chapter 1 discusses the issue of how to describe social systems. It highlights different traditions in the social sciences and the respective advantages and disadvantages of these approaches, stressing their complementary character. The focus of this book, however, will be on simple models of social interactions, explaining various aspects of social self-organization.
- Chapter 2 discusses the method of agent-based computational modeling, how to do it right, and what are future perspectives of this approach. A particular focus is put on the question, how to avoid mistakes and how to derive meaningful results.
- Chapter 3 demonstrates the concept of self-organization for the example of pedestrian crowds. We analyze the spontaneous outbreak of social order under everyday conditions and its breakdown at extreme densities. The chapter illustrates how fundamental research can lead to useful models that enable an understanding of macroscopic outcomes of social interactions. It also summarizes related empirical and experimental work as well as practical applications.
- Chapter 4 provides a second example of agent-based modeling, namely a continuous opinion formation model. It shows why previous models did not solve the puzzle why one finds global diversity in spite of local convergence. Strikingly, it is the tendency of individualization which promotes pluralism through the self-organization of groups.
- Chapter 5 turns the attention from mobility in opinion space to mobility in geographical space. Assuming social interactions with neighboring locations, where the outcome of these interactions is quantified by "payoffs" as this is common in game theory, we find the self-organization of spatiotemporal patterns, when success-driven mobility to neighboring locations occurs. Even when starting with a uniform distribution in space, we observe interesting segregation phenomena and different kinds of agglomeration phenomena, depending on the respective payoff structure. These come about when success-driven mobility increases local differences and thereby destabilizes a homogeneous distribution in space.
- Chapter 6 focuses on the problem of cooperation in social dilemma situations, where it appears more advantageous to selfish individuals to exploit others than to cooperate with them. It is discussed in what ways social mechanisms can effectively change the payoff structure and, thereby, the rules and character of the "game" individuals are playing. It is shown that different mechanisms such as repeated interactions, reputation effects, or social networking can imply different routes to cooperation. The chapter also provides a classification of different kinds of transitions to cooperation and shows that adaptive group pressure can promote cooperation in the prisoner's dilemma game even without changing the properties of its equilibrium solutions.
- Chapter 7 combines the elements of Chaps. 5 and 6, i.e., it studies individuals facing social dilemma situations in space, considering success-driven mobility. While one would think that social cooperation and mobility are unrelated, it surprisingly turns out that mobility is an important factor supporting human sociality, and that it promotes a co-evolution of social behavior and social environment. This model may shed new light on a number of fundamental

questions such the following: Why do similarly behaving people tend to agglomerate, e.g., form groups or cities (an observation which is often referred to as "homophily")? How are "social milieus" come about and why do they influence the behavior of people? Why do they persist so long? And why are selfish people more cooperative in reality than expected? What is the role of fluctuations and exploratory behavior for the emergence of cooperation?[1]

- Chapter 8 looks at a further mechanism that has been suggested to promote cooperation, namely "costly punishment." It is shown that the consideration of neighborhood interactions can resolve the so-called second-order free-rider puzzle that wonders why people would invest into the punishment of uncooperative behavior, if they can profit from other people's sanctioning efforts. The chapter also suggests that the spreading of morals and double moral behavior can be understood with concept of evolutionary game theory. The related system dynamics shows quite a number of surprising features.[2]
- Chapter 9 studies effects of network interactions and transaction costs in coordination games. Developing a percolation theoretical description for the related system dynamics allows one to analytically understand the competitive spreading of innovations, opinions, or products from a new scientific angle. Furthermore, we point out that system-wide coordination is a double edge sword.
- Chapter 10 focuses on the implications of heterogeneity in the inclinations of individuals. For this, we study the interaction of several populations with incompatible preferences. This implies a large variety of different system behaviors, such as the outbreak or breakdown of cooperation, the formation of commonly shared norms, the evolution of subcultures, or the occurrence of conflicts which may cause "revolutions". Despite its richness, the model is simple enough to facilitate an analytical understanding of the possible system behaviors.[3] It would be highly desirable to test the predictions of the model by behavioral experiments.[4]

---

[1]The following related paper may interest the reader as well: D. Helbing, W. Yu, and H. Rauhut (2011) Self-organization and emergence in social systems: Modeling the coevolution of social environments and cooperative behavior. *Journal of Mathematical Sociology* **35**,177–208.

[2]The reader may be interested in the related follow up work as well: D. Helbing, A. Szolnoki, M. Perc, and G. Szabó (2010) Punish, but not too hard: how costly punishment spreads in the spatial public goods game. *New Journal of Physics* **12**, 083005; D. Helbing, A. Szolnoki, M. Perc, and G. Szabó (2010) Defector-accelerated cooperativeness and punishment in public goods games with mutations. *Physical Review E* **81**(5), 057104.

[3]D. Helbing and A. Johansson (2010) Evolutionary dynamics of populations with conflicting interactions: Classification and analytical treatment considering asymmetry and power. *Physical Review E* **81**, 016112.

[4]Recently, the emergence of social norms has been further investigated, and the related study may be interesting for the reader as well: D. Helbing, W. Yu, K.-D. Opp, and H. Rauhut (2011) The emergence of homogeneous norms in heterogeneous populations. Santa Fe Working Paper 11-01-001, see http://www.santafe.edu/media/workingpapers/11-01-001.pdf.

- Chapter 11, therefore, discusses ways in which social experiments should and could be done in the future, using the help of computers.
- Chapter 12 focuses on the example of a route choice experiment, in which a sudden transition to turn-taking behavior is found after many interactions, and it is shown that this transition can be understood with a reinforcement learning model.
- Chapter 13 analyzes the same route choice game for the case of several participating players. It is found that, rather than applying probabilistic strategies, experimental participants develop specialized and almost deterministic strategies over time. There is strong evidence of social differentiation. Furthermore, we show how an individualized information system can be developed that supports social adaptation and avoids a self-defeating prophecy effect.
- Chapter 14 then looks at social systems from a complex systems perspective. In this way, it analyzes systemic socioeconomic risks and the underlying mechanisms. In particular, it discusses factors that have probably contributed to the current financial crisis.
- Chapter 15 addresses the question, how to manage the complexity of social systems, considering that classical control concepts are known to fail.
- Chapter 16 finally tries to identify fundamental and real-world challenges in economics, thereby suggesting questions and approaches for future research.[5]

Although these contributions have originally not been written as chapters of a book, they are largely complementary to each other and follow a common approach that tries to understand macroscopic behavioral patterns from interactions between many individuals. All models pursue an agent-based approach, and for many chapters, supplementary video animations are available at http://www.soms.ethz.ch . Most of the models furthermore follow an evolutionary game theory perspective and are mutually consistent. In fact, the wide spectrum of phenomena that can be described by evolutionary game theoretical models, ranging from coordination and cooperation over social norms and conflict up to revolutions, suggests that this theoretical framework may be flexible enough to form the basis of a future integrated theory of socioeconomic interactions. With this vision in mind, I hope the reader will find this book inspiring.

Zurich *Dirk Helbing*

[5]Beyond proposing research questions, the Visioneer White papers "From Social Data Mining to Forecasting Socio-Economic Crises," "From Social Simulation to Integrative System Design," and "How to Create an Innovation Accelerator" make suggestions how to foster scientific progress in the socioeconomic sciences, see http://www.visioneer.ethz.ch. These are published in *EPJ Special Topics* **195**, 1–186 (2011).

Acknowledgments

The research reported in this book and its production were partially supported by the Future and Emerging Technologies programme FP7-COSI-ICT of the European Commission through the project QLectives (grant no. 231200), through the Coordination and Support Action GSDP - Global System Dynamics and Policy of the European Commission (grant no. 266723), and the FET Flagship Pilot Project FuturICT (grant no. 284709).

Contents

Chapter 1
Modeling of Socio-Economic Systems*

1.1 Introduction

When the "father of sociology", August Comte, came up with the idea of a "social physics", he hoped that the puzzles of social systems could be revealed with a natural science approach [1]. However, progress along these lines was very difficult and slow. Today, most sociologists do not believe in his positivistic approach anymore. The question is whether this proves the failure of the positivistic approach or whether it just shows that social scientists did not use the right methods so far. After all, social scientists rarely have a background in the natural sciences, while the positivistic approach has been most successful in fields like physics, chemistry, or biology.

In fact, recently new scientific communities are developing, and they are growing quickly. They call themselves socio-physicists, mathematical sociologists, computational social scientists, agent-based modelers, complexity or network scientists. Researchers from the social sciences, physics, computer science, biology, mathematics, and artificial intelligence research are addressing the challenges of social and economic systems with mathematical or computational models and lab or web experiments. Will they end up with resignation in view of the complexity of social and economic systems, or will they manage to push our knowledge of social systems considerably beyond what was imaginable even a decade ago? Will August Comte's vision of sociology as "the queen of the sciences" [2] finally become true?

My own judgement is that it is less hopeless to develop mathematical models for social systems than most social scientists usually think, but more difficult than most natural scientists imagine. The crucial question is, how substantial progress in a field as complicated and multi-faceted as the social sciences can be made, and how the current obstacles can be overcome? Moreover, what are these obstacles, after all? The current contribution tries to make the controversial issues better understandable

---

*This chapter reprints a previous publication to be cited as: D. Helbing, Pluralistic Modeling of Complex System. *Science and Culture* **76**(9/10), 399–417 (2010).

D. Helbing (ed.), *Social Self-Organization*, Understanding Complex Systems,
DOI 10.1007/978-3-642-24004-1_1, © Springer-Verlag Berlin Heidelberg 2012

to scientific communities with different approaches and backgrounds. While each of the points may be well-known to *some* scientists, they are probably not so obvious for others. Putting it differently, this contribution tries to build bridges between different disciplines interested in similar subjects, and make thoughts understandable to scientific communities with different points of views.

A dialogue between social, natural and economic sciences seems to be desirable not only for the sake of an intellectual exchange on fundamental scientific problems. It also appears that science is lacking behind the pace of upcoming socio-economic problems, and that we need to become more efficient in addressing practical problems [3]. President Lee C. Bollinger of New York's prestigious Columbia University formulated the challenge as follows: "The forces affecting societies around the world ... are powerful and novel. The spread of global market systems ... are ... reshaping our world ..., raising profound questions. These questions call for the kinds of analyses and understandings that academic institutions are uniquely capable of providing. Too many policy failures are fundamentally failures of knowledge" [4].

The fundamental and practical scientific challenges require from us that we do everything we can to find solutions, and that we do not give up before the limits or failure of a scientific approach have become obvious. As will be argued in the Discussion and Outlook, different methods should be seen complementary to each other and, even when inconsistent, may allow one to get a better picture than any single method can do, no matter how powerful it may seem.

1.2 Particular Difficulties of Modeling Socio-Economic Systems

When speaking about socio-economic systems in the following, it could be anything from families over social groups or companies up to countries, markets, or the world economy including the financial system and the labor market. The constituting system elements or system components would be individuals, groups, or companies, for example, depending on the system under consideration and the level of description one is interested in.

On the macroscopic (systemic) level, social and economic systems have some features that seem to be similar to properties of certain physical or biological systems. One example is the hierarchical organization. In social systems, individuals form groups, which establish organizations, companies, parties, etc., which make up states, and these build communities of states (like the United States or the European Union, for example). In physics, elementary particles form atoms, which create molecules, which may form solid bodies, fluids or gases, which together make up our planet, which belongs to a solar system, and a galaxy. In biology, cells are composed of organelles, they form tissues and organs, which are the constituting parts of living creatures, and these make up ecosystems.

Such analogies are certainly interesting and have been discussed, for example, by Herbert Spencer [5] and later on in systems theory [6]. It is not so obvious, however, how much one can learn from them. While physical systems are often well understood by mathematical models, biological and socio-economic systems are usually not. This often inspires physicists to transfer their models to biological and socio-economic problems (see the discussion in Sect. 1.4.4: "The Model Captures Some Features..."), while biologists, social scientists, and economists often find such attempts "physicalistic" and inadequate. In fact, social and economic systems possess a number of properties, which distinguish them from most physical ones:

1. The number of variables involved is typically (much) larger (considering that each human brain contains about one thousand billion neurons).
2. The relevant variables and parameters are often unknown and hard to measure (the existence of "unknown unknowns" is typical).
3. The time scales on which the variables evolve are often not well separated from each other.
4. The statistical variation of measurements is considerable and masks laws of social behavior, where they exist (if they exist at all).
5. Frequently there is no ensemble of equivalent systems, but just one realization (one human history).
6. Empirical studies are limited by technical, financial, and ethical issues.
7. It is difficult or impossible to subdivide the system into simple, non-interacting subsystems that can be separately studied.
8. The observer participates in the system and modifies social reality.
9. The non-linear and/or network dependence of many variables leads to complex dynamics and structures, and sometimes paradoxical effects.
10. Interaction effects are often strong, and emergent phenomena are ubiquitous (hence, not understandable by the measurement and quantification of the individual system elements).
11. Factors such as a large degree of randomness and heterogeneity, memory, anticipation, decision-making, communication, consciousness, and the relevance of intentions and individual interpretations complicate the analysis and modeling a lot.
12. The same applies to human features such as emotions, creativity, and innovation.
13. The impact of information is often more decisive for the behavior of a socio-economic system than physical aspects (energy, matter) or our biological heritage.
14. The "rules of the game" and the interactions in a social or economic system may change over time, in contrast to what we believe to be true for the fundamental laws and forces of physics.
15. In particular, social systems are influenced by normative and moral issues, which are variable.

For such reasons, social systems are the most complex systems we know. They are certainly more complex than physical systems are. As a consequence, a considerable fraction of sociologists thinks that mathematical models for social systems are destined to fail, while most economists and many quantitatively oriented social scientists seem to believe in models with many variables. Both is in sharp contrast to the often simple models containing only a few variables that physicists tend to propose. So, who is right? The following discussion suggests that this is the wrong question. We will therefore analyze why different scientists, who apparently deal with the same research subject, come to so dramatically different conclusions.

It is clear that this situation has some undesirable side effects: Scientists belonging to different schools of thought often do not talk to each other, do not learn from each other, and probably reject each others' papers and project proposals more frequently. It is, therefore, important to make the approach of each school understandable to the others.

1.3 Modeling Approaches

1.3.1 Qualitative Descriptions

Many social scientists think that the 15 challenges listed above are so serious that it is hopeless to come up with mathematical models for social systems. A common view is that all models are wrong. Thus, a widespread approach is to work out narratives, i.e. to give a qualitative (non-mathematical and non-algorithmic) description of reality that is as detailed as possible. This may be compared with a naturalist painting.

Narratives are important, as they collect empirical evidence and create knowledge that is essential for modelers sooner or later. Good models require several steps of intellectual digestion, and the first and very essential one is to create a picture of the system one is interested in and to make sense of what is going on in it. This step is clearly indispensable. Nevertheless, the approach is sometimes criticized for reasons such as the following:

- Observation, description, and interpretation are difficult to separate from each other, since they are typically performed by the same brain (of a single scientist). Since these processes strongly involve the observer, it is hard or even impossible to provide an objective description of a system at this level of detail. Therefore, different scientists may analyze and interpret the system in different, subjective ways. What is an important aspect for one observer may be an irrelevant detail for another one, or may even be overlooked. In German, there is a saying that "one does not see the forest amongst all the trees", i.e. details may hide the bigger picture or the underlying mechanisms. In the natural sciences, this problem has been partially overcome by splitting up observation, description, and interpretation into separate processes: measurements, statistical analysis, and modeling attempts. Many of these steps are supported by technical instruments, computers, and software tools to reduce the individual element and subjective influence. Obviously, this method can not be easily transferred to the study of social systems, as individuals and subjective interpretations *can* have important impacts on the overall system.

- Despite its level of detail, a narrative is often not suited to be translated into a computer program that would reproduce the phenomena depicted by it. When scientists try to do so, in many cases it turns out that the descriptions are ambiguous, i.e. still not detailed enough to come up with a unique computer model. In other words, different programmers would end up with different computer models, producing different results. Therefore, Joshua Epstein claims: "If you didn't grow it, you didn't explain it" [7] (where "grow" stands here for "simulate in the computer"). For example, if system elements interact in a non-linear way, i.e. effects are not proportional to causes, there are many different possibilities to specify the non-linearity: is it a parabola, an exponential dependence, a square root, a logarithm, a power law, ...? Or when a system shows partially random behavior, is it best described by additive or multiplicative noise, internal or external noise? Is it chaotic or turbulent behavior, or are the system elements just heterogeneous? It could even be a combination of several options. What differences would these various possibilities make?

1.3.2 Detailed Models

In certain fields of computational social science or economics, it is common to develop computer models that grasp as many details as possible. They would try to implement all the aspects of the system under consideration, which are known to exist. In the ideal case, these facts would be properties, which have been repeatedly observed in several independent studies of the kind of system under consideration, preferably in different areas of the world. In some sense, they would correspond to the overlapping part of many narratives. Thus, one could assume that these properties would be *characteristic* features of the kind of system under consideration, not just properties of a single and potentially quite particular system.

Despite it sounds logical to proceed in this way, there are several criticisms of this approach:

- In case of many variables, it is difficult to specify their interdependencies in the right way. (Just remember the many different possibilities to specify non-linear interactions and randomness in the system.)
- Some models containing many variables may have a large variety of different solutions, which may be highly dependent on the initial or boundary conditions, or the history of the system. This particularly applies to models containing non-linear interactions, which may have multiple stable solutions or non-stationary ones (such as periodic or non-periodic oscillations), or they may even show chaotic behavior. Therefore, depending on the parameter choice and the initial condition, such a model could show virtually *any* kind of behavior. While one

may think that such a model would be a flexible world model, it would in fact be just a fit model. Moreover, it would probably not be very helpful to understand the mechanisms underlying the behavior of the system. As John von Neumann pointed out: "With four parameters I can fit an elephant and with five I can make him wiggle his trunk." This wants to say that a model with many parameters can fit anything and explains nothing. This is certainly an extreme standpoint, but there is some truth in it.

- When many variables are considered, it is hard to judge which ones are independent of each other and which ones not. If variables are mutually dependent, one effect may easily be considered twice in the model, which would lead to biased results. Dependencies among variables may also imply serious problems in the process of parameter calibration. The problem is known, for example, from sets of linear equations containing collinear variables.
- Models with many variables, particularly non-linear ones, may be sensitive to the exact specification of parameters, initial, or boundary conditions, or to small random effects. Phenomena like hysteresis (history-dependence) [8], phase transitions [9] or "catastrophes" [10], chaos [11], or noise-induced transitions [12] illustrate this clearly.
- Parameters, initial and boundary conditions of models with many variables are hard to calibrate. If small (or no) data sets are available, the model is under-specified, and the remaining data must be estimated based on "expert knowledge", intuition or rules of thumb, but due to the sensitivity problem, the results may be quite misleading. The simulation of many scenarios with varying parameters can overcome the problem in part, as it gives an idea of the possible variability of systemic behaviors. However, the resulting variability can be quite large. Moreover, a full exploration of the parameter space is usually not possible when a model contains many parameters, not even with supercomputers.
- In models with many variables, it is often difficult to identify the mechanism underlying a certain phenomenon or system behavior. The majority of variables may be irrelevant for it. However, in order to understand a phenomenon, it is essential to identify the variables and interactions (i.e. the interdependencies among them) that matter.

1.3.3 Simple Models

Simple models try to avoid (some of) the problems of detailed models by restricting themselves to a minimum number of variables needed to reproduce a certain effect, phenomenon or system behavior. They are aiming at a better understanding of so-called "stylized facts", i.e. simplified, abstracted, or "ideal-typical" observations ("the essence"). For example, while detailed descriptions pay a lot of attention to the particular content of social norms or opinions and how they change over time in relation to the respective cultural setting, simple models abstract from the content of social norms and opinions. They try to formulate general rules of how social norms come about or how opinions change, independently of their content, with the aim of understanding *why* these processes are history-dependent ("hysteretic") and in what way they dependent on microscopic and macroscopic influences.

It is clear that simple models do not describe (and do not even *want* to describe) all details of a system under consideration, and for this reason they are also called minimal or toy models sometimes. The approach may be represented by a few quotes. The "KISS principle" for building a model demands to "*k*eep *i*t *s*imple and *s*traightforward" [13]. This is also known as Occam's (or Ockham's) razor, or as principle of parsimony. Albert Einstein as well demanded [14]: "Make everything as simple as possible, but not simpler".

A clear advantage of simple models is that they may facilitate an analytical treatment and, thereby, a better understanding. Moreover, it is easy to extend simple models in a way that allows one to consider a heterogeneity among the system components. This supports the consideration of effects of individuality and the creation of simple "ecological models" for socio-economic systems. Nevertheless, as George Box puts it: "Essentially, all models are wrong, but some are useful" [15].

The last quote touches an important point. The choice of the model and its degree of detail should depend on the purpose of a model, i.e. its range of application. For example, there is a large variety of models used for the modeling and simulation of freeway traffic. The most prominent model classes are "microscopic" car-following models, focussing on the interaction of single vehicles, "mesoscopic" gas-kinetic models, describing the change of the velocity distribution of vehicles in space and time, "macroscopic" fluid-dynamic models, restricting themselves to changes of the average speed and density of vehicles, and cellular automata, which simplify microscopic ones in favor of simulation speed. Each type of model has certain ranges of application. Macroscopic and cellular automata models, for example, are used for large-scale traffic simulations to determine the traffic situation on freeways and perform short-term forecasts, while microscopic ones are used to study the interaction of vehicles and to develop driver assistance systems. For some of these models, it is also known how they are mathematically connected with each other, i.e. macroscopic ones can be derived from microscopic ones by certain kinds of simplifications (approximations) [16, 17].

The main purpose of models is to guide people's thoughts. Therefore, models may be compared with city maps. It is clear that maps simplify facts, otherwise they would be quite confusing. We do not want to see any single detail (e.g. each tree) in them. Rather we expect a map to show the facts we are interested in, and depending on the respective purpose, there are quite different maps (showing streets, points of interest, topography, supply networks, industrial production, mining of natural resources, etc.).

One common purpose of models is *prediction*, which is mostly (mis)understood as "forecast", while it often means "the identification of implications regarding how a system is expected to behave under certain conditions". It is clear that, in contrast to the motion of a planet around the sun, the behavior of an individual can hardly be forecasted. Nevertheless, there are certain tendencies or probabilities of doing certain things, and we usually have our hypotheses of what our friends, colleagues, or family members would do in certain situations.

Moreover, it turns out that, when many people interact, the aggregate behavior can sometimes be quite predictable. For example, the "wisdom of crowds" is based on the statistical law of large numbers [18], according to which individual variations (here: the independent estimation of facts) are averaged out.

Furthermore, interactions between many individuals tend to restrict the degree of freedom regarding what each individual can or will do. This is, why the concept of "social norms" is so important. Another example is the behavior of a driver, which is constrained by other surrounding vehicles. Therefore, the dynamics of traffic flows can be mathematically well understood [17, 19]. Nevertheless, one cannot exactly forecast the *moment* in which free traffic flow breaks down and congestion sets in, and therefore, one cannot forecast travel times well. The reason for this is the history-dependent dynamics, which makes it dependent on random effects, namely on the size of perturbations in the traffic flow. However, what can be predicted is what are the possible traffic states and what are conditions under which they can occur. One can also identify the probability of traffic flows to break down under certain flow conditions, and it is possible to estimate travel times under free and congested flow conditions, given a measurement of the inflows. The detail that cannot be forecasted is the exact moment in which the regime shift from free to congested traffic flow occurs, but this detail has a dramatic influence on the system. It can determine whether the travel time is 5 or 40 min.

However, it is important to underline that, in contrast to what is frequently stated, the purpose of developing models is not only prediction. Joshua Epstein, for example, discusses 16 other reasons to build models, including explanation, guiding data collection, revealing dynamical analogies, discovering new questions, illuminating core uncertainties, demonstrating tradeoffs, training practitioners, and decision support, particularly in crises [20].

Of course, not everybody favors simple models, and typical criticisms of them are:

- It is usually easy to find empirical evidence, which is not compatible with simple models (even though, to be fair, one would have to consider the purpose they have been created for, when judging them). Therefore, one can say that simple models tend to over-simplify things and leave out more or less important facts. For this reason, they may be considered inadequate to describe a system under consideration.
- Due to their simplicity, it may be dangerous to take decisions based on their implications.
- It may be difficult to decide, what are the few relevant variables and parameters that a simple model should consider. Scientists may even disagree about the stylized facts to model.
- Simple models tend to reproduce a few stylized facts only and are often not able to consistently reproduce a large number of observations. The bigger picture and the systemic view may get lost.

- Making simple models compatible with a long list of stylized facts often requires to improve or extend the models by additional terms or parameter dependencies. Eventually, this improvement process ends up with detailed models, leaving one with the problems specified in the related section.
- Certain properties and behaviors of socio-economic systems may not be understandable with methods that have been successful in physics: Subdividing the system into subsystems, analyzing and modeling these subsystems, and putting the models together may not lead to a good description of the overall system. For example, several effects may act in parallel and have non-separable orders of magnitude. This makes it difficult or impossible to start with a zeroth or first order approximation and to improve it by adding correction terms (as it is done, for example, when the falling of a body is described by the effect of gravitational acceleration plus the effect of air resistance). Summing up the mathematical terms that describe the different effects may not converge. It is also not clear whether complex systems can be always understood via simple principles, as the success of complexity science might suggest. Some complex systems may require *complex* models to explain them, and there may even be phenomena, the complexity of which is irreducible. Turbulence [21] could be such an example. While it is a long-standing problem that has been addressed by many bright people, it has still not been explained completely.

It should be added, however, that we do not know today, whether the last point is relevant, how relevant it is, and where. So far, it is a *potential* problem one should be aware of. It basically limits the realm, in which classical modeling will be successful, but we have certainly not reached these limits, yet.

1.3.4 Modeling Complex Systems

Modeling socio-economic systems is less hopeless than many social scientists may think [22]. In recent years, considerable progress has been made in a variety of relevant fields, including:

- Experimental research [23–25]
- Data mining [26]
- Network science [27]
- Agent-based modeling [7, 28]
- The theory of complex systems (including emergent and self-organized phenomena, or chaos) [29]
- The theory of phase transitions [9] ("catastrophes" [10]), critical phenomena [30], and extreme events [31]
- The engineering of intelligent systems [32, 33]

These fields have considerably advanced our understanding of complex systems. In this connection, one should be aware that the term "complexity" is used in many different ways. In the following, we will distinguish three kinds of complexity:

1. structural,
2. dynamical, and
3. functional complexity

One could also add algorithmic complexity, which is given by the amount of computational time needed to solve certain problems. Some optimization problems, such as the optimization of logistic or traffic signal operations, are algorithmically complex [34].

Linear models are not considered to be complex, no matter how many terms they contain. An example for *structural* complexity is a car or airplane. They are constructed in a way that is dynamically more or less deterministic and well controllable, i.e. dynamically simple, and they also serve relatively simple functions (the motion from a location A to another location B). While the acceleration of a car or a periodic oscillation would be an example for a simple *dynamics*, examples for complex dynamical behavior are non-periodic changes, deterministic chaos, or history-dependent behaviors. Complex *dynamics* can already be produced by simple sets of non-linearly coupled equations. While a planet orbiting around the sun follows a simple dynamics, the interaction of three celestial bodies can already show a chaotic dynamics. Ecosystems, the human body or the brain would be *functionally* complex systems. The same would hold for the world wide web, financial markets, or running a country or multi-national company.

While the interrelation between function, form and dynamics still poses great scientific challenges, the understanding of structurally or dynamically complex systems has significantly progressed. Simple agent-based models of systems with a large number of interacting system elements (be it particles, cars, pedestrians, individuals, or companies) show properties, which remind of socio-economic systems. Assuming that these elements mutually adapt to each other through non-linear or network interactions (i.e. that the elements are influenced by their environment while modifying it themselves), one can find a rich, history-dependent system behavior, which is often counter-intuitive, hardly predictable, and seemingly uncontrollable. These models challenge our common way of thinking and help to grasp behaviors of complex systems, which are currently a nightmare for decision-makers.

For example, complex systems are often unresponsive to control attempts, while close to "critical points" (also known as "tipping points"), they may cause sudden (and often unexpected) phase transition (so-called "regime shifts"). These correspond to discontinuous changes in the system behavior. The breakdown of free traffic flow would be a harmless example, while a systemic crisis (such as a financial collapse or revolution) would be a more dramatic one. Such systemic crises are often based on cascade spreading through network interactions [35]. Hence, complex adaptive systems allow one to understand extreme events as a result of strong interactions in a system (rather than as externally caused shocks). Furthermore, the interaction of many system elements may give rise to interesting self-organization phenomena and emergent properties, which cannot be understood from the behaviors of the single elements or by adding them up. Typical examples are collective patterns of motion in pedestrian crowds or what is sometimes called "swarm intelligence" [36].

Considering this, it is conceivable that many of today's puzzles in the social sciences may one day be explained by simple models, namely as *emergent* phenomena resulting from interactions of many individuals and/or other system elements. It is important to note that emergent phenomena cannot be explained by *linear* models (which are most common in many areas of quantitative empirical research in the social sciences and economics).

Unfortunately, there is no standard way to set up models of emergent phenomena. On the one hand, there are many possible kinds of non-linear functional dependencies ("interactions") (see the end of the section on "Qualitative Descriptions"). On the other hand, model assumptions that appear plausible do often not produce the desired or expected effects.

In spite of these difficulties, taking into account time-dependent change, a non-linear coupling of variables, spatial or network interactions, randomness, and/or correlations (i.e. features that many social and economic models currently do not consider to the necessary extent), can sometimes deliver unexpected solutions of long-standing puzzles. For example, it turns out that representative agent models (which are common in economics) can be quite misleading, as the same kinds of interactions among the system components can imply completely different or even opposite conclusions, when interactions take place in a socio-economic *network* rather than with average (or randomly chosen) interaction partners [37]. Therefore, models often produce counter-intuitive results, when spatio-temporal or network interactions are relevant. A simple non-linear model may explain phenomena, which complicated linear models may fail to reproduce. In fact, this generally applies to systems that can show *several* possible states (i.e. systems which do not have just one stable equilibrium). Out-of-equilibrium models are also required for the description of systemic crises such as the current financial crisis [35].

1.4 Challenges of Socio-Economic Modeling

Many people before and after Popper have been thinking about the logic of scientific discovery [38]. A wide-spread opinion is that a good model should be applicable to measurements of many systems of a certain kind, in particular to measurements in different areas of the world. The more observations a model can explain and the less parameters it has, the more powerful it is usually considered to be.

Models with a *few* parameters can often be easier to calibrate, and cause-and-effect relationships may be better identified, but one can usually not expect that these models would provide an *exact* description of reality. Nevertheless, a good model should make predictions regarding some possible, but previously unobserved system behaviors. In this connection, prediction does not necessarily mean the *forecast* of a certain event at a specific future point in time. It means a specific system behavior that is expected to occur (or to be possible) under certain conditions (e.g. for certain parameter combinations or certain initial conditions). When such conditions apply and the system shows the expected behavior, this would be considered to verify the model, while the model would be falsified or seriously questioned, if the predicted system behavior would not occur. By experimentally challenging models based on their predictions (implications), it has been possible in the natural sciences to rate alternative models based on their quality in reproducing and predicting measured data. Unfortunately, it turns out that this approach is less suited to identify "the right model" of a social or economic system under consideration. As we will discuss in the following, this is not only due to the smaller amount of data available on most aspects of social and economic systems and due to experimental limitations for financial, technical and ethical reasons...

1.4.1 Promises and Difficulties of the Experimental Approach

So far, it is very expensive to carry out social and economic experiments, for example in the laboratory. While the study of human behavior under controlled conditions has become a common research method not only in psychology, but also in experimental economics and in sociology, the number of individuals that can be studied in such experiments is limited. This implies a large degree of statistical variation, which makes it difficult to determine behavioral laws or to distinguish between different models. The statistical noise creates something like a foggy situation, which makes it difficult to see what is going on. In physics, this problem can be usually solved by better measurement methods (apart from uncertainty that results from the laws of quantum mechanics). In social systems, however, there is an irreducible degree of randomness. The behavior varies not only between individuals due to their heterogeneity (different "personality"). It also varies from one instance to another, i.e. the decision-making of an individual is usually not deterministic. This could be due to various reasons: unknown external influences (details attracting the attention of the individual) or internal factors (exploration behavior, memory effects, decisions taken by mistake, etc.).

The large level of behavioral variability within and between individuals is probably not only due to the different histories individuals have, but also due to the fact that exploration behavior and the heterogeneity of behaviors are beneficial for the learning of individuals and for the adaptability of human groups to various environmental conditions. Applying a theory of social evolution would, therefore, suggest that randomness is significant in social and economic systems, because it increases system performance. Besides, heterogeneity can also have *individual* benefits, as differentiation facilitates specialization. The benefit of a variation *between* individuals is also well-known from ecological systems [39].

Besides impeding the discovery of behavioral laws, the limited number of participants in laboratory experiments also restricts the number of repetitions and the number of experimental settings or parameter combinations that can be studied. Scanning parameter spaces is impossible so far, while it would be useful to detect different system behaviors and to determine under which conditions they occur. It can be quite tricky to select suitable system parameters (e.g. the payoff matrix in a game-theoretical experiment).

Computer simulations suggest that one would find interesting results mainly, if the parameters selected in different experimental setups imply different system behaviors, i.e. if they belong to different "phases" in the parameter space (see Fig. 1.1). To identify such parameter combinations, it is advised to perform computer simulations *before*, in order to determine the phase diagram for the system under consideration [25]. The problem, however, is that the underlying model is unlikely to be perfect, i.e. even a good social or economic model is expected to make predictions which are only approximately valid. As a consequence, the effect one likes to show may appear for (somewhat) different parameter values, or it may not occur at all (considering the level of randomness) [40].

1.4.2 Several Models Are Right

The above mentioned properties of socio-economic systems imply that it is difficult to select the "right" model among several alternative ones. For an illustration, let us take car-following models, which are used for the simulation of urban or freeway traffic. Thanks to radar sensors, it has become possible to measure the acceleration of vehicles as a function of the typical variables of car-following models, which are the distance to the car ahead, the own speed, and the speed difference. When fitting the parameters of various car-following models to data of such measurements, it turns out that the remaining error between computer simulations and measurements is about the same for most of the models. The calibration error varies between 12 and 17%, and according to the authors, "no model can be denoted to be the best" [41]. When the error of different models (i.e. the deviation between model and data) is determined for a *new* data set (using the model parameters determined with the *previous* data set), the resulting validation error usually varies between 17 and 22% (larger validation errors mainly result, when the calibration data set is overfitted) [41]. Again, the performance of the different models is so similar that it would not be well justified to select one of them as the "correct" model and exclude all the others. A closer analysis shows that the parameters of the car-following dynamics vary among different drivers, but the behavior of specific drivers also vary over time [42]. We have to assume that the same applies to basically all kinds of behavior, not only for car driving. Moreover, it is likely that many behaviors (such as decision-making behaviors) vary even more than car-following behavior does. As a consequence, it would be even more difficult to distinguish between different models by means of empirical or experimental data, which would mean that we may have to accept several models to be (possibly) "right", even when they are not consistent with each other. In other words, the question "What is the best model?" or "How to choose the model?" may not be decidable in a reasonable way, as is also suggested by the

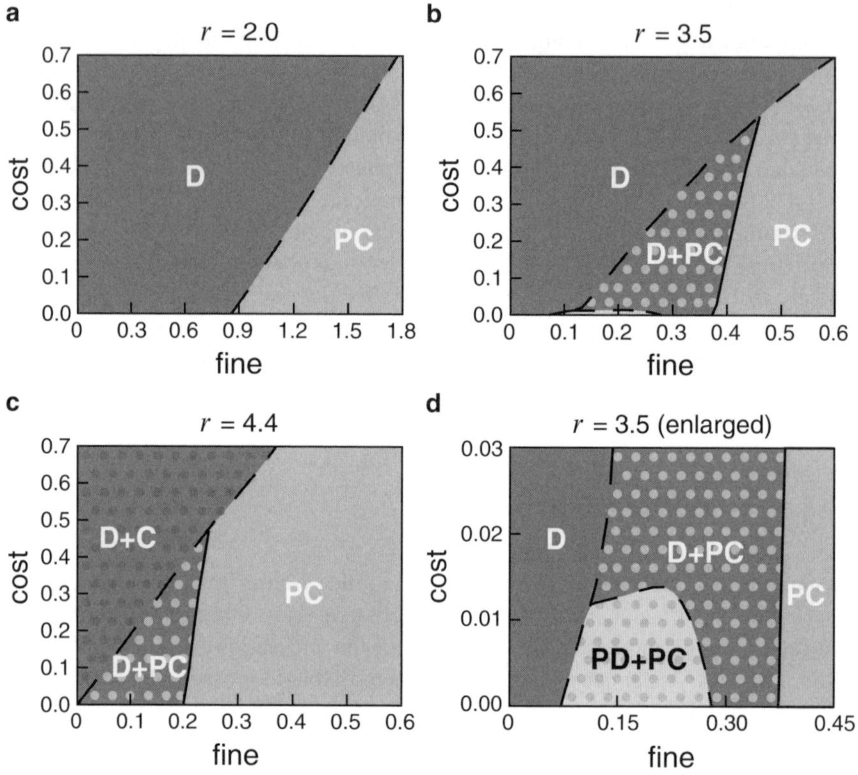

Fig. 1.1 So-called "phase diagram", showing the finally remaining strategies in the spatial public goods game with cooperators (C), defectors (D), cooperators who punish defectors (PC) and hypocritical punishers (PD), who punish other defectors while defecting themselves (after [37]). Initially, each of the four strategies occupies 25% of the sites of the two-dimensional lattice, in which individuals interact, and their distribution is uniform in space. However, due to their evolutionary competition, two or three strategies die out after some time. The finally resulting state depends on the punishment cost, the punishment fine, and the synergy r of cooperation (the factor by which cooperation increases the sum of investments). The displayed phase diagrams are for (**a**) $r = 2.0$, (**b**) $r = 3.5$, and (**c**) $r = 4.4$. (**d**) Enlargement of the small-cost area for $r = 3.5$. *Solid separating lines* indicate that the resulting fractions of all strategies change continuously with a modification of the punishment cost and punishment fine, while *broken lines* correspond to discontinuous changes. All diagrams show that cooperators and defectors cannot stop the spreading of costly punishment, if only the fine-to-cost ratio is large enough (see the *PC area*). Note that, in the absence of defectors, the spreading of punishing cooperators is extremely slow and follows a voter model kind of dynamics. A small level of strategy mutations (which continuously creates a small number of strategies of all kinds, in particular defectors) can largely accelerate the spreading of them. Furthermore, there are parameter regions where punishing cooperators can crowd out "second-order free-riders" (non-punishing cooperators) in the presence of defectors (D+PC). Finally, for low punishment costs, but moderate punishment fines, it may happen that "moralists", who cooperate and punish non-cooperative behavior, can only survive through an "unholy alliance" with "immoral", hypocritical punishers (PD+PC). For related videos, see http://www.soms.ethz.ch/research/secondorder-freeriders or http://www.matjazperc.com/games/moral.html

next section. This situation reminds a bit of Gödel's Undecidability Theorem [43], which relates to the (in)completeness of certain axiom systems.

It may be tempting to determine the best model as the one which is most *successful*, for example in terms of the number of citations it gets. However, success is not necessarily an indicator of a good model. Let us take models used for stock trading as an example. Clearly, even if the stock prices vary in a perfectly random manner and if the average success of each model is the same over an infinite time period; when different traders apply different trading models, they will be differently successful at any chosen point in time. Therefore, one would consider some models more successful than others, while this would be only a matter of luck. At other points in time, different models would be the most successful ones.

Of course, if behaviors are not just random, some models should be better than others, and it should eventually be possible to separate "good" from "bad" models through the "wisdom of crowds" effect. However, the "wisdom of crowds" assumes independent judgements, while scientists have repeated interactions. It has be shown experimentally that this tends to create consensus, but that this consensus will often deviate from the truth [44]. The problem results from social influence, which creates a herding effect that can undermine the "wisdom of crowds". Of course, this mainly applies, when the facts are not sufficiently obvious, which however is the case in the social sciences due to the high variability of observations, while the problem is less pressing in the natural sciences thanks to the higher measurement precision. Nevertheless, the physicist Max Planck is known for the quote: "Science progresses funeral by funeral" [45]. Thomas Kuhn's study of scientific revolutions [46] suggests as well that scientific progress is not continuous, but there are sudden paradigm shifts. This reveals the problem of herding effects. Even a collective agreement is no guarantee for the correctness of a model, as the replacement of classical mechanics by relativistic quantum theory shows. In other words, success is no necessarily an indicator for good models. It may just be an indicator for what model is most fashionable at a given time. The problem becomes worse by the academic selection process that decides, what scientists make a career and which ones not. It creates a considerable inertia in the adjustment to new knowledge, i.e. scientific trends are likely to persist longer than what is justified by facts.

1.4.3 No Known Model Is Right

A typical approach in the natural sciences is to verify or falsify previously untested predictions (implications) of alternative models by sometimes quite sophisticated experiments. Only in the minority of cases, two alternative theories turn out to be the same, like the wave and the particle picture of quantum mechanics. In most cases, however, two theories A and B are non-identical and inconsistent, which means that they should make different predictions in particular kinds of situations. Experiments are performed to find out whether theory A or theory B is right, or whether both of them deviate from the measurements. If the experimental data confirm theory

A and are not compatible with theory B (i.e. deviate significantly from it), one would discard theory B forever. In this way, experiments are thought to narrow down the number of alternative theories, until a single theory remains, which is considered to be "true".

When social or economic systems are modeled, the following situation is not unlikely to happen: Scientists identify mutually incompatible predictions of theories A and B, and it turns out that an experiment supports theory A, but not theory B. One day, another scientist identifies a different set of incompatible predictions, and another experiment supports theory B, but not theory A. Due to the inherent simplifications of socio-economic models, for *any* model it should be easy to find empirical evidence that contradicts it. What should one do in such cases? Giving up on modeling would probably not be the best idea. Generalizing a model is always possible, but it will usually end up with detailed models, which implies a number of problems that have been outlined in the related section. One could also stay with many particular models and determine their respective ranges of validity. This, however, will never result in a holistic or systemic model. A possible way out would be the approach of pluralistic modeling outlined in the Summary and Outlook.

Modeling in modern physics seems to face similar problems. While one would expect that each experiment narrows down the number of remaining, non-falsified models, one actually observes that, after each experiment, scientists come up with a number of new models. As people say: "Each answered question raises ten new ones." In fact, there is an abundance of elementary particle models, and the same applies to cosmological models. Many models require to assume the existence of factors that have never been measured and perhaps *will* never be measured, such as Higgs bosons, dark matter, or dark energy. We will probably have to live with the fact that models are just *models* that never grasp all details of reality.

Moreover, as has been pointed out, understanding elementary particles and fundamental forces in physics does not explain at all what is happening in the world around us [47, 48]. Many emergent phenomena that we observe in the biological, economic and social world will never be derived from elementary particle physics, because emergent properties of a system cannot be understood from the properties of its system components alone. They usually come about by the interaction of a large number of system components. Let us be honest: Our textbooks do not even explain the particular properties of water, as simple as H_2O molecules may be. (Of course, this does not mean that this situation will remain forever – see e.g. H. Eugene Stanley's related work.)

Generally, there is still a serious lack in understanding the connection between function, dynamics, and form. Emergence often seems to have an element of surprise. The medical effect of a new chemical drug cannot be understood by computer simulation alone. So far, we also do not understand emotions and consciousness, and we cannot calculate the biological fitness of a species in the computer. The most exciting open puzzles in science concern such emergent phenomena. It would be interesting to study, whether social and economic phenomena such as trust, solidarity, and economic value can be understood as emergent phenomena as well [3].

1.4.4 *The Model Captures Some Features, But May Be Inadequate*

Scientists are often prompted to transfer their methods to another areas of application, based on analogies that they see between the behavior of different systems. Systems science is based on such analogies, and physicists generalize their methods as well. The question is how useful a "physicalist approach" can be, which transfers properties of many-particle systems to social or economic systems, although individuals are certainly more intelligent than particles and have many more behavioral degrees of freedom.

Of course, physicists would never claim that particle models could provide an exact description of social or economic systems. Why, then, do they think the models could make a contribution to the understanding of these systems? This is, because they have experience with what can happen in systems characterized by the non-linear interaction of many system components in space and time, and when randomness plays a role. They know how self-organized collective phenomena on the "macroscopic" (aggregate) level can results from interactions on the "microscopic" (individual) level. And they have learned, how this can lead to phase transitions (also called "regime shifts" or "catastrophes"), when a system parameter ("control parameter") crosses a critical point ("tipping point"). Furthermore, they have discovered that, at a critical point, the system typically shows a scale-free behavior (i.e. power laws or other fat-tail distributions rather than Gaussian distributions).

It is important to note that the characteristic features of the system at the critical point tend to be "universal", i.e. they largely do not depend on the details of the interactions. *This* is, why physicists think they can abstract from the details. Of course, details *are* expected to be relevant when the system is *not* close to a critical point. It should also be added, that there are a number of different kinds of universal behavior, so-called universality classes. Nevertheless, many-particle models may allow one to get a better understanding of regime shifts, which are not so well understood by most established models in economics or the social sciences. However, if the tipping point is far away, the usefulness of many-particle models is limited, and more detailed descriptions, as they are favored by economists and social scientists, appear to be more adequate.

Sometimes, it is not so clear how far analogies can carry, or whether they are useful at all. Let us take neural network models. In a certain sense, they can be used to model learning, generalization, and abstraction. However, the hope that they would explain the functioning of the brain has been largely disappointed. Today, we know that the brain works quite differently, but neural network theory has given birth to many interesting engineering applications that are even commercially applied.

Let us consider models of cooperation based on coupled oscillators as a second example. Without any doubt, the synchronization of cyclical behavior is among the most interesting collective phenomena we know of, and models allow one to study if and how groups of oscillators will coordinate each other or fall apart into subgroups (which are not synchronized among each other, while the oscillators in each of them are) [49]. Despite this analogy to group formation and group dynamics, it is not clear, what we can learn from such models for social systems.

A similar point is sometimes raised for spin models, which have been proposed to describe opinion formation processes or the emergence of cooperation in social dilemma situations. In this connection, it has been underlined that social interactions cannot always be broken down into binary interactions. Some interactions involve three or more individuals at the same time, which may change the character of the interaction. Nevertheless, similar phenomena have been studied by overlaying binary interactions, and it is not fully clear how important the difference is.

Let us finally ask whether unrealistic assumptions are generally a sign of bad models? The discussion in the section on "Simple Models" suggests that this is not necessarily so. It seems more a matter of the purpose of a model, which determines the level of simplification, and a matter of the availability of better models, i.e. a matter of competition. Note, however, that a more realistic model is not necessarily more useful. For example, many car-following models are more realistic than fluid-dynamic traffic models, but they are not suited to simulate large-scale traffic networks in real-time. For social systems, there are a number of different modeling approaches as well, including the following:

- *Physical(istic) modeling approach:* Socio- and econo-physicists often abstract social interactions so much that their models come down to multi-particle models (or even spin models with two behavioral options). Such models focus on the effect of non-linear interactions and are a special case of bounded rationality models, sometimes called zero-intelligence models [50]. Nevertheless, they may display features of *collective* or *swarm* intelligence [36]. Furthermore, they may be suited to describe regime shifts or situations of routine choice [51], i.e. situations where individuals react to their environment in a more or less subconscious and automatic way. Paul Omerod, an economist by background, argues as follows [52]: "In many social and economic contexts, self-awareness of agents is of little consequence... No matter how advanced the cognitive abilities of agents in abstract intellectual terms, it is as if they operate with relatively low cognitive ability within the system... The more useful 'null model' in social science agent modelling is one close to zero intelligence. It is only when this fails that more advanced cognition of agents should be considered."
- *Economic modeling approach:* Most economists seem to have quite the opposite approach. Their concept of "homo economicus" (the "perfect egoist") assumes that individuals take strategic decisions, choosing the optimal of their behavioral options. This requires individuals with infinite memory and processing capacities. Insofar, one could speak of an infinite-intelligence approach. It is also known as rational choice approach and has the advantage that the expected behaviors of individuals can be axiomatically derived. In this way, it was possible to build the voluminous and impressive theory of mainstream economics. Again, the reliability of this theory depends, of course, on the realism of its underlying assumptions.

- *Sociological modeling approach:* Certain schools of sociologists use rational choice models as well. In contrast to economists, however, they do not generally assume that individuals would radically optimize their own profit. Their models rather consider that, in social systems, exchange is more differentiated and multi-faceted. For example, when choosing their behavior, individuals may not only consider their own preferences, but the preferences of their interaction partner(s) as well. In recent years, "fairness theory" has received a particular attention [53] and often been contrasted with rational choice theory. These social aspects of decision-making are now eventually entering economic thinking as well [54].
- *Psychological modeling approach:* Psychologists are perhaps least axiomatic and usually oriented at empirical observations. They have identified behavioral paradoxies, which are inconsistent with rational choice theory, at least its classical variant. For example, it turns out that most people behave in a risk-averse way. To account for their observations, new concepts have been developed, including prospect theory [55], satisficing theory [56], and the concept of behavioral heuristics [57]. In particular, it turns out that individual decisions depend on the respective *framing*. In his Nobel economics lecture, Daniel Kahneman put it this way: "Rational models are psychologically unrealistic... the central characteristic of agents is not that they reason poorly, but that they often act intuitively. And the behavior of these agents is not guided by what they are able to compute, but by what they happen to see at a given moment." Therefore, modern research directions relate to the cognitive and neurosciences. These results are now finding their way into economics via the fields of experimental, behavioral, and neuro-economics.

In summary, there is currently no unified approach that scientists generally agree on. Some of the approaches are more stylized or axiomatic. Others are in better quantitative agreement with empirical or experimental evidence, but mathematically less elaborated. Therefore, they are theoretically less suited to derive implications for the behavior in situations, which have not been studied so far. Consequently, all models have their strengths and weaknesses, no matter how realistic they may be. Moreover, none of the mathematical models available so far seems to be sophisticated enough to reflect the full complexity of social interactions between many people.

1.4.5 Different Interpretations of the Same Model

A further difficulty of modeling socio-economic systems is that scientists may not agree on the interpretation of a model. Let us discuss, for example, the multi-nomial logit model, which has been used to model decision-making in a large variety of contexts and awarded with the Nobel prize [58]. This model can be derived in a utility-maximizing framework, assuming perfectly rational agents deciding under conditions of uncertainty. The very same model, however, can also be derived in other ways. For example, it can be linked to psychological laws or to distributions of statistical physics [59]. In the first case, the interpretation is compatible with the infinite-intelligence approach, while in the last case, it is compatible with the zero-intelligence approach, which is quite puzzling. A comparison of these approaches is provided by [59].

1.5 Discussion and Outlook

1.5.1 Pluralistic or Possibilistic Modeling and Multiple World Views: The Way Out?

Summarizing the previous discussion, it is quite unlikely that we will ever have a single, consistent, complete, and correct model of socio-economic systems. Maybe we will not even find such a grand unified theory in physics. Recently, doubts along these lines have even been raised by some particle physicists [60, 61]. It may be the time to say good bye to a modeling approach that believes in the feasibility of a unique, general, integrated and consistent model. At least there is no theoretical or empirical evidence for the possibility of it.

This calls for a paradigm shift in the modeling approach. It is important to be honest that each model is limited, but most models are useful for something. In other words, we should be tolerant with regard to each others' models and see where they can complement each other. This does not mean that there would be separate models for non-overlapping parts of the system, one for each subsystem. As has been pointed out, it is hard to decide whether a particular model is valid, no matter how small the subsystem is chosen. It makes more sense to assume that each model has a certain validity or usefulness, which may be measured on a scale between 0 and 1, and that the validity furthermore depends on the part or aspect of the system addressed. This validity may be quantified, for example, by the goodness of fit of a given system or the accuracy of description of another system of the same kind.

As there are often *several* models for each part or aspect of a system, one could consider all of them and give each one a weight according to their respective validity, as determined statistically by comparison with empirical or experimental data. Analogously to the "wisdom of crowds" [18], which is based on the law of large numbers, this should lead to a better quantitative fit or prediction than most (or even each) model in separation, despite the likely inconsistency among the models. Such an approach could be called a *pluralistic* modeling approach [62], as it tolerates and integrates multiple world views. It may also be called a *possibilistic* approach [63], because it takes into account that each model has only a certain likelihood to be valid, i.e. each model describes a possible truth. However, this should not be misunderstood as an appeal for a subjectivistic approach. The pluralistic modeling approach still assumes that there is some underlying reality that some, many, or all of us share (depending on the aspect one is talking about).

As shocking as it may be for many scientists and decision-makers to abandon their belief in the existence of a unique, true model, solution, or policy, the pluralistic modeling approach is already being used. Hurricane prediction and climate modeling are such examples [64]. Even modern airplanes are controlled by multiple computer programs that are run in parallel. If they do not agree with each other, a majority decision is taken and implemented. Although this may seem pretty scary, this approach has apparently worked surprisingly well so far. Moreover, when crash tests of newly developed cars are simulated in the computer, the simulations are again performed with *several* models, each of which is based on different approximation methods. Therefore, it is plausible to assume that pluralistic modeling will be much more widely used in future, whenever a complex system shall be modeled.

1.5.2 Where Social Scientists and Natural Scientists or Engineers Can Learn from Each Other

It has been argued that all modeling approaches have their strengths and weaknesses, and that they should be considered complementary to each other rather than being each others' enemies. This also implies that scientists of different disciplines may profit and learn from each other. Areas of fruitful multi-disciplinary collaboration could be:

- The modeling of socio-economic systems themselves.
- Understanding the impacts that engineered systems have on the socio-economic world.
- The modeling of the social mechanisms that drive the evolution and spreading of innovations, norms, technologies, products etc.
- Scientific challenges related to the questions how to manage complexity or to design better systems.
- The application of social coordination and cooperation mechanisms to the creation of self-organizing technical systems (such as decentralized traffic controls or peer-to-peer systems).
- The development of techno-social systems [65], in which the use of technology is combined with social competence and human knowledge (such as Wikipedia, prediction markets, recommender systems, or the semantic web).

Given the large potentials of such collaborations, it is time to overcome disciplinary boundaries. They seem to make less and less sense. It rather appears that multi-disciplinary, large-scale efforts are needed to describe and understand socio-economic systems well enough and to address practical challenges of humanity (such as the financial and economic crisis) more successfully [66].

Acknowledgements The author is grateful for support by the ETH Competence Center "Coping with Crises in Complex Socio-Economic Systems" (CCSS) through ETH Research Grant CH1-01 08-2 and by the Future and Emerging Technologies programme FP7-COSI-ICT of the European Commission through the project Visioneer (grant no.: 248438).

References

1. A. Comte, *Social Physics: From the Positive Philosophy* (Calvin Blanchard, New York, 1856)
2. A. Comte, *Course on Positive Philosophy* (1830–1842)
3. D. Helbing, Grand socio-economic challenges, Working Paper, ETH Zurich (2010)
4. L.C. Bollinger, Announcing the Columbia committee on global thought, see http://www.columbia.edu/content/announcing-columbia-committee-global-thought.html, last accessed on March 6, 2012
5. H. Spencer, *The Principles of Sociology* (Appleton, New York, 1898; the three volumes were originally published in serial form between 1874 and 1896)
6. L. von Bertalanffy, *General System Theory: Foundations, Development, Applications* (George Braziller, New York, 1968)
7. J.M. Epstein, *Generative Social Science. Studies in Agent-Based Computational Modeling* (Princeton University, 2006), p. 51
8. I.D. Mayergoyz, *Mathematical Models of Hysteresis and their Applications* (Academic Press, 2003)
9. H.E. Stanley *Introduction to Phase Transitions and Critical Phenomena* (Oxford University, 1987)
10. E.C. Zeeman (ed.), *Catastrophe Theory* (Addison-Wesley, London, 1977)
11. H.G. Schuster, W. Just, *Deterministic Chaos* (Wiley-VCH, Weinheim, 2005)
12. W. Horsthemke, R. Lefever, *Noise-Induced Transitions: Theory and Applications in Physics, Chemistry, and Biology* (Springer, Berlin, 1983)
13. "KISS principle" at Wikipedia.org, see http://en.wikipedia.org/wiki/KISS_principle, last accessed on March 6, 2012
14. A. Einstein, "On the Method of Theoretical Physics". The Herbert Spencer Lecture, delivered at Oxford (10 June 1933); also published in *Philosophy of Science* **1**(2), p. 165 (April 1934).
15. G.E.P. Box, N.R. Draper, *Empirical Model-Building and Response Surfaces* (Wiley, NJ, 1987), pp. 74+424
16. D. Helbing, Derivation of non-local macroscopic traffic equations and consistent traffic pressures from microscopic car-following models. Eur. Phys. J. B **69**(4), 539–548 (2009), see also http://www.soms.ethz.ch/research/traffictheory
17. D. Helbing, Traffic and related self-driven many-particle systems. Rev.Mod. Phys. **73**, 1067–1141 (2001)
18. F. Galton, Vox populi. Nature **75**, 450–451 (1907)
19. D. Helbing, et al., see collection of publications on analytical traffic flow theory at http://www.soms.ethz.ch/research/traffictheory, last accessed on March 6, 2012
20. J.M. Epstein, Why model? J Artif. Soc. Soc. Simulat. **11**(4), 12 (2008). see http://jasss.soc.surrey.ac.uk/11/4/12.html
21. P.A. Davidson, *Turbulence* (Cambridge University, Cambridge, 2004)
22. W. Weidlich, *Sociodynamics: A Systemic Approach to Mathematical Modelling in the Social Sciences* (Dover, 2006)
23. J.H. Kagel, A.E. Roth, *The Handbook of Experimental Economics* (Princeton University, Princeton, NJ, 1995)
24. F. Guala, *The Methodology of Experimental Economics* (Cambridge University Press, New York, 2005)
25. D. Helbing, W. Yu (2010) The future of social experimenting. *Proceedings of the National Academy of Sciences USA (PNAS)* **107**(12), 5265–5266; see also http://www.soms.ethz.ch/research/socialexperimenting
26. O. Maimon, L. Rokach, *The Data Mining and Knowledge Discovery Handbook* (Springer, Berlin, 2005)
27. M.O. Jackson, *Social and Economic Networks* (Princeton University, Princeton, 2008)
28. N. Gilbert (ed.), *Computational Social Science* (Sage, CA, 2010)

29. J.H. Miller, S.E. Page, *Complex Adaptive Systems: An Introduction to Computational Models of Social Life* (Princeton University, Princeton, NJ, 2007)
30. D. Sornette, *Critical Phenomena in Natural Sciences. Chaos, Fractals, Selforganization and Disorder: Concepts and Tools* (Springer, Berlin, 2006)
31. S. Albeverio, V. Jentsch, H. Kantz (eds.), *Extreme Events in Nature and Society* (Springer, Berlin, 2005)
32. D. Floreano, C. Mattiussi, Bio-Inspired Artificial Intelligence: Theories, Methods, and Technologies (MIT, Cambridge, MA, 2008)
33. S. Nolfi, D. Floreano, *Evolutionary Robotics : The Biology, Intelligence, and Technology of Self-Organizing Machines* (MIT, Cambridge, MA, 2000)
34. D. Helbing, A. Deutsch, S. Diez, K. Peters, Y. Kalaidzidis, K. Padberg, S. Lämmer, A. Johansson, G. Breier, F. Schulze, M. Zerial, Biologistics and the struggle for efficiency: Concepts and perspectives. Adv. Complex Syst. **12**(6), 533–548 (2009)
35. D. Helbing, System risks in society and economics. Sante Fe Institute Working Paper #09-12-044 (2009). See http://www.santafe.edu/media/workingpapers/09-12-044.pdf, last accessed on March 6, 2012
36. M. Moussaid, S. Garnier, G. Theraulaz, D. Helbing, Collective information processing and pattern formation in swarms, flocks, and crowds. Top. Cognit. Sci. **1**(3), 469–497 (2009)
37. D. Helbing, A. Szolnoki, M. Perc, G. Szabó, Evolutionary establishment of moral and double moral standards through spatial interactions. PLoS Comput. Biol. **6**(4), e1000758 (2010)
38. K.R. Popper, *The Logic of Scientific Discovery* (Hutchinson, 1959); original German version: *Logik der Forschung* (Springer, Vienna, 1935)
39. D. Tilman, D. Wedin, J. Knops, Productivity and sustainability influenced by biodiversity in grassland ecosystems. Nature **379**, 718–720 (1996)
40. A. Traulsen, D. Semmann, R.D. Sommerfeld, H.-J. Krambeck, M. Milinski, Human strategy updating in evolutionary games. Proc. Natl. Acad. Sci. USA (PNAS) **107**(7), 2962–2966 (2010)
41. E. Brockfeld, R.D. Kühne, P. Wagner, Calibration and validation of microscopic traffic flow models. Transport. Res. Board **1876**, 62–70 (2004)
42. A. Kesting, M. Treiber, Calibrating car-following models by using trajectory data: Methodological study. Transport. Res. Record **2088**, 148–156 (2008)
43. K. Gödel, *On Formally Undecidable Propositions of Principia Mathematica and Related Systems* (Basic, New York, 1962)
44. J. Lorenz, H. Rauhut, F. Schweitzer, D. Helbing, How social influence undermines the wisdom of crowds. Proc. Natl. Acad. Sci. USA (PNAS) **108**(22), 9020–9025 (2011)
45. M. Planck: "An important scientific innovation rarely makes its way by gradually winning over and converting its opponents, but rather because its opponents eventually die, and a new generation grows up that is familiar with it."
46. T.S. Kuhn, *The Structure of Scientific Revolutions* (University of Chicago, Chicago, 1962)
47. T. Vicsek, The bigger picture. Nature **418**, 131 (2002)
48. L. Pietronero, Complexity ideas from condensed matter and statistical physics. europhysicsnews **39**(6), 26–29
49. A.S. Mikhailov, V. Calenbuhr, *From Cells to Societies. Models of Complex Coherent Action* (Springer, Berlin, 2002)
50. R.A. Bentley, P. Ormerod, Agents, intelligence, and social atoms, in *Creating Consilience: Integrating the Sciences and the Humanities*, ed. by M. Collard, E. Slingerland (Oxford University Press, 2011)
51. H. Gintis, *The Bounds of Reason: Game Theory and the Unification of the Behavioral Sciences* (Princeton University, Princeton, 2009)
52. P. Omerod, What can agents know? The feasibility of advanced cognition in social and economic systems. In *Proceedings of the AISB 2008 Convention on Communication, Interaction and Social Intelligence* **6** (Aberdeen, Scotland, 2008), pp. 17–20
53. E. Fehr, K.M. Schmidt, A theory of fairness, competition, and cooperation. Q. J. Econ. **114**(3), 817–868 (1999)

54. B. Frey, *Economics as a Science of Human Behaviour: Towards a New Social Science Paradigm* (Kluwer Academics, Dordrecht, 1999)
55. D. Kahneman, A. Tversky, Prospect theory: An analysis of decision under risk. Econometrica **47**(2), 263–291 (1979)
56. H.A. Simon, A behavioral model of rational choice. Q. J. Econ. **69**(1), 99–118 (1955)
57. G. Gigerenzer, P.M. Todd, and the ABC Research Group, *Simple Heuristics That Make Us Smart* (Oxford University, New York, 2000)
58. D. McFadden, Conditional logit analysis of qualitative choice behaviour, in *Frontiers of Econometrics*, ed. by P. Zarembka (Academic Press, New York, 1974), pp. 105–142
59. D. Helbing, *Quantitative Sociodynamics. Stochastic Methods and Models of Social Interaction Processes* (Kluwer Academic, Dordrecht, 1995)
60. P. Woit, *Not Even Wrong: The Failure of String Theory and the Search for Unity in Physical Law for Unity in Physical Law* (Basic, New York, 2006)
61. L. Smolin, *The Trouble With Physics: The Rise of String Theory, The Fall of a Science, and What Comes Next* (Mariner, Boston, 2007)
62. J. Rotmans, M.B.A. van Asselt, Uncertainty management in integrated assessment modeling: Towards a pluralistic approach. Environ. Monit. Assess. **69**(2), 101–130 (2001)
63. D. Dubois, H. Prade, Possibilistic logic: a retrospective and prospective view. Fuzzy Set. Syst. **144**(1), 3–23 (2004)
64. V. Lucarini, Towards a definition of climate science. Int. J. Environ. Pollut. **18**(5), 413–422 (2002)
65. A. Vespignani, Predicting the behavior of techno-social systems. Science **325**, 425–428 (2009)
66. D. Helbing, The FuturIcT knowledge accelerator: Unleashing the power of information for a sustainable future, Project Proposal (2010), see http://arxiv.org/abs/1004.4969 and http://www.futurict.eu, last accessed on March 6, 2012

Chapter 2
Agent-Based Modeling*

Since the advent of computers, the natural and engineering sciences have enormously progressed. Computer simulations allow one to understand interactions of physical particles and make sense of astronomical observations, to describe many chemical properties ab initio, and to design energy-efficient aircrafts and safer cars. Today, the use of computational devices is pervasive. Offices, administrations, financial trading, economic exchange, the control of infrastructure networks, and a large share of our communication would not be conceivable without the use of computers anymore. Hence, it would be very surprising, if computers could not make a contribution to a better understanding of social and economic systems. While relevant also for the statistical analysis of data and data-driven efforts to reveal patterns of human interaction [1], we will focus here on the prospects of computer simulation of social and economic systems. More specifically, we will discuss the techniques of agent-based modeling (ABM) and multi-agent simulation (MAS), including the challenges, perspectives and limitations of the approach. In doing so, we will discuss a number of issues, which have not been covered by the excellent books and review papers available so far [2–10]. In particular, we will describe the different steps belonging to a thorough agent-based simulation study, and try to explain, how to do them right from a scientific perspective. To some extent, computer simulation can be seen as experimental technique for hypothesis testing and scenario analysis, which can be used complementary and in combination with experiments in real-life, the lab or the Web.

---

*This chapter has been prepared by D. Helbing and *S. Balietti* under the project title "How to Do Agent-Based Simulations in the Future: From Modeling Social Mechanisms to Emergent Phenomena and Interactive Systems Design".

D. Helbing (ed.), *Social Self-Organization*, Understanding Complex Systems,
DOI 10.1007/978-3-642-24004-1_2, © Springer-Verlag Berlin Heidelberg 2012

2.1 Why Develop and Use Agent-Based Models?

2.1.1 Potential of Computer Simulation in the Socio-Economic Sciences

It is well-known that the ways in which social scientists analyze human behavior, social interactions, and society vary largely. The methods range from qualitative to quantitative ones, and among the quantitative ones, some communities prefer detailed models with many variables and parameters, while others prefer simple or simplified models with a few variables and parameters only. Reference [11] discusses these different types of system description and their respective advantages and disadvantages. Overall, each method has its justification, and the choice of the proper method very much depends on the respective purpose. For example, the elaboration of applications such as new systems designs often requires a quite realistic and, hence, detailed description of all relevant aspects. In contrast, simple models may be used to get a better understanding of how social mechanisms work. They serve to reduce the complexity of a given system to an extent that allows to guide our thinking and provide an intuition how certain changes in the system would affect its dynamics and outcome.

The application of computational models is currently not common in the social and economic sciences. This is perhaps because many people consider them as intransparent and unreliable (as compared to analytical methods) and/or as unsuitable for prediction. These points will be addressed later on. In fact, if properly done, computer simulations *can* deliver reliable results beyond the range of analytical tractability (see Sect. 2.3.6). Moreover, we will show in Sect. 2.4.1 that prediction is not generally impossible for socio-economic systems and that it is even not necessary to improve a system (e.g. to reduce instabilities or vulnerabilities), see Sect. 2.4.2. Besides, the benefit of computational models is not restricted to prediction. Joshua Epstein, for example, discusses 16 other reasons to build models, including explanation, guiding data collection, revealing dynamical analogies, discovering new questions, illuminating core uncertainties, demonstrating tradeoffs, training practitioners, and last but not least decision support, particularly in crisis situations [12].

In fact, computer models can naturally complement classical research methods in the socio-economic sciences. For example, they allow one to test whether mechanisms and theories used to explain certain observed phenomena are sufficient to understand the respective empirical evidence, or whether there are gaps or inconsistencies in the explanation. Moreover, they allow one to study situations, for which analytical solutions cannot be found anymore, and to go beyond the idealizations and approximations of simple models. Without the exploration of model behaviors that can only be numerically determined, scientific analysis is often restricted to unrealistic models and to situations, which may be of little relevance for reality. For example, the financial crisis may have been the result of approximations and simplifications of economic models, which were not sufficiently justified (for a more detailed discussion of this point see [13]).

2.1.2 Equation-Based Versus Agent-Based Approach

Today, computer-simulation in the natural sciences and engineering mostly rely on equation-based modeling (e.g. of the dynamics of gases, fluids, or solid bodies). Such an approach would certainly be hard to transfer to the social sciences, as most system behaviors have not been formalized mathematically. A method that seems to be more suited for the computer simulation of socio-economic systems is agent-based modeling (ABM) [2–6]. The corresponding computational technique is called multi-agent simulation (MAS) or agent-based computational modeling ("ABC modeling"). Depending on the problem of interest, agents may for example represent individuals, groups, companies, or countries and their interactions. The behaviors and interactions of the agents may be formalized by equations, but more generally they may be specified through (decision) rules, such as if-then kind of rules or logical operations. This makes the modeling approach much more flexible. Besides, it is easily possible to consider individual variations in the behavioral rules ("heterogeneity") and random influences or variations ("stochasticity").

To give a clearer picture, let us provide below a list of properties, which may be given to an agent representing an individual:

- Birth, death, and reproduction
- Individual needs of resources (e.g. to eat and drink)
- Competition and fighting ability
- Toolmaking ability (e.g. the possibility to grow food, hunt etc.)
- Perception
- Curiosity, exploration behavior, ability for innovation
- Emotions
- Memory and future expectations
- Mobility and carrying capacity
- Communication
- Learning and teaching ability
- The possibility of trading and exchange
- The tendency to have relationships with other agents (e.g. family or friendship ties etc.)

Agent-based computational models also appear ideal to study interdependencies between different human activities (both, symbiotic and competitive relationships) [14]. Therefore, they can shed new light on social and economic systems from an "ecological" perspective [15]. In fact, evolutionary ecological models can also reflect the feature of steady innovation, which is typical for socio-economic systems. Moreover, such models appear particularly suited to study the sustainability and resilience of systems. Finally, they can be well combined with other simulation methods used in the natural and engineering sciences, including statistical physics, biology, and cybernetics.

2.1.3 Scientific Agent-Based Models Versus Computer Games

Agent-based computational modeling is also well suited for visualization. Here, it makes sense to distinguish agent-based models in science from computer games. While the latter may actually be *based* on agent-based models, they are often oriented at *believability*, i.e. at appearing as realistic as possible. However, computer games usually do not care too much about making realistic *assumptions*. That is, while the outcome may look realistic, the underlying mechanisms may not be well justified. As a consequence, they may not be suited to understand the outcome of the simulation, to draw conclusions, or to make predictions. Implications outside the exact settings that the game was prepared for are likely to be unreliable. Therefore, in many cases, such computer simulations will not produce useful knowledge beyond what has been put into the model.

Scientific agent-based models, in contrast, do often not invest into believability. That is, they intentionally make simplifications and may, for example, represent people by circles, or purposefully restrict themselves to very few properties from the list given in Sect. 2.1.2. Instead of focusing on a plausible appearance, they try to represent a few characteristic features (such as certain kinds of interaction mechanisms) more realistically. Such computer simulations should enable one to draw conclusions about previously unexperienced scenarios, i.e. they should be in reasonable agreement with later empirical observations or experimental results. In other words, scientific simulations are more focused on getting the processes rather than the visual representation right. They are interested in explanatory power.

Finally, agent-based simulations for engineering applications are often located somewhere between the two archetypical cases discussed above. However, rather than on the basis of the level of detail and believability, it also makes sense to classify models as follows:

- *Physical models* assume that individuals are mutually reactive to current (and/or past) interactions.
- *Economic models* assume that individuals respond to their future expectations and take decisions in a selfish way.
- *Sociological models* assume that individuals respond to their own and *other* people's future expectations (and their past and current experiences as well).

2.1.4 Advantages of Agent-Based Simulations

Agent-based simulations are suited not only to reflect interactions between different individuals (and other entities). They allow one to start off with the descriptive power of verbal argumentation and to determine the implications of different hypotheses. From this perspective, computer simulation can provide "an orderly formal framework and explanatory apparatus" [16]. Other favorable features of agent-based simulations are [17]: modularity, great flexibility, large expressiveness, and the possibility to execute them in a parallelized way.

Agent-based models can be combined well with other kinds of models. For example, when simulating the interaction with the environment, the environment may be represented by a discrete or continuous field. Such an approach is pursued within the framework of active walker models [18, 19]. One can easily couple agent-based models with continuum models, such as gas-kinetic or fluid-dynamic models. Such an approach is, for example, used to simulate the evacuation of people in scenarios where poisonous gas spreads in the environment [20, 21]. A similar approach would be applied, when weather, environmental, or climate simulations would be combined with models of human response to the respective external conditions.

In certain contexts, for reasons of computational efficiency it may also be reasonable to replace an agent-based by an aggregate ("macroscopic") simulation approach. For example, traffic flows can not only be well represented by a car-following (agent-based) model, but also by a fluid-dynamic one [22]. It is even possible to relate the car-following models with fluid-dynamic ones in an analytical way [23]. In other words, it is possible to construct a mathematical bridge between the micro- and macro-level of description [24] – something which would be very nice to have for economics and other fields as well.

In the economic sciences, multi-agent computer simulations make it possible to overcome limitations of the theoretical concept of *homo economicus* (the "perfect egoist") [25], by relaxing idealized assumptions that are empirically not well enough supported. They also offer a possibility to go beyond the representative agent models of macroeconomics [26], and to establish a natural link between the micro- and macro-level description, considering heterogeneity, spatio-temporal variability, and fluctuations, which are known to change the dynamics and outcome of the system sometimes dramatically [13].

Finally, agent-based simulations are suited for detailed hypothesis-testing, i.e. for the study of the consequences of ex-ante hypotheses regarding the interactions of agents. Insofar, one could say that they can serve as a sort of magnifying glass or telescope ("socioscope"), which may be used to understand our reality better. It is usually no problem to apply methods from statistics and econometrics to simulation outcomes and to compare simulation results with actual data (after processing them in a way reflecting the measurement process). Moreover, by modeling the relationships on the level of individuals in a rule-based way, agent-based simulations allow one to produce characteristic features of the system as emergent phenomena without having to make a priori assumptions regarding the aggregate ("macroscopic") system properties.

2.1.5 Understanding Self-organization and Emergence

Agent-based simulations are a suitable tool to study complex systems. Complex systems are systems with many interacting entities and non-linear interactions among them. Such systems may behave in a number of interesting and often unexpected (sometimes even paradoxical) ways, which justifies to call them complex:

- They may have several stationary states (a phenomenon known as "multi-stability"), and the resulting outcome may depend on the previous history (such as the size of occurring perturbations, the "initial state", etc., and such history-dependencies are often called "hysteresis effect") (see Figs. 2.1 and 2.3).
- They may be "out of equilibrium" and behave in non-stationary ways.
- They may "self-organize", showing periodic or non-periodic oscillations, "chaotic" or "turbulent" behavior, or spatio-temporal pattern formation (such as stop-and-go waves in traffic flows).
- They are often robust to small perturbations, i.e. "relax" to their previous behavior (the "stable attractor").
- Consequently, they often resist external manipulation or control attempts.
- However, at so-called "tipping points", small influences may cause a sudden and often unexpected "systemic shift" ("phase transition"), after which the system behaves substantially different (see Fig. 2.2).
- More generally, they may show new, "emergent" properties, which cannot be understood from the properties of their system elements ("the system is more than the sum of its parts") (see Fig. 2.1).
- Correlations may determine the system dynamics, and neglecting them may lead to completely wrong conclusions.
- During systemic shifts (so-called "phase transitions") or due to a phenomenon called "self-organized criticality" (SOC), cascading effects on all scales (i.e. of any size) may occur, so that local factors may have a systemic ("global") impact ("critical phenomena").
- Therefore, "extreme events" may happen with probabilities much higher than expected according to a normal distribution, and are distributed according to "(truncated) power laws" or other "fat tail distributions".
- The system may have features such as reproduction, innovation, reinforcement learning, an expectation-related dynamics, etc.
- There may be singularities after a finite time.

Many of the above features are results of strong interactions in real or abstract space, or of network interactions within the system. Such interactions can often lead to counter-intuitive behaviors [29, 30]. Here are a number of examples (most of them related to traffic systems, as they are well-known to everybody):

- Even when all drivers try to drive fluently, vehicles are sometimes stopped by "phantom traffic jams" (i.e. jams without an obvious reason such as an accident or bottleneck) [31, 32].
- Stop-and-go traffic may occur despite the best attempts of drivers to move ahead smoothly [33, 34].
- Even when the maximum road capacity is not reached, a temporary reduction in the traffic flow can cause a lasting traffic jam [35].
- Traffic jams do not occur in the bottleneck area, but upstream of it.
- Under certain conditions, speed limits can speed up traffic [31]. Similar "slower-is-faster effects" occur in urban traffic control, logistic systems, administrative processes, etc., i.e. delays at appropriate times and of suitable durations may reduce overall waiting times [36, 37].

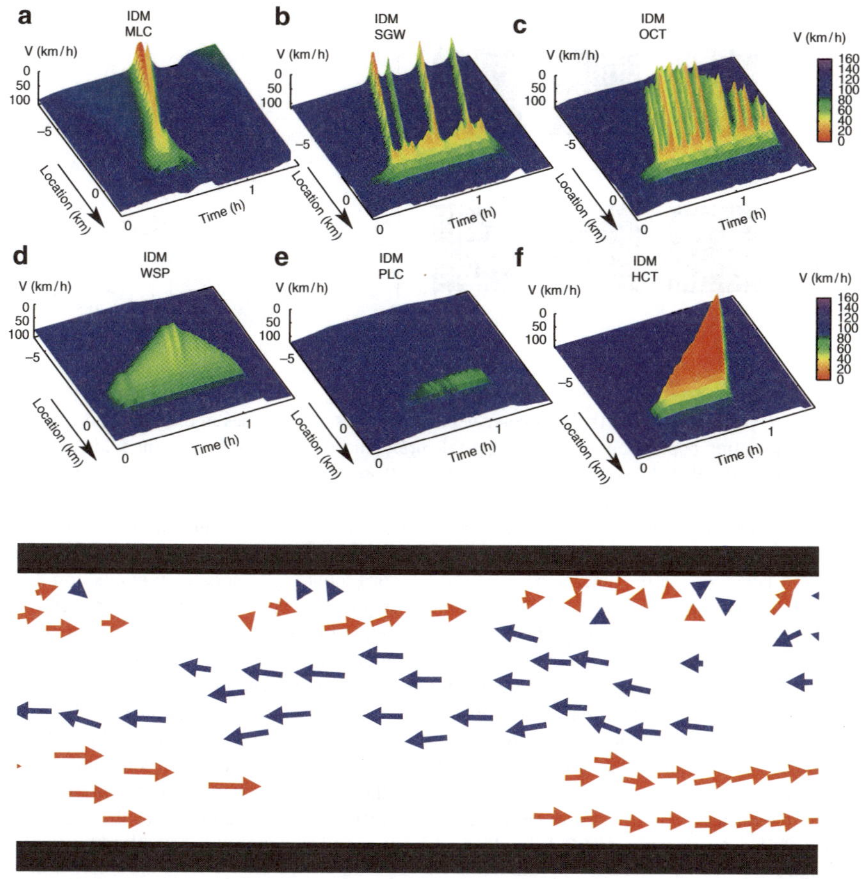

Fig. 2.1 *Top*: Freeway traffic constitutes a dynamically complex system, as it involves the non-linear interaction of many independent driver-vehicle units with a largely autonomous behaviour. Their interactions can lead to the emergence of different kinds of traffic jams, depending on the traffic flow on the freeway, the bottleneck strength, and the initial condition (after [27]): a moving cluster (MC), a pinned localized cluster (PLC), stop-and-go waves (SGW), oscillating congested traffic (OCT), or homogeneous congested traffic (HCT). The different traffic patterns were produced by computer simulation of a freeway with an on-ramp at location $x = 0\,\text{km}$ using the intelligent driver model (IDM), which is a particular car-following model. The velocity as a function of the freeway location and time was determined from the vehicle trajectories (i.e. their spatio-temporal movement). During the first minutes of the simulation, the flows on the freeway and the on-ramp were increased from low values to their final values. The actual breakdown of free traffic flow was triggered by additional perturbations of the ramp flow. *Bottom*: In pedestrian counterflows one typically observes a separation of the opposite walking directions. This "lane formation" phenomenon has been reproduced here with the social force model [28]

- While pedestrians moving in opposite directions normally organize into lanes, under certain conditions a "freezing-by-heating" effects or intermittent flows may occur [38].
- The maximization of system efficiency may lead to a breakdown of capacity and throughput [30].

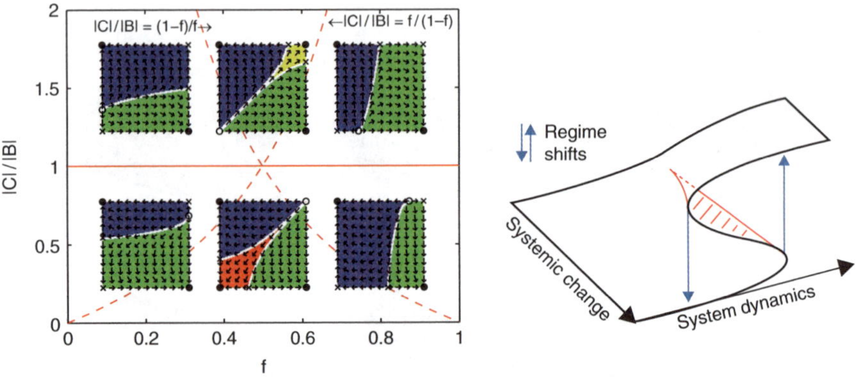

Fig. 2.2 *Left*: Illustration of the parameter-dependent types of outcomes in the social norms game of two interacting populations with incompatible preferences (after [151]). f is the relative size of population 1 and $B = b - c < 0$ and $C = b + c > 0$ are model parameters, which depend on the benefit b of showing the individually preferred behavior, while c is the reward of conforming with the behavior of the respective interaction partner. *Small arrows* illustrate the vector field $(dp/dt, dq/dt)$ as a function of the fraction p of individuals in population 1 showing their preferred behavior (on the horizontal axis) and the corresponding fraction q in population 2 (on the vertical axis). *Empty circles* stand for unstable fix points (repelling neighboring trajectories), *black circles* represent stable fix points (attracting neighboring trajectories), and *crosses* represent saddle points (i.e. they are attractive in one direction and repulsive in the other). The basins of attraction of different stable fix points are represented in different shades of *grey* (colors) [*green* = population 1 sets the norm, *blue* = population 2 sets the norm, *yellow* = each population does what it prefers, *red* = nobody shows the preferred behavior]. The *solid red lines* indicates the threshold at which a continuous phase transition takes place, *dashed lines* indicate discontinuous phase transitions. *Right*: When a complex system is manipulated (e.g. by external control attempts), its system parameters, stability, and dynamics may be affected. This figure illustrates the occurrence of a so-called "cusp catastrophe". It implies a discontinuous transition ("regime shift") in system dynamics

- Sometimes, the combination of two dissatisfactory solutions can be the best solution [39,40].
- There is clearly a global diversity of opinions and behaviors, despite a rather strong tendency of local convergence [41].
- In repeated interactions, the "wisdom of crowds" mechanism may lead to collective error [42].
- One can often find cooperation among selfish individuals in social dilemma situations (where it seems more profitable to exploit the cooperativeness of others) [39].
- People often do not show the behavior they like and often show behaviors they do not like [43].

Generally, there is *still* a serious lack of understanding regarding the connections between the structure, dynamics, and function of complex networks such as techno-socio-economic-environmental systems. Therefore, emergence often has an element

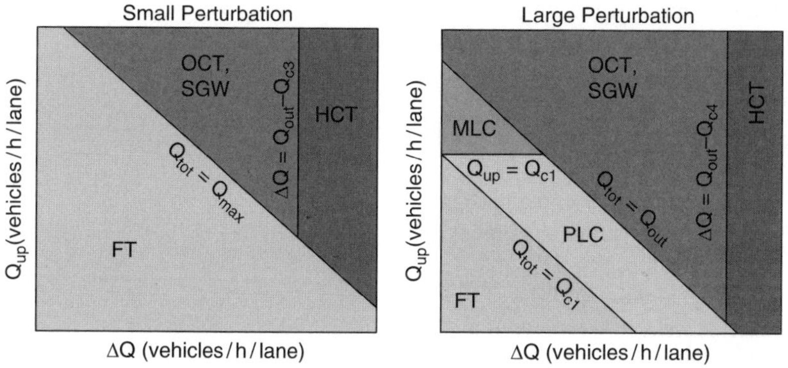

Fig. 2.3 Schematic (idealized) phase diagrams of the traffic patterns expected for a particular traffic model as a function of the freeway traffic flow Q_{up} upstream of a ramp bottleneck and the on-ramp flow ΔQ (after [27]). The *left figure* is for negligible, the *right figure* for large perturbations in the traffic flow. The situation for medium-sized perturbations can lie anywhere between these two extremes. Different colors correspond to different possible traffic states (FT = free traffic flow, PLC = pinned localized cluster, MLC = moving localized cluster, OCT = oscillating congested traffic, SGW = stop-and-go waves, HCT = homogenous congested traffic, see Fig. 2.1). The equations next to the separating lines can be analytically calculated, but are not important here. For details see [27])

of surprise. So far, for example, we do not understand emotions and consciousness, and we cannot even calculate the "fitness" of a behavioral strategy in the computer. The most exciting open puzzles in science concern such emergent phenomena. It would be interesting to study, whether social and economic phenomena such as trust, solidarity, and economic value can be understood as emergent phenomena as well [13]. Agent-based simulations appear to be a promising approach to make scientific progress in these areas.

2.1.6 Examples of Agent-Based Models

The method of agent-based modeling is very versatile. It may be applied, for example, to the following problems:

- Social influence and opinion formation [41,44]
- Coalition formation [45,46]
- Collective intelligence [47]
- Social networks [48–50]
- Group dynamics [51]
- Social cooperation [52,53]
- Social norms [14,54,55]
- Social conflict [56,57]

- Financial markets [58–60]
- Competition and cooperation between firms [61,62]
- Micro-economic models [63,64]
- Macro-economic models [65,66]
- Organization and managerial decisions [67]
- Migration [68]
- Agglomeration and segregation [69,70]
- Urban and regional development [71–73]
- Traffic dynamics [74,75]
- Crowd dynamics [76,77]
- Systemic risks in socio-economic systems [78,79]

and more [80–84].

2.1.7 Social Super-Computing

Multi-agent simulations are well suited for parallelization. Therefore, using super-computing power, it is possible in principle to run agent-based simulations with millions (or, in future, even billions) of agents. The following examples give an idea of the state-of-the art:

- A first successful application area was large-scale traffic simulation. The TRANSIMS project [85–87] of the Los Alamos National Institute (LANL), for example, has created agent-based simulations of whole cities such as Dallas [88] or Portland [89]. The approach has been recently extended to the simulation of the travel behavior of the 7.5 million inhabitants of Switzerland [90, 91]. These simulations are obviously based on parallel computing. They generate realistic individual activity patterns according to detailed statistical panel data ("travel diaries") [92, 93], which are nowadays complemented by GPS data and public mobility data (e.g. from Greater London Area's OYSTER Card). Other extensions look at interconnections between the traffic system and regional development [72,73].
- Recent applications are studying contingency plans for large-scale evacuations of cities [94,95]. The key aspect here is understanding the interdependency of infrastructure systems [96,97] and their vulnerabilities through natural disasters, terrorist attacks, accidents, and other incidents. For example, the Los Alamos National Laboratories have already established a Critical Infrastructure Protection Decision Support System [98]. Its advanced simulation capabilities have already been extensively used during past emergencies.
- Large-scale simulations have also been applied to study and predict the spreading of diseases. While previous epidemic spreading models such as the SIR model [99–101] have neglected spatial interaction effects, recent models [102] take into account effects of spatial proximity [20], air traffic [103] and land transport [104], using TRANSIMS and other traffic simulation models. The current scientific

development also tries to take into account behavioral changes, which may reduce the spreading rate of diseases.

- Furthermore, there are attempts to model financial markets with agent-based simulations. Two examples for this are the Santa-Fe Stock Market Simulator [105] and U-mart [106]. Recent attempts are even heading towards the simulation of the whole economic system (see for example the EU project EURACE [107]). Other simulation studies are trying to understand the evolution of cooperation [39], social norms [14, 54], language [108–110], and culture [3]. Such simulations explore the conditions under which trust, cooperation and other forms of "social capital" can thrive in societies (see also [13, 111]). They also show that the crust of civilization is disturbingly vulnerable. Such simulations can reveal possible causes of breakdowns of social order. Examples for events where this actually happened are as diverse as the war in former Yugoslavia, lootings after earthquakes or other natural disasters, or the violent demonstrations we have recently seen in some European countries.

It appears logical that supercomputing will be ultimately moving on from applications in the natural and engineering sciences to simulations of social and economic systems, as more and more complex systems become understandable and the required data become available. And it is obvious that virtual three-dimensional worlds (such as Google Earth) are waiting to be filled with realistic life.

2.2 Principles of Agent-Based Modeling

After describing the potential of multi-agent simulations, let us now discuss the principles of how to craft agent-based models. A thorough scientific study involves a number of steps:

- First, one should clearly describe the evidence to be explained by the respective study. What are the empirical or experimental data or observations to be reproduced, or what are the "stylized facts", i.e. the simplified, idealized properties of the system under consideration?
- Second, one should explain what is the purpose of the simulation? To understand a phenomenon? To get a more accurate description? To make predictions? To develop an application (e.g. a new traffic control)? In the social sciences, it is common to formulate a scientific puzzle, i.e. to describe a problem that is hard to understand and why. This could be an unexpected or even paradoxical individual or system behavior. Emergent system behaviors are particularly interesting candidates for the formulation of such a puzzle ("scientific mystery").
- Next, one needs to decide how to choose the agents in the model. For example, when the competition of companies shall be studied, it may not be necessary to simulate all employees of all companies. It may be sufficient to choose the companies as the agents of the model. In fact, it can be shown mathematically (e.g. by eigenvalue analysis) that mutually coupled agents may jointly behave

like one entity, i.e. one agent. An example for this is the quasi-species concept in the theory of evolution [112].

- After specifying the agents, one should formulate hypotheses regarding the underlying socio-economic processes or fundamental mechanisms leading to the particular system behavior that needs to be explained. Ideally, these mechanisms should be sociologically or economically justified, i.e. there should be some empirical evidence for the mechanisms on which the agent-based model is based. The transfer of models from other sciences (such as spin, epidemic, or synchronization models) requires particular justification beyond saying that the resulting system behavior is reminiscent of features that have been observed elsewhere.
- When specifying the mechanisms underlying the multi-agent simulation, one should not put into the model assumptions what one wants to explain. The mechanisms on which the multi-agent simulations are based should be (at least) one level more elementary than the evidence to be understood. For example, the rich-gets-richer effect [113] may be used as an ingredient, if class formation shall be described. Moreover, "homophily" [114] may be assumed in models of coalition formation or solidarity. Moreover, social network characteristics may be used to explain the spreading of behaviors [115–117].

  However, if the income distribution is to be explained, it is favorable not to start with the rich-gets-richer effect, but instead with a mechanism that is purely random and not biased in favor of anybody in the beginning. Moreover, even if this is not realistic, it would be interesting to start the computer simulation with identical wealth of everybody [118]. Furthermore, if social segregation is to be explained, one should not assume "homophily" already, but to let it evolve in a system that starts off with identical and non-segregated individuals [39, 119]. Finally, if group formation is to be explained, social network characteristics should not be assumed as an input [41]. They should, for example, result from certain rules regarding the formation and deletion of social ties [120, 121].
- Last but not least, one should compare the computer simulation results with the empirical evidence. Here, one should avoid to be selective, i.e. one should state what are the features that are correctly reproduced, and which ones are not. Pointing out the limitations of a model is equally important as underlining its explanatory power.

Note that, even though linear models may be sensitive to parameter variations or perturbations, they cannot explain self-organization or emergent phenomena in socio-economic systems. This underlines that the consideration of non-linear interaction mechanisms is crucial to understand many observations in social and economic systems [13].

When specifying the properties of agents and their interactions, it makes sense to select some from the list given in Sect. 2.1.2, but (by far) not all. As was pointed out before, the goal of scientific computer-simulation is usually not a realistically looking computer game, but the explanation of a set of observations from a minimum set of assumptions.

It is definitely not obvious how each of the points on the list in Sect. 2.1.2 is modeled best. Typically, there are a number of plausible alternatives. To gain an understanding and intuition of the studied system, simple assumptions (even if idealized) are often preferable over detailed or complicated ones. For example, rather than assuming that individuals would strictly optimize over all options when decision are taken (as "homo economicus" would do), it seems justified to use simple decision heuristics, as evidence from social psychology suggests [122]. However, it would obviously be interesting to compare the implications of both modeling approaches (the classical economics and the heuristics-based one).

2.2.1 Number of Parameters and Choice of Model

When formulating the agent-based model, a number of aspects should be considered, as discussed in the following. The adequate number of model parameters and variables depends on the purpose of the model and its required degree of realism and accuracy. A model for the elaboration of practical applications tends to have more parameters than a model aimed at the fundamental understanding of social or economic mechanisms. However, one should keep in mind that parameter-rich models are hard to calibrate and may suffer from over-fitting. Hence, their predictive power may not be higher than that of simple models (see Sect. 2.3.2).

Generally, one can say that

- A model with meaningful parameters (which have a clear interpretation and are measurable) should be favored over a model with meaningless fit parameters.
- The same applies to a model with operational variables as compared to a model containing variables that cannot be measured.
- Given the same number of parameters, an explanatory model should be prefered over a purely descriptive ("fit") model.
- In case of a comparable goodness of fit in the model calibration step, one should choose the model with the better predictive power (i.e. which better matches data sets that have not been used for calibration).
- Given a comparable predictive power of two models, one should select the simpler one (e.g. the one with analytical tractability or with fewer parameters) according to Einstein's principle that *a model should be as simple as possible, but not simpler.*

The goodness of fit should be judged with established statistical methods, for example with the adjusted R-value or similar concepts considering the number of model parameters [123, 124]. This adjusted formula tries to compensate for the fact that it is easier to fit a data set with more parameters. For a further discussion of issues related to the calibration and validation of agent-based models see Sect. 2.3.2.

2.3 Implementation and Computer Simulation of Agent-Based Models

2.3.1 Coding Multi-agent Simulations and Available Software Packages

Agent-based simulations may be programmed from scratch in most computer languages (e.g. C, C++, Java, etc.). This may be favorable from the perspective of computational efficiency. However, there are also more user-friendly ways of specifying agent-based models. This includes Swarm [125] and Repast [126], which provide low-level libraries, as well as graphical modelling environments such as Netlogo [127] or Sesam [128], which are more suited for beginners. Another recent software tool for the simulation of large crowds is MASSIVE [129]. For a comprehensive comparison of the features of available packages for agent-based simulations see [130, 131].

2.3.1.1 Documentation

No matter how multi-agent simulations are implemented, it is crucial to document the code and the simulation scenarios well. Not only would the code be useless for any future collaboration partner, if it is not well documented. It is also quite easy to lose the overview over all the changes one may have experimented with. There are several publications, which report results that do not fit the model discussed in the paper. Such accidents happen often for one of the following reasons:

- The computer code has not been properly debugged – checked for mistakes, see Sect. 2.3.1.3 (e.g. variables are not properly initiated, or local variables are overwritten by global variables, or variables are mixed up, or there is a mismatch of different types of variables, such as integer values and real numbers).
- Changes are made in the code, but not well documented and forgotten.
- Readimade routines from software libraries are applied, but the preconditions for their use are not satisfied.

2.3.1.2 Plausibility Tests

The most famous mistake in a computer code was the explosion of a space shuttle caused by confusing a semicolon and a comma. However, not always do programming mistakes become evident in such a dramatic way. In fact, the danger is that they remain unnoticed. Computer codes do not automatically generate errors, when something is incorrectly implemented. On the contrary, most computer codes produce results independently of how reasonable the code or input is. Therefore, one should better have a healthy distrust towards any computational output. Computer codes always require verification and proper testing, and it is crucial to check the plausibility of the output. Section 2.3.4.2 discusses some ways how this can be done.

2.3.1.3 Common Mistakes and Error Sources

To avoid errors in the computer code, the following precautions should be taken:

- Structuring the computer code into subroutines allows one to keep a better overview (and do a thorough, component-wise testing) as compared to one monolithic code.
- Different variants of the computer code should be distinguished with different version numbers,[1] and the main features, particularly the changes with regard to the last version, should be clearly documented in the beginning of the code and, ideally, shown on the computer display and in the output file, when the code is run.
- One should not include the parameters of the model in the computer code itself ("hardcoding"), but instead they should be read from a separate input file, which should also have an own version number. This parameter file should contain information, with what version of the computer code the parameter file needs to be run. It makes sense to write out the parameters and their values in the output file, to reduce unnoticed errors (such as adding a new parameter, which is erroneously initiated with the value of another parameter).
- Programming languages initiate variables and parameters in different ways (and some of them with random contents), so that it may easily remain unnoticed when a variable or parameter was not set. Therefore, all parameters and variables should be initialized immediately, when they are introduced in the code.
- Moreover, one should not make several significant changes at a time. For example, it is common to change only one parameter value and compare the new result with the previous one, before another parameter is changed. This helps to check the plausibility of the computational results.

2.3.2 Specification of Initial and Boundary Conditions, Interaction Network and Parameters

The simulation of agent-based models requires a number of specifications before the simulations can be run:

---

[1] There exist a number of software packages aimed at supporting developers in versioning the code. They automatise several operations such as assigning sequential versions numbers, comparing different versions of files, undo or merging changes on the same files, etc. For examples, see [132–135].

- It is necessary to define the interaction network. For example, agents may interact in space, and their interaction frequency or strength may depend on their distance. However, in many cases, spatial proximity is just one factor or even irrelevant. In many cases, one needs to consider friendship networks or other interaction networks. It is well-known that the statistical properties of the interaction network may matter for the dynamics of the system dynamics. As a consequence, it may also be needed to run the simulation for different interaction networks. For example, interactions in square grids and in hexagonal grids may sometimes lead to qualitatively different outcomes. Moreover, the number of interaction partners may be relevant for the system dynamics, as may be a heterogeneity in the number of interaction partners or the existence of loops in the interaction network.
- One needs to specify the initial values of all variables (and if memory effects are considered as well, also their previous history). In many cases, it is common to make simplifying assumptions (e.g. all individuals are characterized by identical initial values, or more realistically, the initial variables are assumed to vary according to a particular distribution (e.g. a normal distribution). It is advised to test the sensitivity of the model with respect to the initial condition.
- Furthermore, boundary conditions may have to be specified as well. For example, if a simulation of agents is performed in space, one needs to decide how the rules should look like at the boundary. One may decide to use a finite world such as a chess board. However, this may cause artificial behavior close to the boundaries of the simulation area. Therefore, in many cases one assumes "periodic" boundary conditions, which corresponds to a simulation on the surface of a torus. This is often simpler than the simulation on the surface of a sphere. Note, however, that an improper choice of the boundary conditions can sometimes produce artifacts, and that the simulated system size may affect the result. Therefore, one needs to have an eye on this and may have to test different specifications of the boundary conditions.
- Finally, it is necessary to specify the open model parameters, i.e. to calibrate the model. If the initial and boundary conditions, or the structure of the interaction network cannot be independently measured, they have to be treated as model parameters as well. In case there are enough empirical data, the parameters can be determined by minimizing an "error function", which measures the difference between simulation results and empirical data. Note that the choice of a suitable error function may be relevant, but non-trivial. An improperly chosen function may not be able to differentiate well between good and bad models. For example, minimizing the error between actual and simulated trajectories of pedestrians does not manage to distinguish well between a simple extrapolation model and a circular repulsion model for pedestrians, although the latter avoids collisions and is much more realistic [136]. Or maximizing throughput may produce some unrealistic effects, such as the occurrence of pedestrians or drivers, who get trapped in a certain place [137], while a minimization of delay times ensures that everybody gets forward. Therefore, not only the model, but also the error function must be well chosen.

If there are not enough (or even no) empirical or experimental measurements to calibrate the model with, on can still try to find parameters, which "do the job", i.e. which deliver plausibly looking simulation results. Such a qualitative simulation approach has,for example, been pursued in the early days of pedestrian modeling. Despite the simplicity of this method, it was surprisingly successful and managed to reproduce a number of striking self-organization phenomena observed in pedestrian crowds [136].

Finally note that, even when empirical or experimental data are available, the size of empirical data sets typically does not allow one to determine the model parameters accurately. Particularly, if the model contains many parameters, the reliability of the parameter estimates tends to be poor. In the worst case, this can imply dangerously misleading model predictions. To get an idea of the confidentiality intervals of parameters, one should therefore determine all parameter combinations which are compatible with the error bars of the empirical data to be reproduced.

2.3.2.1 Model Validation

A high goodness of fit during model calibration does not necessarily imply a high predictive power [138, 139], i.e. a good fit of new data sets. In many cases, one faces the problem of over-fitting (i.e. the risk of fitting noise or irrelevant details in the data). Therefore, it is necessary to determine the "predictive power" of a model by a suitable validation procedure.

In the ideal case, the model parameters can be measured independently (or can, at least be estimated by experts). If the parameters have a concrete meaning, it is often possible to restrict the parameters to a reasonable range. In case of non-meaningful fit parameters, however, this is not so simple (or even unfeasible).

One way of validating a model is to subdivide the empirical or experimental data into two non-overlapping parts: a calibration and a validation dataset. The calibration dataset is used to determine the model parameters, and the validation dataset is used to measure the goodness of fit reached with the model parameters determined in the calibration stage. In order to make this calibration and validation procedure independent of the way in which the original dataset is subdivided, the whole procedure should be performed either for all possible subdivisions into calibration and validation datasets or for a representative statistical sample of all possibilities. As each of these subdivisions delivers a separate set of parameters in the calibration step, this results in a distribution of model parameters, which are consistent with the data. From these distributions, one can finally determine average or most likely model parameters as well as confidence intervals. Furthermore, the distribution of goodness-of-fit values reached in the related validation steps reflects the predictive power of the model.

Another way of judging the power of a model is to determine the number of stylized facts that a model can reproduce. It sometimes makes sense to prefer a model that reproduces many different observations qualitatively well (for example, different observed kinds of traffic patterns) over a model whose goodness of fit is quantitatively better (for example, in terms of reproducing measured travel times). This applies in particular, if the model, which appears to be quantitatively superior, is not well consistent with the stylized facts [27]. To distinguish models of different quality, it can also be useful to measure the goodness of fit with several different error functions.

2.3.2.2 Sensitivity Analysis

As empirically or experimentally determined parameter values have a limited accuracy, one should also carry out a sensitivity analysis. For this, the simulation is performed with modified parameters to determine, how robust the simulation results are with respect to the choice of parameters. The sensitivity can be measured with Theil's inequality coefficient. Note that the parameters should be varied at least within the range of the confidence interval that determines the range, within which the actual parameters may vary according to the calibration dataset.

Besides determining the robustness of the model to the parameter specification, it is also recommended to test the robustness to the model assumptions themselves. By simulating variants of the model, one can figure out, which conclusions stay qualitatively the same, and which ones are changing (e.g., if the network characteristics or the system size or the learning rule are modified).

2.3.3 *Performing Multi-agent Simulations*

2.3.3.1 Choice of the Time Discretization

The first problem relates to the choice of the time step Δt. In models that are discrete in time (such as deterministic cellular automata), the time step is assumed to be fixed, and the question may matter or not. When the model contains differential equations, however, the choice of Δt is always relevant. The rule is that large Δt speed up the simulation, but may lead to wrong results. For example, the discrete logistic equation $x_{n+1} = rx_n(1 - x_n)$ may behave very differently from the continuous one $dx/dt = ax(1 - x)$ (the former one may perform a chaotic motion, while the latter one evolves smoothly in time).

To determine a suitable time discretization, Δt is reduced from moderate values (like $\Delta t = 1$ or 0.1) to smaller values, until the results do not change significantly anymore (e.g. less than 0.1% when Δt is chosen five times smaller).

2.3.3.2 Relevance of Considering Fluctuations

A widely recognized fact of socio-economic systems is that they do not behave deterministically. There are always sources of fluctuations ("noise"), such as errors, exploratory behavior, or the influence of unknown factors. For this reason, the probabilistic behavior should be reflected in the model. Neglecting "noise" may lead to misleading conclusions. For example, the zero-noise case may behave completely different from a system with just a little bit of noise, no matter how small it may be [119,140].[2] In such cases, the result of the deterministic model without noise should be considered as artificial and of no practical relevance.

Note that the significance of fluctuations in techno-socio-economic-environmental systems is often wrongly judged. While large noise usually has a destructive influence on a system, as expected, it is quite surprising that small noise intensities can actually *trigger* structure formation or increase system performance [39,141–143].

The implementation of noise in computer simulations is not fully trivial, i.e. mistakes can be easily made. First of all, no random number generator produces random numbers in a strict sense, but rather quasi-random numbers. In other words, there is a certain degree of statistical interdependency between computer-generated random numbers, and this may create artifacts. The quality of random number generators can be very different. Therefore, it makes sense to test the number generator in advance.

Moreover, when adding a Gaussian noise to differential equations (e.g. $dx/dt = f(x,t)$), it must be considered that the variance of the related diffusion process increases linearly with time. This implies that the prefactor of the noise term in the differential equation is not proportional to the time step Δt, but proportional to $\sqrt{\Delta t}$, which can be easily overlooked. This has to do with the particular properties of stochastic equations. (As a consequence, the discretized version of the above differential equation with noise would be $x(t+\Delta t) - x(t) = \Delta t\, f(x,t) + D\sqrt{\Delta t}$, where D allows to vary the noise strength.)

2.3.3.3 Simulating Statistical Ensembles

In case of a model containing noise, one needs to determine statistical distribution(s) (or, more simply, the averages and standard deviations), running the simulation many times with different sets of random numbers (i.e. with a differently initialized random number generator[3]). In other words, running a model with noise a single time is of relatively little meaning. Sometimes, it may serve to illustrate a "typical" system behavior (but then, the simulation run should not be selected by the authors; it should be randomly chosen to avoid a bias through selective presentation). In any case, a scientific paper should present the statistical distributions over sufficiently many (typically at least 100 simulation runs) or the mean value and variability (either one, two or three standard deviations, or quantiles, as it is done by box plots).

---

[2]The reason for this is that deterministic systems may easily get trapped in local optima, which can be overcome by noise [119].

[3]Some random number generators do this automatically by coupling to the clock.

2.3.3.4 Choice of the Discretization Method

Finally, note that appropriate discretization schemes are also not trivial. Depending on how they are done, they may be quite inaccurate or inefficient. For example, the simplest possible discretization of the differential equation $dx/dt = f(x,t)$, namely $x(t + \Delta t) - x(t) = \Delta t\, f(x,t)$, may converge slowly. In case of so-called "stiff" systems of differential equations, the convergence may be so inefficient (due to different time scales on which different variables in the system change) that it may be completely useless. Certain discretizations may even not converge at all towards the correct solution, if Δt is not chosen extremely small. (For example, the solution procedure may face instabilities, which may be recognized by oscillatory values with growing amplitudes.) In such cases, it might be necessary to choose particular solution approaches. The situation for partial differential equations (which contain derivatives in several variables, such as space and time), is even more sophisticated. Normally, the trivial discretization does not work at all, and particular procedures such as the upwind scheme may be needed [144].

2.3.3.5 Performance and Scalability

Multi-agent simulations may require a considerable computational power for the following reasons:

- The dependence of the system dynamics on random fluctuations requires many simulation runs.
- Multi-agent simulations may involve a large number of agents.
- The simulation of rational behavior (i.e. systematic optimization over future developments resulting from behavioral alternatives) or of human cognitive and psychological processes (such as personality, memory, strategic decision-making, reflexivity, emotions, creativity etc.) is potentially quite resource demanding.
- The calibration of model parameters to empirical data also requires many simulation runs for various parameter combinations to determine the parameter set(s) with the minimum error.
- The parameter space needs to be scanned to determine possible system behaviors (see Sect. 2.3.4.5).
- Performing scenario analyses (with changed model assumptions) requires many additional simulation runs. A typical example is the method of "sensitivity analysis" to determine the robustness of a model (see Sect. 2.3.4.6).
- The visualization of simulation scenarios may require further substantial computer power (see Sect. 2.3.4.4).

For the above reasons, the performance of a multi-agent simulations can matter a lot. Unfortunately, this often makes it advantageous to write a specialized own computer program rather than using a general-purpose agent-based simulation platform. However, the performance can sometimes be increased by a factor of 100, 1,000, or more by a number of measures such as

- Using suitable compiler options to optimize the executable file of the computer program.
- Avoiding output on a computer display (for example, by writing the numerical results into a file every 100 or 1,000 time steps and visualizing the results afterwards).
- Avoiding multiple entangled loops and performing loops in the right order (as is favorable for read/write operations).
- Avoiding exponential functions, logarithms, and exponents, where possible.
- Applying an efficient numerical integration method together with a proper time discretization (see Sect. 2.3.3.4).
- Using appropriate parameter values (for example, dividing by small numbers often causes problems and, considering limitations of numerical accuracy, may create almost any output).
- Starting with well-chosen initial conditions (e.g. an approximate analytical solution).
- Considering that there are simple ways of determining certain quantities (e.g. the standard deviation can be easily determined from the sum and sum of squares of data values; a moving average can be determined by adding a new value and subtracting the last value; or an exponential average can be determined by multiplying the previous value of the average with a factor $q < 1$ and adding $(1-q)$ times the new value).

The performance is often measured by the analysis of how the required computational time increases with the number N of agents in the system. In the ideal case, such a *scalability analysis* gives a linear (or constant) dependency on the "system size" N. In many cases, however, the computational time scales like a polynomial, or, what is even worse, like an exponential function. Hence, evaluating how a computer code scales with system size allows one to distinguish efficient implementations from inefficient ones. In computer science, the performance of algorithms is expressed by the "complexity" of an algorithm. For example, the term NP-hard refers to an algorithm which does not scale polynomially, which means that the computer time required for simulations explodes with the system size. In such cases, only moderate system sizes are numerically tractable on a PC. Larger systems may still be treated by parallelization of the computer code and parallel processing on dozens or hundreds of processors, but NP-hard problems can be too demanding even for the biggest supercomputers. However, it is sometimes possible to reduce the complexity considerably by applying reasonable approximations. For example, the simulation of pedestrian crowds can be significantly accelerated by assuming that pedestrians do *not* interact if their distance is greater than a certain value. Moreover, many optimization problems can be approximately solved by using suitable "heuristics" [122, 145–147].

2.3.4 *Presentation of Results*

2.3.4.1 Reproducibility

As it is customary in other scientific areas, the result of multi-agent simulations must be presented in a way that allows other scientists to reproduce the results without having to ask the authors for details. In the ideal case, the source code underlying the computer simulations is published as supplementary information in a well documented form.

In order to be reproducable, a publication must contain all the information discussed in Sect. 2.3.2 (including the initial and boundary condition, kind of interaction network, and model parameters). Furthermore, it must be specified how the noise was implemented. The update rule (such as parallel or random sequential update) and the order of update steps must be provided as well as the full set of rules underlying the agent-based model. Any relevant approximations must be pointed out, and it may make sense to specify the numerical solution method and the way, in which random numbers and statistical distributions were produced.

For the sake of comfort, one should consider to provide parameter values in tables or figure captions. Moreover, besides specifying them, it is desirable to use meaningful names for each parameter and to explain the reasons, why the rules underlying the multi-agent simulation, certain initial and boundary conditions, particular network interactions, etc. were used.

2.3.4.2 Plausibility Considerations

As underlined before, there are quite a number of mistakes that can be made in multi-agent simulations (see Sects. 2.3.1.3 and 2.3.2). Therefore, the computer code and its single subroutines should be carefully checked. Moreover, it should be described what plausibility checks have been performed. For example, as has been pointed out in Sect. 2.3.1.2, the model may have exact or approximate solutions in certain limiting cases (e.g. when setting certain parameters to 0, 1 or very large values).

Furthermore, the computational results should have the right order of magnitude, and change in a plausible way over time or when the model parameters are modified. In addition, one should take into account that some variables are restricted to a certain range of values. For example, probabilities must always be between 0 and 1. Besides, there may be so-called "constants of motion". For example, probabilities must add up to 1 at any point in time, or the number of vehicles in a closed road network must stay the same. There may also be certain quantities, which should develop monotonously in time (e.g. certain systems have an entropy or Lyapunov function). All these features can be used to check the plausilibilty of simulation results. It may be enough to determine the values of such quantities every 1,000 time steps.

Finally, any unexpected results must be tracked back to their origin, to make sure they are well understood. Seemingly paradoxical results must be carefully studied, and their origins and underlying mechanisms must be clearly identified and explained.

2.3.4.3 Error Bars, Statistical Analysis, and Significance of Results

Like empirical data, the data resulting from multi-agent simulations should be subject to statistical analysis. In particular, it is not sufficient to present single simulation runs or mean values of many simulation runs. No matter how far apart two mean values may be, this does not necessarily mean that the difference is statistically significant. The judgment of this requires a proper statistical analysis (such as a variance analysis) [148].

A minimum requirement to judge the significance of results is the presentation of error bars (and it should be stated whether they display one, two, or three standard deviations). Box plots (i.e. presenting the median, minimum and maximum value, and quantiles) are likely to give a better picture. Based on error bars or box plots, the significance of simulation results (and differences between different simulation settings) can be often visually assessed, but a thorough statistical analysis is clearly favorable.

When performing statistical analyses, it must be taken into account that the frequency distributions resulting in multi-agent simulations may not be of Gaussian type. One may find multi-modal distributions or strongly skewed, "fat-tail" distributions such as (truncated) power laws. We also point out that fitting power laws is tricky and that straight-forward fitting approaches may easily lead to wrong exponents and confidence intervals. Besides, one speaks of a power law only if there is a linear relationship in a log-log plot over at least two orders of magnitude (i.e. over a range of the horizontal axis that spans a factor 100). Deviations from power laws may be as meaningful as the power laws themselves and should be pointed out in the related discussion of results.

2.3.4.4 Visualization

Besides presenting tables with statistical analyses, visualization is a useful way of presenting scientific data, which is widely applied in the natural and engineering sciences. Some classical ways of reprensation are

- Time-dependent plots
- Two- or three-dimensional plots indicating the interdependencies (or correlations) between two or three variables (which may require to project the full parameter space into two or three dimensions)
- Pie charts or frequency distributions
- Representations of relative shares (percentages) and their changes over time

- Snapshots of spatial distributions or videos of their development in time
- Illustrations of network dependencies

In some sense, visualization is the art of transferring relevant information about complex interdependencies to the reader quickly and in an intuitive way. Today, large-scale data mining and massive computer simulations steadily create a need for new visualization techniques and approaches (see [111] for a more detailed discussion).

2.3.4.5 Scanning of Parameter Spaces and Phase Diagrams

An important way of studying the behavior of socio-economic systems with agent-based models is the scanning of the parameter space. As was pointed out in Sect. 2.1.5, systems composed of many non-linearly interacting agents are likely to produce a number of self-organized or emergent phenomena. It is, therefore, interesting to determine the conditions under which they occur. In particular, it is relevant to identify the separating lines (or hypersurfaces) in the parameter space that separate different system behaviors ("system states") from each other, and to determine the character of the related "phase transitions": One typically checks whether it is a continuous ("second-order") transition, at which systemic properties start to change smoothly, or a discontinuous ("first-order") transition, where a sudden regime shift occurs (see Fig. 2.2). The latter transitions occur, for example, in cases of hysteresis ("history dependence"), and their possible types have been studied by "catastrophe theory" [149] (see Fig. 2.2). Scientific paradigm shifts, for example, are typically first-order transitions [150], as are revolutions in societies [151, 152].

The parameter dependencies of the different kinds of system states ("phases") and their separating lines or hypersurfaces are usually represented by "phase diagrams". A particularly interesting case occurs, if the system displays multi-stability, i.e. were different stable states are possible, depending on the respective initial condition or history. For example, in traffic theory, various kinds of congested traffic states may result, depending on the size of perturbations in the traffic flow (see Fig. 2.3). That circumstance allows one to understand systems, which show apparently inconsistent behaviors. In fact, it is quite common for social systems that a certain characteristic system behavior is reported in one part of the world, while another one is observed elsewhere [153]. There may nevertheless be a common theory explaining both system behaviors, and history-dependence may be the reason for the different observations.

Scanning parameter spaces typically requires large computational resources. Even when using a computer pool, it may require weeks to determine a two-dimensional phase diagram at reasonable accuracy. Varying more parameters will consume even more time. However, one way of determining interesting areas of the parameter space is to use the "overlay method", which simulates interactions in two-dimensional space, but additionally varies the parameter values in horizontal

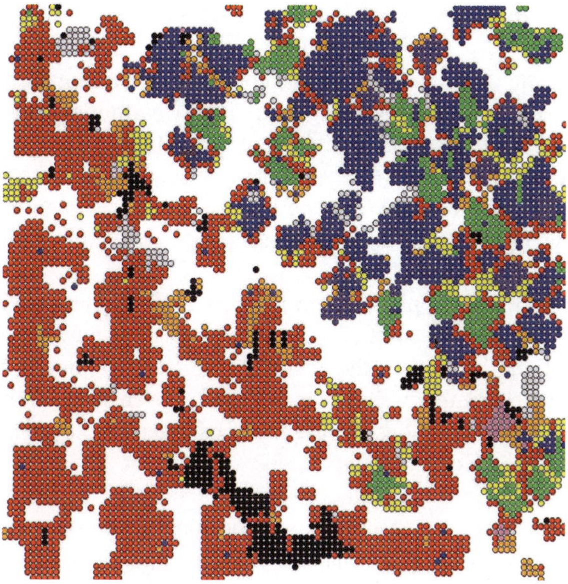

Fig. 2.4 Overlay phase diagram for an evolutionary public goods game with success-driven migration as well as prosocial and antisocial punishment, representing different behaviors by different colors (recent work with Wenjian Yu; see Chap. 8 for an explanation of the different behaviors such as 'moralists' and 'immoralists'). The prevalent behavior depends on the size of the model parameters (here: the punishment cost and fine), which is varied along the axes. One can see that people showing the same behavior tend to form clusters (a phenomenon called "homophily"). Moreover, cooperators (mainly *green* moralists and a few *blue* non-punishing cooperators) spread above a hyperbolic kind of line. Below it, defectors (*red or black*) florish. The spreading of moralists above a certain punishment level gets rid of the conventional free-rider and the second-order free-rider problem. Mobility speeds up the convergence to the finally resulting strategy distribution. It also increases the green area of moralists, i.e. it pushes the hyperbolic separation line to lower punishment values. Defectors who punish non-punishers (*grey*) occur around the separating line. Defectors who punish defectors (*yellow* immoralists) occur in separation from each other (as loners). They require enough space, which they mainly find at low densities, or when mobility creates areas of low density. In the mixed phase of *black* and *red*, and in the mixed phase of *blue* and *green*, there is only a slow logarithmic coarsening, because the payoffs are the same. (This looks like a coexistence of two strategies, if the simulations are not run long enough.) The occurence of *black* defectors who punish cooperators can explain the existence of antisocial punishers. Black antisocial punishers can exist basically at all punishment levels, if they can cluster together and are not in direct neighborhood to moralists

and vertical direction. In this way, one may get a quick impression of the spatio-temporal dynamics in different areas of the considered two-dimensional parameter space (see Fig. 2.4). After identifying the approximate location of separating lines between different phases (i.e. qualitatively different system behaviors), a fine-grained analysis (e.g. scaling analysis) can be made to reveal the detailed behavior next to the phase separation lines (or hysersurfaces).

2.3.4.6 Sensitivity Analysis

Preparing phase diagrams already provide good hints regarding the sensitivity of an agent-based model to parameter changes. Generally, within a given phase, there is not much variation of the system behavior, given the chosen parameter combinations are not too close to the lines (or hypersurfaces) separating it from other phases. However, when a transition line (or hypersurface) is crossed, significant changes in the system behavior are expected, particularly if the transition is of first order (i.e. discontinuous).

Beyond determining the phase diagram, the sensitivity can also be measured in terms of Theil's inequality coefficient [154]. It measures how different two time-dependent solutions are, when the model parameters (or initial conditions) are slightly changed. In a similar way, one may study how sensitive the model is towards the consideration of fluctuations ("noise"). A characterization can also be made by determining the Lyapunov exponents [155].

However, a multi-agent simulation may not only be sensitive to parameter changes. It may also be sensitive to minor modifications of the agent-based model itself. For example, slight changes of the interaction network (by adding or subtracting nodes or links) may impact the system behavior. Analyses of failure and attack tolerance demonstrate this very well [156]. To investigate so-called k-failures, one randomly removes k agents or links from the system and studies changes in the system performance. Similarly, one may investigate the impact of adding k links or agents. The method is capable to reveal certain kinds of "structural instabilities". A further kind of structural instabilities may be caused by modifications in the rules determining the interactions of agents. Such modifications may reflect innovations, but also inaccuracies of the model as compared to reality. For example, "unknown unknowns" are factors overlooked in the model, but they may be discovered to a certain extent by varying the model assumptions. They may also be identified by comparing models of different researchers, focusing on their incompatible features.

2.3.5 Identification of the Minimum Model Ingredients

One important part of scientific analysis is the identification of the minimum set of rules required to explain certain empirical observations or "stylized facts" derived from them (i.e. simplified, idealized, characteristic features). Particularly in models with many parameters, it is sometimes difficult to understand the exact mechanism underlying a certain phenomenon. Therefore, one should attempt to successively reduce the model to a simpler one with less terms, parameters and/or variables, in order to find out under what conditions the phenomenon of interest disappears. It is clear that simplifications of the model will often reduce the level of quantitative agreement with empirical data. However, in many cases one is mainly interested in questions such as:

- Does the system have multiple stable states?
- Does the model behave in a history-dependent way?

- Does it produce an oscillatory behavior or a stable equilibrium?
- Is the statistical distribution Gaussian, bimodal, multi-modal or heavily skewed, e.g. a (truncated) power law?
- What kinds of observed patterns can the model reproduce?
- Is a linear or an equilibrium model sufficient to reproduce the observations, or does it miss out important facts?
- Are spatial (neighborhood) interactions important or not?
- Is a heterogeneity of agent properties relevant for the explanation or not?
- Does small or moderate noise have a significant influence on the system behavior or not?
- Are correlations important, or is a mean field approximation ("representative agent approach"), assuming well-mixed interactions good enough?
- How important are the degree distribution or other specifics of the interaction network?

From statistical physics, it is known that all these factors *may* play a significant role for the system behavior [13], but this is not always the case. Therefore, the required ingredients of a model and appropriate level of sophistication very much depend on the phenomena to be described, on the purpose of the model, and the desired accuracy.

2.3.6 Gaining an Analytical Understanding

Besides providing a clearer understanding and intuition, simplifications also have another advantage: they can make models mathematically better tractable. In order to derive the essential phenomena in an analytical way (e.g. by means of stability or perturbation analyses), a radical simplification may be needed, and as a consequence, the resulting model will usually not reproduce empirical details, but just the stylized facts (such as the occurrence of certain kinds of instabilities, patterns, phase transitions/regime shifts, or phase diagrams with certain topological features) (see [27] for an example).

One advantage of analytical tractability is the circumstance that one can often derive parameter dependencies or scaling relationships. Frequently, such parameter dependencies are not obvious, and even a numerical analysis may not give a good picture, if several parameters are involved.

Analytical treatments often allow one to determine the location of stationary points and their stability properties. From this information, one can derive the fundamental features of the phase diagram, which gives a pretty good picture of the possible system behaviors. Therefore, the fundamental properties of a system may indeed be analytically understood. This is nicely illustrated by the example of multi-population evolutionary games [151]. Also the properties of freeway traffic flow and the possible congestion patterns have been analytically understood, despite their complexity [27].

2.3.7 *Some Problems and Limitations of Computational Modeling*

Despite all the virtues of mathematical modeling, one should not forget some possible problems. So far, it is not known what phenomena can be understood by agent-based models, and what are the fundamental limits of this approach. It is conceivable that there exist phenomena, which are irreducibly complex [11]. For example, the method of physics to reduce most observations to the behavior of individual particles and pair interactions may not be fully appropriate in socio-economic systems. Some phenomena require a more integrated treatment of the interactions between many agents. Public good games are just one example [157]. Recent models of pedestrian interactions are also turning away from pair interaction approaches in favor of heuristics that respond to an integrated visual pattern [158]. The corresponding behaviors can still be treated by agent-based models, but one must be aware that they may have fundamental limitations as well.

Overestimating the power of models can be quite harmful for society, as the financial crisis has shown. For this reason, it is important to state known limitations of a model or, in other words, its range of validity. It should be made clear what the purpose of a particular model is, e.g. whether it serves to understand stylized facts or scientific puzzles better, or whether the model aims at predictions or real-world applications. When it comes to "predictions", it should be said whether they are meant to be "forecasts" (in time) or "model implications" (in the sense of system states that are expected to occur when model parameters or initial conditions etc. are changed in a certain way).

In order to assess the reliability of a model, it is favorable to derive a number of predictions (or implications) of a model, which may later be verified or falsified. It is also good to point out advantages *and* disadvantages with regard to other existing models. This can give a better picture of the strengths and weaknesses of a certain approach, and guide further research aimed at developing models that can consistently explain a large set of empirical observations. It also helps to be aware of crucial modeling assumptions, on which the validity of the model implications depends. This is particularly important for models, which are practically applied.[4]

It must be underlined that specifying a model correctly is not simple. In many cases, different plausible mechanisms may exist that promise to explain the same observations [159]. For example, there are many possible explanation of power laws [160]. Moreover, if empirical data vary considerably (as it is common for socio-economic data), it may be difficult to decide empirically, which of the proposed models is the best [138, 139]. It may very well be that all known models are wrong, or that their parameter specification is wrong (as it happened in financial risk models that were calibrated with historical data). Actually, due to the implicit

[4]One should be aware that this may sooner or later happen to any model, if it promises to be useful to address real-world phenomena.

simplifications and approximations of most models, this is true most of the time [161], and this is why it is so important to state what a model is aimed to be useful for, and what are its limits. In other words, usually, models contain a grain of truth, but are not valid in every single respect. Consequently, a pluralistic modeling approach makes sense [11], and overlaying the implications of several models may give better results than the best available model itself (due to the "wisdom of crowds" effect or the law of large numbers).

2.4 Practical Application of Agent-Based Models: Potentials and Limitations

Having discussed potential problems and limitations of computational modeling in several passages of Sect. 2.3.3, and in particular Sect. 2.3.7, we should not forget to point out, where agent-based models of socio-economic systems may actually be more powerful than one would think.

2.4.1 Stylized Facts and Prediction in Socio-Economic Systems

Some people think that the lack of collections of stylized facts in the socio-economic sciences may be a result of the non-existence of such facts due to the great degree of flexibility of social and economic systems. However, there *are* actually a number of stylized facts with a surprisingly large range of validity:

1. The Fisher equation of financial mathematics (which determines the relationship between nominal and real interest rates under inflation) [162]
2. The fat tail character of many financial and economic distributions [163, 164]
3. The Matthew effect (i.e. the rich-gets-richer effect) [113, 160]
4. Dunbar's number (limiting the number of people one can have stable social relations with) [109]
5. Pareto's principle (according to which roughly 80% of an effect comes from about 20% of the causes) [165]
6. Zipf's law (determining the distribution of city rank sizes and many other things) [166]
7. The gravity law (describing the distribution of trade flows and migration) [167–169]
8. Goodhart's law (according to which any observed statistical regularity, e.g. a risk model, breaks down once pressure is placed upon it for control purposes) [170, 171]

Complementary, a list of stylized facts regarding social norms can be found in [43].

Even when accepting that the above "laws" *tend* to apply,[5] many people doubt the possibility to predict the behavior of socio-economic systems, while they believe that this is a precondition for crisis relief. However, both is not exactly true:

1. As pointed out before, there are many other purposes of modeling and simulation besides prediction [12]. For example, models may be used to get a picture of the robustness of a system to perturbations (i.e. applied to perform simulated "stress tests", considering the effect of interactions among the entities constituting the system).
2. "Model predictions" should be better understood as "model implications" rather than "model forecasts", e.g. a statements that say what systemic outcomes (e.g. cooperation, free-riding, or conflict) are expected to occur for certain (regulatory) boundary conditions [14,54] (see Sect. 2.4.2.2).
3. It is important to recognize that forecasts *are* possible *sometimes*. A famous example is Moore's law regarding the performance of computer chips (which is ultimately a matter of the innovation rate [172]). Moreover, while the detailed ups and downs of stock markets are hard to predict, the manipulation of interest rates by central banks leads, to a certain extent, to foreseeable effects. Also the sequence in which the real-estate market in the US affected the banking system, the US economy, and the world economy was quite logical. That is, even though the exact timing can often not be predicted, causality networks allow one to determine likely courses of events (see Fig. 2.5). In principle, this enables us to take counter-actions in order to avoid or mitigate the further spreading of the crisis [173].
4. As weather forecasts show, even unreliable short-term forecasts can be useful and of great economic value (e.g. for agriculture). Another example illustrating this is a novel self-control principle for urban traffic flows, which was recently invented [40, 174]. Although its anticipation of arriving platoons is of very short-term nature, it manages to reduce travel times, fuel consumption, and vehicle emissions.

In conclusion, prediction is limited in socio-economic systems, but more powerful than many people believe. Moreover, in certain contexts, it is not necessary to forecast the course of events. For example, in order to reduce problems resulting from bubbles in financial markets, it is not necessary to predict the time of their bursting or even to know about their concrete existence. The reason for this will be explained in Sect. 2.4.2.3. Furthermore, Sec.2.4.2.2 shows how the often raised problem of self-destroying prophecies can be avoided.

[5] Probably, nobody would claim that they are always true.

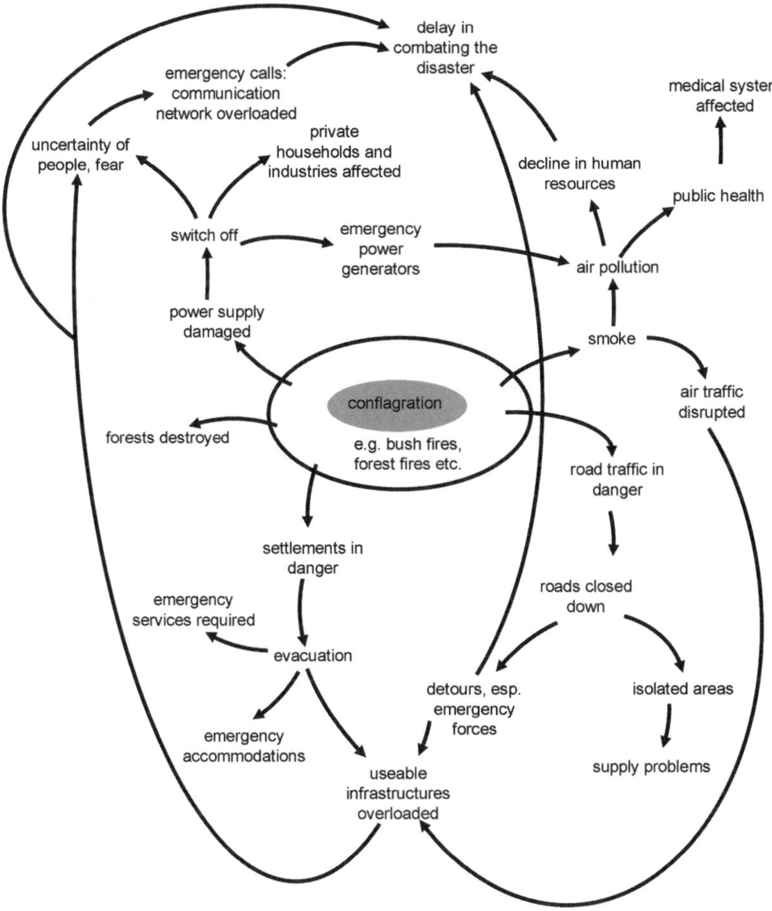

Fig. 2.5 Illustration of cascading effects in techno-socio-economic systems triggered by forest fires (after [173]). Note that the largest damage of most disasters is caused by cascading effects, i.e. the systemic impact of an over-critical local perturbation

2.4.2 Possibilities and Limitations in the Management of Socio-Economic Systems

2.4.2.1 Paradigm Shift from Controlling Systems to Managing Complexity

When trying to improve socio-economic systems, first of all, it must be stressed that the idea to *control* socio-economic systems is not only inadequate – it is also not working well. As socio-economic systems are complex systems, cause and effect are usually not proportional to each other. In many cases, complex systems tend to resist manipulation attempts (cf. "Goodhart's law"), while close to so-called "tipping points" (or "critical points"), unexpected "regime shifts" ("phase transitions", "catastrophes") may happen. Consequently, complex systems cannot be controlled like a technical system (such as a car) [29].

The above property of systemic resistance is actually a result of the fact that complex systems often self-organize, and that their behavior is robust to not-too-large perturbations. While forcing complex systems tends to be expensive (if systemic resistance is strong) or dangerous (if an unexpected systemic shift is caused), the alternative to support the self-organization of the system appears to be promising. Such an approach "goes with the flow" (using the natural tendencies in the system) and is resource-efficient. Therefore, a reasonable way to manage complexity is to guide self-organization or facilitating coordination [29, 175].

In a certain sense, this self-organization or self-control approach moves away from classical regulation to mechanism design [176]. Regulation often corresponds to changing the boundary conditions, while mechanism design changes the interactions in the system in a way that reduces instabilities (e.g. due to delays) and avoids that the system is trapped in local optima (and continues its evolution to a system-optimal state). For example, slightly modifying the interaction of cars by special driver assistant systems can stabilize traffic flows and avoid bottleneck effects to a certain degree [177].

2.4.2.2 Self-destroying Prophecies and Proper Design of Information Systems

It is often pointed out that socio-economic systems would not be predictable, because the reaction of people to information about the system would destroy the validity of the forecast. If done in the wrong way, this is actually true. Let us illustrate this by an example: Assume that all car drivers are given the same information about existing traffic jams. Then, drivers may most likely over-react, i.e. more drivers may use an alternative road than is required to reach a system-optimal distribution of traffic flows [178].

However, as has been shown by laboratory route choice experiments, an almost optimal route choice behavior may be reached by an information system that gives user-specific recommendations. In other words, some drivers would be asked to stay on the congested road, and others to leave it. When the recommendation system compensates for the fact that not everyone follows the recommendations, one can avoid over- or under-reactions of drivers in congestion scenarios [178].

One crucial issue of such individualized recommender systems, however, is their reliability. An unreliable system or one that is systematically biased, will be only poorly followed, and people will eventually compensate for biases [178]. Therefore, it is also essential to design the information system in a way that is fair to everyone. That is, nobody should have a *systematic* advantage. Nevertheless, the system should be flexible enough to allow a trading of temporary (dis)advantages. For example, somebody who was asked to take the slower road on a given day (and who would have a right to use the faster road on another day), may still use the faster road. However, he or she would have to pay a fee for this, which would be earned by somebody else, who would exchange his or her "ticket" for the faster road for a "ticket" for the slower road. In such a way, the system optimum state could still be maintained [178].

In summary, the above described information system would cheat nobody, and it would be flexible and fair. Only people who use the faster road more often than average would have to pay a road usage fee. A normal driver would either pay nothing on average: While he or she will pay on some days (when being under a pressure of time, while the recommendation asks to take the slower road), the same amount of money can be earned on other days by taking the slower road. In other words, fair usage would be free of charge on average, and drivers would still have a freedom of route choice. The primary goal of the system would not be to suppress traffic flows through road pricing, but the pricing scheme would serve to reach a system optimal traffic state.

2.4.2.3 New Approaches and Designs to Manage Complexity

It is quite obvious that there is no single scheme, which allows one to manage *all* kinds of complex systems optimally, independently of their nature. The success of a management concept very much depends on the characteristics of the system, e.g. its degree of predictability. The systems design must account for this.

If *long-term forecasts* are possible, there must obviously be an almost deterministic relationship between input- and output-variables, which allows one to change the temporal development and final outcome of the system. If long-term predictability is *not* given, management attempts must be oriented at a sufficiently frequent re-adjustment, which requires a suitable monitoring of the system.

As the example of weather forecasts shows, even unreliable *short-term forecasts* can be very useful and economically relevant (e.g. for agriculture).[6] The success principle in case of short-term forecasts is the flexible adjustment to the local conditions, while well predictable systems often perform well with a fixed (rather than variable) organization principle.

In systems where *no forecasts* over time are possible at all, it may still be feasible to improve the system behavior by modifying the statistical distribution. Taking the time-dependencies at stock markets for illustration, introducing a Tobin tax would reduce excessive levels of speculation (e.g. high-frequency trading). Moreover, introducing "noise" (further sources of unpredictability), could destroy undesirable correlations and impede insider trading [179].

[6] Another example is the "self-control" of urban traffic flows, which is based on a special, traffic-reponsive kind of decentralized traffic light control [40], see Sect. 2.4.2.1.

These are just a few examples illustrating that there actually *are* possibilities to influence systems involving human behavior in a favorable way. A more detailed discussion of the issue of managing complexity is given in [29, 180–183].

2.4.2.4 Changing the Rules of the Game and Integrative Systems Design

Generally, there are two ways of influencing the dynamics and outcome of a system by changing the "rules of the game". If the interactions in the system are weak, the system dynamics can be well influenced by modifying the boundary conditions of the system (i.e. by regulatory measures). However, if the interactions are strong, as in many social and economic processes, the self-organization of the system dominates the external influence. In this case, a modification of interactions in the system ("mechanism design") seems to be more promising. (A good example for this is a traffic assistance system that reduces the likelihood of congestion by special driver assistance systems [177].) Of course, regulatory measures and mechanism design may also be combined with each other.

While mechanism design is relatively common in computer science and some other areas (e.g. in evolutionary game theory, mathematics, and partly in physics), it seems that these methods have not been extensively applied in the social sciences so far. For example, there are many different mechanisms to match supply and demand, and it would be interesting to know what systemic properties they imply (such as the level of stability, the efficiency of markets, the resulting wealth distribution, the creation of investment opportunities, etc.). It is also not clear how to reach the best combination of top-down and bottom-up elements in decision-making processes, and how to find the best balance between centralized and decentralized coordination approaches. All this poses interesting and practically relevant challenges that determine the prosperity and well-being of societies (see also [111]).

Moreover, in the past, mechanism design has been applied basically to subsystems, i.e. parts of the complex overall system we are living in. However, due to the interconnection of all sorts of (sub-)systems (e.g. of the traffic, supply, industrial, environmental, health and social systems), measures in one (sub)system may have undesired side effects on other (sub)systems. In fact, for fundamental reasons, it is quite frequent that taking the best action in one (sub)system affects another (sub)system in a negative way. Such partial improvements will usually not promote an optimal state of society (whatever the optimal state of society may be). A good example for this is the poor performance of traffic light controls that optimize locally (without a coordination with neighboring intersections) [177]. Unfortunately, such kinds of problems are not at all restricted to traffic systems. Undesirable feedback effects (like spill-over effects and mutual obstructions) are quite common for many networked systems, such as logistic or production systems, or even administrative processes [174].

2.4.2.5 Choice of the Goal Function

Improving a socio-economic system is far from trivial. Even if one would have a perfectly realistic and predictive model, the result may largely depend on the chosen goal function. An improper choice of the goal function can cause more harm than benefit. For example, maximizing the efficiency of a system may make it vulnerable to breakdowns [30]. Besides, there is a danger of misusing models to promote individual interests that are not compatible with human well-being. However, the following goals appear to be widely acceptable:

- Increase the self-awareness of society
- Reduce vulnerability and risk
- Increase resilience (the ability to absorb societal, economic, or environmental shocks)
- Avoid loss of control (sudden, large and unexpected systemic shifts)
- Develop contingency plans
- Explore options for future challenges and opportunities
- Increase sustainability
- Facilitate flexible adaptation
- Promote fairness
- Increase social capital and the happiness of people
- Support social, economic and political inclusion and participation
- Balance between central and decentral (global and local) control
- Protect privacy and other human rights, pluralism and socio-bio-diversity
- Support collaborative forms of competition and vice versa ("coopetition")
- Promote human well-being

If several goals shall be promoted at the same time, the question arises how to perform such multi-goal optimization. Most optimization methods used today eliminate heterogeneity in the system, i.e. there is one optimal solution which is applied to everyone. For socio-economic systems, this appears to be particularly problematic, as it tends to reduce socio-diversity and innovation. Besides, it is promoting average performance rather than individual strengths. A way to overcome this problem is suggested in the following.

The crucial question in this connection is, how to translate the performance values X_{ij} of the alternative systems i (measured on multiple scales j by goal functions g_j) into one scale. Traditionally, this is done by weighting each criterion j or goal function with a certain factor w_j. This results in the overall individual performance

$$x_i = \sum_j w_j x_{ij}, \tag{2.1}$$

where $x_{ij} = X_{ij}/\langle X_{ij}\rangle_i$ is the value X_{ij}, scaled by the average performance $\langle X_{ij}\rangle_i$ of all alternatives i. The overall individual performance values x_i can ordered on a one-dimensional scale, i.e. ranked. Such an approach, however, promotes average performance rather than excellence, since excellence is typically characterized by extreme values on one or a few rating scales, but not on all of them.

In order to reward individual strengths of alternative system designs, one may proceed as follows: Political decision-makers could choose the weight they would like to attribute to each criterion or goal function, say $w_1 = 0.35$, $w_2 = 0.25$, $w_3 = 0.25$, and $w_4 = 0.15$ (assuming only four relevant goals in this example). An index that is favorable with respect to individual strengths, would for example be

$$y_i = \sum_j w_j x_{ij} + 0.1(y_{i1} + y_{i2} - y_{i3} - y_{i4}), \qquad (2.2)$$

where the values y_{ij} correspond to the values x_{ij}, sorted according to their size in descending order. This formula overweights the particular strengths of each individual system i, and it is possible that different alternative systems perform equally well. Putting this into the context of the European Union for illustration, each country could choose a systemic design which fits the respective national strengths best. Hence, the "pluralistic" goal function (2.2) overcomes a number of problems of the optimization methods that are predominantly used today (namely, one-dimensional ranking scales, which measure the performance in an individually non-differentiated way, which typically creates one winner and many losers).

2.5 Summary, Discussion, and Outlook

In this contribution, we have presented an overview, how to do agent-based modeling (ABM) and multi-agent simulations (MAS) properly and how to avoid a number of traps associated with this research approach. In particular, we have discussed the potentials, limitations, and possible problems. Multi-agent simulations can be used for hypothesis testing and to get a better understanding of complex systems. They are flexible and allow one to reflect many characteristic features of techno-socio-economic-environmental systems in a natural way (including heterogeneity and network interactions). As a result, one can expect insights into the different possible "states" or behaviors of a system and the preconditions for their occurrence. In particular, phase diagrams facilitate the representation of different characteristic phases and of transitions between them. Considering the possibility of multi-stability and history-dependence, phase diagram are also a promising approach to make sense of seemingly inconsistent empirical evidence.

We have underlined how essential it is to proceed carefully when modeling and simulating socio-economic systems. In particular, a publication should clearly describe:

- The research question (challenge, "puzzle", "mystery") addressed, including the purpose of the model
- The research methodology/approach used
- The assumptions underlying the agent-based model
- The current empirical or experimental evidence

- Implications or predictions that allow others to assess the explanatory power (e.g. through lab or Web experiments)
- The expected range of validity and limitations of the approach

Agent-based models can be published in a number of journals (searching for "agent-based model" or "multi-agent simulation" at http://scholar.google.com will give a good overview). However, studies presenting multi-agent simulations are currently still hard to publish in mainstream economic or social science journals, possibly because most authors do not back up their computational results with analytical ones. Proceeding more carefully when performing multi-agent simulations, as suggested in this contribution, will most likely increase the interest in this approach over time, particularly as it can produce significant results beyond the range of phenomena that are understandable through analytical methods. Finally, multi-agent simulations can also be useful to identify interesting experimental setups [142], considering in particular that experiments are costly and restricted to a small number of conditions and repetitions.

2.5.1 Future Prospects and Paradigm Shifts

Great prospects for agent-based modeling do not only result from the experience gained with multi-agent simulations, the availability of user-friendly simulation platforms, greater computer power, and improved visualization techniques. We also expect a number of paradigm shifts:

- The social sciences are currently experiencing a transition from a data-poor to a data-rich situation. This allows one to verify or falsify models, calibrate their parameters, and to move to data-driven modeling approaches [1, 158, 184]. Moreover, it will be possible to improve the level of detail, accuracy, and scale of agent-based models by orders of magnitude. At the same time, thanks to the availability of user-friendly simulation tools, the development times for multi-agent simulations will shrink dramatically.
- The application of methods from statistical physics and the theory of complex systems to socio-economic data generates a chance of moving beyond descriptive (fit) models towards explanatory models. The improved data situation supports the comprehension of the inner relationships and significant patterns of complex systems.
- New possibilities to mine real-time data (e.g. text mining of news, blogs, twitter feeds, etc.) create the opportunity to move from measurements with a delay (such as the classical ways of determining the gross national product or the number of people, who have got the flu) towards reliable real-time estimates ("nowcasting") [185, 186]. Furthermore, using particular properties of spreading processes in networks, it seems even possible to achieve two-week forecasts (based on the method of "health sensors") [187]. More generally, "reality mining" will facilitate multi-agent simulations of realistic scenarios, the determination of

model parameters (and other relevant model inputs) on the fly, and the timely determination of advance warning signs. It will also help to avoid destabilizing delays and to increase the efficiency of crisis response measures. (Delays are a major problem in the efficiency of disaster response management and mitigation [188].)

- Multi-agent simulations will integrate measurement-based data-mining and model-based simulation approaches. This approach goes beyond feeding in real-time inputs (such as initial and boundary conditions, parameters, and network characteristics) into multi-agent simulations: it performs a data-driven pattern recognition and modeling in parallel to related computer simulations and, thereby, combines the strengths of both methods to reach the optimum accuracy and predictability. For example, integrating two incomplete sets of traffic information (cross-sectional measurements and floating car data) and a fluid-dynamic real-time traffic simulation facilitates to reduce the delay times between the formation and detection of traffic jams by 50%, and to double the reliability of such traffic information.
- It will be possible to move from a batch processing of alternative simulation scenarios to interactive real-time specifications and scenario analyses. This will facilitate to explore policy options and "parallel worlds" (i.e. possible futures), as the situation evolves and pressing decisions must be taken. For example, evacuation scenarios in response to certain disasters have to be developed and evaluated quickly. More generally, interactive supercomputing would facilitate more flexible contingency plans that are tailored to the actual situation of crisis.
- Multi-agent simulations could be directly coupled with lab and web experiments. In fact, the decisions of agents in computer simulations could be taken by real people. Serious multi-player online-games provide the opportunity of involving a large number of people into the analysis of complex data and the exploration of realistic decision-making scenarios in virtual worlds, which realistically map possible future worlds. In this way, agent-based simulation approaches may be applied for crowd sourcing and eGovernance applications, to make use of the "wisdom of crowds". For example, one could populate the three-dimensional virtual model of a new shopping center, railway station, or airport in order to find out, how well the architectures would fulfil its function, and to determine which design is favored by the future users.
- In the medium-term future, one can expect a confluence of real and virtual worlds. For example, Google Earth and similar virtual representations of the real world could be populated with simulated people or real ones. In fact, people agreeing to share their GPS coordinates could be represented in these worlds directly, to the level of detail they like. An augmented reality approach would allow people to share information about their interests, backgrounds, values, etc. The amount of information shared may be decided interactively or by the kinds of interaction partners (e.g. people are expected to share private information more openly with people they consider to be their friends). Such augmented reality tools will be able to serve as "translator" or "adaptor" for people with different languages or cultural backgrounds, helping them to make themselves

understandable to each other. The resulting techno-social systems would also offer many new opportunities for social and economic participation, both in the virtual and in the real world.

Given this development, we envision a new way of performing socio-economic research, which may be called "Social Supercomputing". This approach would facilitate the integration of different kinds of data (e.g. demographic, socio-economic, and geographic data) and different kinds of simulation approaches (e.g. agent-based and equation-based ones), at an unprecedented scale and level of detail. It would enable simulations and interactivity on all scales. In fact, the computer simulation of techno-socio-economic systems on a global scale is a vision that appears to become feasible within the next 10 to 15 years, with unprecedented opportunities for societies and economies, if done in the right way. Some of the challenges for computer scientists and other researchers on the way are described in the Visioneer White papers (particularly Sects. 3.3 and 3.4.2 of [189]).

References

1. D. Helbing, S. Balietti, From social data mining to forecasting socio-economic crisis. visioneer white paper (2010) http://www.visioneer.ethz.ch.
2. N. Gilbert, K.G. Troitzsch, *Simulation for the Social Scientist*. (Open University Press, England, 2005)
3. J.M. Epstein, R. Axtell, *Growing artificial societies*. (Cambridge, MA, 1996)
4. J.M. Epstein, *Generative Social Science: Studies in Agent-Based Computational Modeling*. Princeton Studies in Complexity. (Princeton University Press, Princeton, 2007)
5. N.R. Jennings, On agent-based software engineering. Artif. Intell. **117**(2), 277–296 (2000)
6. A.M. Uhrmacher, D. Weyns, *Multi-Agent Systems: Simulation and Applications*. (CRC Press, Inc., Boca Raton, FL, USA, 2009)
7. N. Gilbert, S. Bankes, Platforms and methods for agent-based modeling. Proc. Natl. Acad. Sci. USA **99**(Suppl 3), 7197 (2002)
8. M.W. Macy, R. Willer, From factors to actors: computational sociology and agent-based modeling. Annu. Rev. Sociol. **28**(1), 143–166 (2002)
9. J.P. Davis, K.M. Eisenhardt, C.B. Bingham, Developing theory through simulation methods. Acad. Manag. Rev. **32**(2), 480 (2007)
10. R.A. Bentley, P. Ormerod, Agents, intelligence, and social atoms, in *Creating Consilience: Integrating the Sciences and the Humanities*, ed. by M. Collard, E. Slingerland (Oxford University Press, 2011)
11. D. Helbing, Pluralistic modeling of complex systems (2010) CCSS-10-009
12. J.M. Epstein, Why model? J. Artif. Soc. Soc. Simulat. **11**(4), 12 (2008). http://jasss.soc.surrey.ac.uk/11/4/12.html
13. D. Helbing, S. Balietti, Fundamental and real-world challenges in economics. Sci. Cult. **76**, 399–417 (2010). Special Issue: 15 year of econophysics research
14. D. Helbing, A. Szolnoki, M. Perc, G. Szabó, Evolutionary establishment of moral and double moral standards through spatial interactions. PLoS Comput. Biol. **6**(4), e1000758, 04 (2010)
15. R. May, A. McLean, *Theoretical Ecology: Principles and Applications*, 3rd edn. (Oxford University Press, USA, 2007)
16. I. Foster, A two-way street to science's future. Nature **440**, 419 (2006)
17. C.S. Taber, R.J. Timpone, *Computational Modelign*. (Sage, London, 1996)

18. F. Schweitzer, *Brownian Agents and Active Particles. On the Emergence of Complex Behavior in the Natural and Social Sciences*. (Springer, Berlin, 2003)
19. D. Helbing, J. Keltsch, P. Molnar, Modelling the evolution of human trail systems. Nature **388**(6637), 47–50 (1997)
20. J.M. Epstein, Modelling to contain pandemics. Nature **460**(7256), 687 (2009)
21. J. Parker, J. Epstein, A global-scale distributed agent-based model of disease transmission, ACM Transactions on Modelling and Computer Simulation (TOMACS) **22**(1), (2011)
22. M. Treiber, A. Hennecke, D. Helbing, Congested traffic states in empirical observations and microscopic simulations. Phys. Rev. E **62**, 1805 (2000)
23. D. Helbing, Derivation of non-local macroscopic traffic equations and consistent traffic pressures from microscopic car-following models. Eur. Phys. J. B **69**(4), 539–548 (2009)
24. R.K. Sawyer, Artificial Societies. Socio. Meth. Res. **31**(3), 325 (2003)
25. H.J. Aaron, Distinguished lecture on economics in government: Public policy, values, and consciousness. J. Econ. Perspect. **8**(2), 3–21 (1994). http://ideas.repec.org/a/aea/jecper/v8y1994i2p3-21.html
26. A.P. Kirman, Whom or what does the representative individual represent? *J. Econ. Perspect.* **6**(2), 117–36 (1992). http://ideas.repec.org/a/aea/jecper/v6y1992i2p117-36.html
27. D. Helbing, M. Treiber, A. Kesting, M. Schönhof, Theoretical vs. empirical classification and prediction of congested traffic states. Eur. Phys. J. B **69**(4), 583–598 (2009)
28. D. Helbing, *Verkehrsdynamik*. (Springer, Berlin, 1997)
29. D. Helbing, *Managing Complexity: Insights, Concepts, Applications*, 1st edn. (Springer, Berlin, 2007)
30. D. Helbing, Systemic risks in society and economics (2009). Sante Fe Institute, working paper 09-12-044.
31. R. Kuehne, Freeway control using a dynamic traffic flow model and vehicle reidentification techniques. Transport. Res. Record **1320**, 251–259 (1991)
32. B.S. Kerner, P. Konhäuser, Cluster effect in initially homogeneous traffic flow. Phys. Rev. E **48**(4), 2335+ (1993)
33. R.E. Chandler, R. Herman, E.W. Montroll, Traffic dynamics: studies in car following. Oper. Res., 165–184 (1958)
34. R. Herman, E.W. Montroll, R.B. Potts, R.W. Rothery, Traffic dynamics: analysis of stability in car following. Oper. Res. **7**(1), 86–106 (1959)
35. D. Helbing, M. Treiber, Critical discussion of "synchronized flow". Cooper. Transport. Dyn. **1**, 2.1–2.24 (2002)
36. D. Helbing, T. Seidel, S. Laemmer, K. Peters, *Econophysics and Sociophysics - Trends and Perspectives*, chapter Self-organization principles in supply networks and production systems (Wiley, Weinheim, 2006), pp. 535–558
37. D. Helbing, A. Mazloumian, Operation regimes and slower-is-faster effect in the control of traffic intersections. Eur. Phys. J. B **70**(2), 257–274 (2009)
38. Dirk Helbing, Illés J. Farkas, and Tamás Vicsek, Freezing by Heating in a Driven Mesoscopic System. *Phys. Rev. Lett.* **84**(6), 1240–1243 (2000)
39. D. Helbing, W. Yu, The outbreak of cooperation among success-driven individuals under noisy conditions. Proc. Natl. Acad. Sci. **106**(10), 3680–3685 (2009)
40. S. Laemmer, D. Helbing, Self-control of traffic lights and vehicle flows in urban road networks. JSTAT (2008). P04019
41. Michael Mäs, Andreas Flache, and Dirk Helbing, Individualization as Driving Force of Clustering Phenomena in Humans. PLoS Comput. Biol. **6**(10), e1000959+ (2010)
42. J. Lorenz, H. Rauhut, F. Schweitzer, D. Helbing, How social influence undermines the wisdom of crowds. Proc. Natl. Acad. Sci. USA (PNAS) **108**(22), 9020–9025 (2011)
43. D. Helbing, W. Yu, K.-D. Opp, H. Rauhut, The emergence of homogeneous norms in heterogeneous populations. Santa Fe Working Paper 11-01-001 (2011), see http://www.santafe.edu/media/workingpapers/11-01-001.pdf, last accessed on March 6, 2012
44. C. Nardini, B. Kozma, A. Barrat, Who's talking first? consensus or lack thereof in coevolving opinion formation models. Phys. Rev. Lett. **100**(15), 158701 (2008)

45. J.S. Sichman, Depint: Dependence-based coalition formation in an open multi-agent scenario. JASSS J. Artif. Soc. Soc. Simulat. **1**(2) (1998)
46. M.E. Gaston, M. des Jardins, Agent-organized networks for dynamic team formation. In *Proceedings of the fourth international joint conference on Autonomous agents and multiagent systems*, AAMAS '05, pages 230–237, New York, NY, USA, 2005. ACM
47. E. Bonabeau, M. Dorigo, G. Theraulaz, Inspiration for optimization from social insect behaviour. Nature **406**(6791), 39–42 (2000)
48. G. Szabo, G. Fath, Evolutionary games on graphs. Phys. Rep. **446**(4-6), 97–216 (2007)
49. P.J. Carrington, J. Scott, S. Wasserman, *Models and Methods in Social Network Analysis.* (Cambridge University Press, Cambridge, 2005)
50. P. Holme, G. Ghoshal, Dynamics of networking agents competing for high centrality and low degree. Phys. Rev. Lett. **96**(9), 098701 (2006)
51. M. Sierhuis, J.M. Bradshaw, A. Acquisti, R. van Hoof, R. Jeffers, A. Uszok, Human-Agent Teamwork and Adjustable Autonomy in Practice. In *Proceedings of the seventh international symposium on artificial intelligence, robotics and automation in space (I-SAIRAS)*, 2003
52. Robert Axelrod, *The Complexity of Cooperation: Agent-Based Models of Competition and Collaboration*, 1st printing edn. (Princeton University Press, Princeton, 1997)
53. S. Bowles, H. Gintis, The evolution of strong reciprocity: cooperation in heterogeneous populations. Theor. Popul. Biol. **65**(1), 17–28 (2004)
54. D. Helbing, A. Johansson, Cooperation, norms, and conflict: A unified approach (2009) SFI Working Paper
55. Heiko Rauhut and Marcel Junker, Punishment deters crime because humans are bounded in their strategic decision-making. J. Artif. Soc. Soc. Simulat. **12**(3), 1 (2009)
56. L-E. Cederman, Modeling the size of wars: From billiard balls to sandpiles. Am. Polit. Sci. Rev. **97**, 135–150 (2003)
57. L-E. Cederman, *Emergent Actors in World Politics: How States and Nations Develop and Dissolve*. (Princeton University Press, Princeton, NJ, 1997)
58. C. Hommes, Chapter 23 Heterogeneous Agent Models in Economics and Finance, in *Handbook of Computational Economics*, vol. 2 (Elsevier, 2006), pp. 1109–1186
59. B. LeBaron, Building the santa fe artificial stock market. *Physica A*, 2002. Working Paper, Brandeis University.
60. M. Raberto, S. Cincotti, S.M. Focardi, M. Marchesi, Agent-based simulation of a financial market. Phys. Stat. Mech. Appl. **299**(1-2), 319–327 (2001)
61. J. Zhang, Growing Silicon Valley on a landscape: an agent-based approach to high-tech industrial clusters. *Entrepreneurships, the New Economy and Public Policy*, pp. 71–90 (2005)
62. Robert Axtell, The emergence of firms in a population of agents. Working Papers 99-03-019, Santa Fe Institute, March 1999
63. T. Kaihara, Multi-agent based supply chain modelling with dynamic environment. Int. J. Prod. Econ. **85**(2), 263–269 (2003)
64. C. Preist, Commodity trading using an agent-based iterated double auction, in *Proceedings of the third annual conference on Autonomous Agents* (ACM 1999), pp. 131–138.
65. Lebaron, Blake, Tesfatsion, Leigh, Modeling macroeconomies as open-ended dynamic systems of interacting agents. Am. Econ. Rev. **98**(2), 246–250 (2008)
66. Leigh Tesfatsion, Agent-based computational economics: growing economies from the bottom up. Artif. Life **8**(1), 55–82 (2002)
67. J.R. Harrison, Z. Lin, G.R. Carroll, K.M. Carley, Simulation modeling in organizational and management research. Acad. Manag. Rev. **32**(4), 1229 (2007)
68. B.S. Onggo, Parallel discrete-event simulation of population dynamics. In *Proceedings of Winter Simulation Conference 2008* (Miami, FL, USA), pp. 1047–1054
69. Y. Mansury, M. Kimura, J. Lobo, T.S. Deisboeck, Emerging patterns in tumor systems: simulating the dynamics of multicellular clusters with an agent-based spatial agglomeration model. J. Theor. Biol. **219**(3), 343–370 (2002)
70. D. Osullivan, J.M Macgill, C. Yu, Agent-based residential segregation: a hierarchically structured spatial model. *Agent 2003 Challenges in Social Simulation* (2003)

71. M. Batty, *Cities and Complexity: Understanding Cities with Cellular Automata, Agent-Based Models, and Fractals*. (The MIT Press, Cambridge, 2007)
72. D. Helbing, K. Nagel, The physics of traffic and regional development. Contemp. Phys. **45**(5), 405–426 (2004)
73. V. Killer, K.W. Axhausen, D. Guth, C. Holz-Rau, Understanding regional effects of travel times in switzerland and germany 1970-2005. Jonkoping, Sweden, August 2010. 50th European Regional Science Association (ERSA)
74. C.R. Binder, C. Hofer, A. Wiek, R.W. Scholz, Transition towards improved regional wood flows by integrating material flux analysis and agent analysis: the case of appenzell ausserrhoden, switzerland. Ecol. Econ. **49**(1), 1–17 (2004)
75. H. Dia, An agent-based approach to modelling driver route choice behaviour under the influence of real-time information. Transport. Res. C Emerg. Tech. **10**(5-6), 331–349 (2002)
76. C.M. Henein, T. White, Agent-based modelling of forces in crowds, in *Multi-agent and Multi-agent-Based Simulation* (Springer, Berlin, 2005), pp. 173–184
77. M. Batty, Agent-based pedestrian modelling, in *Advanced spatial analysis: the CASA book of GIS*, page 81 (2003)
78. D. Delli Gatti, *Emergent macroeconomics : an agent-based approach to business fluctuations*. (Springer, Berlin, 2008)
79. M. Aoki, H. Yoshikawa, *Reconstructing Macroeconomics: A Perspective from Statistical Physics and Combinatorial Stochastic Processes (Japan-US Center UFJ Bank Monographs on International Financial Markets)*, 1st edn. (Cambridge University Press, Cambridge, 2006)
80. R. Conte, R. Hegselmann, P. Terna, *Simulating Social Phenomena*. (Springer, Berlin, 1997)
81. R. Sun (ed.), *Cognition and Multi-Agent Interaction: From Cognitive Modeling to Social Simulation*, 1st edn. (Cambridge University Press, Cambridge, 2008)
82. N. Gilbert, *Agent-Based Models*. (Sage Publications, Inc, 2007)
83. V. Grimm, E. Revilla, U. Berger, F. Jeltsch, W.M. Mooij, S.F. Railsback, H.H. Thulke, J. Weiner, T. Wiegand, D.L. DeAngelis, Pattern-oriented modeling of agent-based complex systems: lessons from ecology. Science **310**(5750), 987 (2005)
84. R. Axelrod, Advancing the art of simulation in the social sciences, in *Handbook of Research on Nature Inspired Computing for Economy and Management*, ed. by Jean-Philippe Rennard (Hersey, PA, 2005)
85. K. Nagel, Parallel implementation of the transims micro-simulation. Parallel Comput. **27**(12), 1611–1639 (2001)
86. M. Rickert, K. Nagel, Dynamic traffic assignment on parallel computers in transims. Future Generat. Comput. Syst. **17**(5), 637–648 (2001)
87. K. Nagel, R.J. Beckman, C.L. Barrett, Transims for urban planning. Los Alamos Unclassified Report (LA-UR) 98-4389, Los Alamos National Laboratory (1999)
88. K. Nagel, C.L. Barrett, Using microsimulation feedback for trip adaptation for realistic traffic in Dallas. Santa Fe Working Paper 97-03-028 (1997), see http://www.santafe.edu/media/workingpapers/97-03-028.pdf, last accessed on March 6, 2012
89. P.M. Simon, J. Esser, and Nagel K. Simple queueing model applied to the city of portland. Int. J. Mod. Phys. C **10**(5), 941–960 (1999)
90. B. Raney, N. Cetin, A. Vollmy, M. Vrtic, Axhausen. K., K. Nagel, An agent-based microsimulation model of swiss travel: First results. Network Spatial. Econ. **3**, 23–41 (2003)
91. M. Balmer, K.W. Axhausen, K. Nagel, An agent based demand modeling framework for large scale micro-simulations (2005) Working paper, 329, Institute for Transport Planning and Systems (IVT), ETH Zurich, Switzerland.
92. D. Charypar, K. Nagel, Generating complete all-day activity plans with genetic algorithms. Transportation **32**(4), 369–397 (2005)
93. K.W. Axhausen, T. Garling, Activity-based approaches to travel analysis: conceptual frameworks, models, and research problems. Transport Rev. **12**(4), 323–341 (1992)
94. G. Laemmel, D. Grether, K. Nagel, The representation and implementation of time-dependent inundation in large-scale microscopic evacuation simulations. Transport. Res. C Emerg. Tech. **18**(1), 84–98 (2010)

95. H. Taubenboeck, N. Goseberg, N. Setiadi, G. Laemmel, F. Moder, M. Oczipka, H. Klupfel, Roland Wahl, T. Schlurmann, G. Strunz, J. Birkmann, K. Nagel, F. Siegert, F. Lehmann, S. Dech, A. Gress, R. Klein, "last-mile" preparation for a potential disaster - interdisciplinary approach towards tsunami early warning and an evacuation information system for the coastal city of padang, indonesia. Nat. Hazards Earth Syst. Sci. **9**(4), 1509–1528 (2009). http://www.nat-hazards-earth-syst-sci.net/9/1509/2009/
96. J. Min, J.E. Beyler, Brown T.H., Y. Jun Son, A.T. Jones, Toward modeling and simulation of critical national infrastructure interdependencies. IIE Trans., 57–71 (2007) special issue on Industrial Engineering and Operations Research in Homeland Security.
97. C. Barrett, S. Eubank, M. Marathe, Modeling and simulation of large biological, information and socio-technical systems: An interaction based approach, in *Interactive Computation*, ed. by D. Goldin, S.A. Smolka, P. Wegner (Springer, Berlin, 2006), pp. 353–392. 10.1007/3-540-34874-3_14
98. National Infrastructure Simulation and Analysis Center (NISAC), see http://www.lanl.gov/programs/nisac/, last accessed on March 6, 2012
99. W. Kermack, A. Mckendrick, Contributions to the mathematical theory of epidemics–i. Bull. Math. Biol. **53**(1), 33–55 (1991)
100. W. Kermack, A. Mckendrick, Contributions to the mathematical theory of epidemics–ii. the problem of endemicity. Bull. Math. Biol. **53**(1), 57–87 (1991)
101. W. Kermack, A. McKendrick, Contributions to the mathematical theory of epidemics–iii. further studies of the problem of endemicity. Bull. Math. Biol. **53**(1), 89–118 (1991)
102. V. Colizza, A. Barrat, M. Bartholemy, A. Vespignani, The role of the airline transportation network in the prediction and predictability of global epidemics. Proc. Natl. Acad. Sci. USA **103**(7), 2015–2020 (2006)
103. L. Hufnagel, D. Brockmann, T. Geisel, Forecast and control of epidemics in a globalized world. Proc. Natl. Acad. Sci. USA **101**(42), 15124–15129 (2004)
104. S. Eubank, H. Guclu, Anil K.V.S., M.V. Marathe, A. Srinivasan, Z. Toroczkai, N. Wang, Modelling disease outbreaks in realistic urban social networks. Nature **429**(6988), 180–184 (2004)
105. Artificial stock market, see http://sourceforge.net/projects/artstkmkt/, last accessed on March 6, 2012
106. U-mart. http://www.u-mart.org/, last accessed on March 6, 2012
107. Eurace. http://www.eurace.org/, last accessed on March 6, 2012
108. A. Cangelosi, D. Parisi, *Simulating the evolution of language*. (Springer, Berline, 2002)
109. R. Dunbar, *Grooming, gossip, and the evolution of language*. (Harvard University Press, Cambridge, 1998)
110. M.A. Nowak, D.C. Krakauer, The evolution of language. Proc. Natl. Acad. Sci. USA **96**(14), 8028 (1999)
111. D. Helbing, S. Balietti, How to create an innovation accelerator. Eur. Phys. J. Special Topics **195**, 101–136 (2011)
112. M. Eigen, P. Schuster, The Hypercycle. Naturwissenschaften **65**(1), 7–41 (1978)
113. R.K. Merton, The Matthew effect in science. The reward and communication systems of science are considered. Science (New York, N.Y.) **159**(810), 56–63 (1968)
114. M. Mcpherson, L. Smith-Lovin, J.M. Cook, Birds of a Feather: Homophily in Social Networks. Annu. Rev. Sociol. **27**, 415–444 (2001)
115. N.A. Christakis, J.H. Fowler, The spread of obesity in a large social network over 32 years. New Engl. J. Med. **357**(4), 370–379 (2007)
116. N.A. Christakis, J.H. Fowler, The Collective Dynamics of Smoking in a Large Social Network. New Engl. J. Med. **358**(21), 2249–2258 (2008)
117. K.P. Smith, N.A. Christakis, Social Networks and Health. Annu. Rev. Sociol. **34**(1), 405–429 (2008)
118. T. Chadefaux, D. Helbing, How wealth accumulation can promote cooperation. PLoS One **5**(10), e13471, 10 (2010)

119. D. Helbing, W. Yu, H. Rauhut, Self-organization and emergence in social systems: Modeling the coevolution of social environments and cooperative behavior. J. Math. Sociol. **35**, 177–208 (2011)
120. A.P. Fiske, *Structures of Social Life: The Four Elementary Forms of Human Relations: Communal Sharing, Authority Ranking, Equality Matching, Market Pricing*. (Free Press, 1991)
121. Mark Granovetter, The strength of weak ties: a network theory revisited. Sociol. Theor. **1**, 201–233 (1982)
122. G. Gigerenzer, P.M. Todd, ABC Research Group, *Simple Heuristics That Make Us Smart* (Oxford University Press, New York, 2000)
123. W.H. Greene, *Econometric Analysis*. (Prentice Hall, Upper Saddle River, NJ., 2008). Particularly Chap. 7.4: Model selection criteria
124. F. Diebold, *Elements of forecasting*, 3rd edn. (Thomson/South-Western, Mason, Ohio, 2003)
125. Swarm, http://www.swarm.org/, last accessed on March 6, 2012
126. Repast, recursive porous agent simulation toolkit. http://repast.sourceforge.net/, last accessed on March 6, 2012
127. Netlogo, http://ccl.northwestern.edu/netlogo/, last accessed on March 6, 2012
128. Sesam, shell for simulated agent systems. http://www.simsesam.de/, last accessed on March 6, 2012
129. Massive software. http://www.massivesoftware.com/, last accessed on March 6, 2012
130. Wikipedia: Comparison of agent-based modeling software. http://en.wikipedia.org/wiki/Comparison_of_agent-based_modeling_software, last accessed on March 6, 2012
131. S.F. Railsback, S.L. Lytinen, S.K. Jackson, Agent-based simulation platforms: review and development recommendations. Simulation **82**(9), 609–623 (2006)
132. CVS Concurrent Version System. http://www.nongnu.org/cvs/, last accessed on March 6, 2012
133. Subversion, http://subversion.apache.org/, last accessed on March 6, 2012
134. Bazaar, http://bazaar.canonical.com/en/, last accessed on March 6, 2012
135. Git - fast version control system. http://git-scm.com/
136. D. Helbing, A. Johansson, Pedestrian, crowd and evacuation dynamics, in *Encyclopedia of Complexity and Systems Science* (Springer, Berlin, 2010), pp. 6476–6495
137. D. Helbing, A. Johansson, L. Buzna, *Modelisation Du Traffic - Actes du groupe de travail*, chapter New design solutions for pedestrian facilities based on recent empirical results and computer simulations, pp. 67–88. (2003) Actes No. 104
138. E. Brockfeld, R.D. Kuehne, P. Wagner, Toward benchmarking of microscopic traffic flow models. Transport. Res. Record **1852**, 124–129 (2004)
139. E. Brockfeld, R.D. Kuehne, P. Wagner, Calibration and validation of microscopic traffic flow models. Transport. Res. Record **1876**, 62–70 (2004)
140. D. Helbing, M. Schreckenberg, Cellular automata simulating experimental properties of traffic flows. Phys. Rev. E **59**, R2505–R2508 (2010)
141. A.G. Hoekstra, J. Kroc, P.M.A. Sloot (eds.), chapter Game theoretical interactions of moving agents, in *Simulating Complex Systems by Cellular Automata* (Springer, Berlin, 2010), pp.219–239
142. D. Helbing, W. Yu, The future of social experimenting. PNAS **107**(12), 5265–5266 (2010)
143. D. Helbing, T. Platkowski, Self-organization in space and induced by fluctuations. *Int. J. Chaos Theor. Appl.* **5**(4), (2000)
144. R.J. LeVeque, *Numerical Methods for Conservation Laws*. (Birkhauser, Basel, 1992)
145. K. De Jong, Evolutionary computation: a unified approach, in *GECCO '08: Proceedings of the 2008 GECCO conference companion on Genetic and evolutionary computation* (ACM, New York, NY, USA, 2008), pp. 2245–2258
146. P. Ciaccia, M. Patella, P. Zezula, M-tree: An Efficient Access Method for Similarity Search in Metric Spaces. VLDB J. 426–435 (1997)
147. L.A. Wolsey, An analysis of the greedy algorithm for the submodular set covering problem. Combinatorica **2**(4), 385–393 (1982)

148. D. Freedman, R. Pisani, R. Purves, *Statistics*. (W. W. Norton & Company, New York, 1997)
149. E.C. Zeeman, *Catastrophe Theory: Selected Papers, 1972–1977* (Addison-Wesley, 1980)
150. T.S. Kuhn, *The Structure of Scientific Revolutions*. (University Of Chicago Press, Chicago, 1962)
151. D. Helbing, A. Johansson, Evolutionary dynamics of populations with conflicting interactions: Classification and analytical treatment considering asymmetry and power. Phys. Rev. E **81**, 016112+ (2010)
152. D. Helbing, A. Johansson, Cooperation, Norms, and Revolutions: A Unified Game-Theoretical Approach. PLoS ONE **5**, e12530 (2010)
153. Benedikt Herrmann, Christian Thöni, and Simon Gächter. Antisocial Punishment Across Societies. Science **319**(5868), 1362–1367 (2008)
154. D. Helbing, *Quantitative Sociodynamics: Stochastic Methods and Models of Social Interaction Processes*. (Springer, Berlin, 2010)
155. V. Nemytsky, V. Stepanov, *Qualitative theory of differential equations*. (Princeton University Press, Princeton NJ, 1960)
156. R. Albert, H. Jeong, A.-L. Barabasi, Error and attack tolerance of complex networks. Nature **406**, 378–382 (2000)
157. M. Olson, *The Logic of Collective Action : Public Goods and the Theory of Groups*. Harvard economic studies, v. 124. (Harvard University Press, Cambridge, MA, 1971)
158. M. Moussaïd, D. Helbing, S. Garnier, A. Johansson, M. Combe, G. Theraulaz, Experimental study of the behavioural mechanisms underlying self-organization in human crowds. Proc. Biol. Sci. B: Biol. Sci. **276**(1668), 2755–2762 (2009)
159. M. Treiber, A. Kesting, D. Helbing, Three-phase traffic theory and two-phase models with a fundamental diagram in the light of empirical stylized facts. Transport. Res. B: Methodological, **44**(8–9), 983–1000 (2010)
160. A.-L. Barabási, R. Albert, Emergence of Scaling in Random Networks. Science **286**(5439), 509–512 (1999)
161. John P. Ioannidis, Why Most Published Research Findings Are False. PLoS Med. **2**(8), e124+ (2005)
162. I. Fisher, *The Theory of Interest*. (Augustus M. Kelley Publishers, NJ, 1930)
163. B.B. Mandelbrot, The variation of certain speculative prices. J. Bus. **36**, 394–419 (1963)
164. R.N. Mantegna, E.H. Stanley, *An Introduction to Econophysics: Correlations and Complexity in Finance*. (Cambridge University Press, New York, 1999)
165. V. Pareto, *Translation of Manuale di economia politica ("Manual of political economy")*. (A.M. Kelley, New york, 1971)
166. George Zipf, *The economy of geography* (Addison-Wesley Publishing Co. Inc, Cambridge, 1949), pp.347–415
167. Ravenstein E. The birthplaces of the people and the laws of migration. *The Geographical Magazine III*, pp. 173–177, 201–206, 229–233 (1876)
168. G.K. Zipf, The p1 p2/d hypothesis: on the intercity movement of persons. Am. Soc. Rev. **11**, 677–686 (1946)
169. J. Tinbergen, *Shaping the world economy: suggestions for an international economic policy*. (Twentieth Century Fund, New York, 1962)
170. C.A.E. Goodhart, Monetary relationships: a view from Threadneedle Street. Papers monetary Econ. **1** (1975)
171. J. Danielsson, The emperor has no clothes: Limits to risk modelling. J. Bank. Finance **26**(7), 1273–1296 (2002)
172. D. Helbing, M. Treiber, N.J. Saam, Analytical investigation of innovation dynamics considering stochasticity in the evaluation of fitness. Phys. Rev. E **71**, 067101 (2005)
173. D. Helbing, H. Ammoser, C. Kuehnert, chapter Disasters as extreme events and the importance of network interactions for disaster response management, in *Extreme Events in Nature and Society* (Springer, Berlin, 2005), pp. 319–348
174. Verfahren zur Koordination von vernetzten Abfertigungsprozessen oder zur Steuerung des Transports von mobilen Einheiten innerhalb eines Netzwerkes, (2010), see http://www.patent-de.com/20100805/DE102005023742B4.html, last accessed on March 6, 2012

175. D. Helbing, A. Deutsch, S Diez, K Peters, Y Kalaidzidis, K. Padberg-Gehle, S. Laemmer, A. Johansson, G. Breier, F. Schulze, M. Zerial, Biologistics and the struggle for efficiency: concepts and perpesctives. Adv. Complex Syst. **12**(06), 533+ (2009)
176. L. Hurwicz, Optimality and informational efficiency in resource allocation processes, in *Mathematical Methods in the Social Sciences*, ed. by K.J. Arrow, S. Karlin, P. Suppes (Stanford University Press, Stanford, CA, 1960), pp. 27–46
177. A. Kesting, M. Treiber, M. Schonhof, D. Helbing, Adaptive cruise control design for active congestion avoidance. Transport. Res. C Emerg. Tech. **16**(6), 668–683 (2008)
178. D. Helbing, chapter Dynamic decision behavior and optimal guidance through information services: Models and experiments, in *Human Behaviour and Traffic Networks* (Springer, Berlin, 2004), pp. 47–95
179. D. Helbing, M. Christen, Physik der Finanzmärkte (Wirtschaftswoche, December 22, 2010)
180. N. Wiener, *Cybernetics, Second Edition: or the Control and Communication in the Animal and the Machine*. (The MIT Press, Cambridge, MA, 1965)
181. B. Fabien, *Analytical System Dynamics: Modeling and Simulation*. (Springer, Berlin, 2008)
182. S. Skogestad, I. Postlethwaite, *Multivariable Feedback Control: Analysis and Design*, 2nd edn. (Wiley-Interscience, New York, 2005)
183. A.L. Fradkov, I.V. Miroshnik, V.O. Nikiforov, *Nonlinear and Adaptive Control of Complex Systems*. (Springer, Berlin, 1999)
184. A. Johansson, D. Helbing, P. Shukla, Specification of the social force pedestrian model by evolutionary adjustment to video tracking data. Adv. Complex Syst. **10**, 271–288 (2007)
185. Google flu trends. http://www.google.org/flutrends/, last accessed on March 6, 2012
186. J. Vernon Henderson, A. Storeygard, D.N. Weil, Measuring economic growth from outer space. Technical Report 15199, National Bureau of Economic Research, July (2009) http://www.nber.org/papers/w15199.
187. N.A. Christakis, J.H. Fowler, Social network sensors for early detection of contagious outbreaks. *PloS one* **5**(9), e12948+, September (2010)
188. L. Buzna, K. Peters, H. Ammoser, C. Kuehnert, D. Helbing, Efficient response to cascading disaster spreading. Phys. Rev. E **75**, 056107 (2006)
189. D. Helbing, S. Balietti, From social simulation to intgrative system design. visioneer white paper, 2010. http://www.visioneer.ethz.ch.

Chapter 3
Self-organization in Pedestrian Crowds*

3.1 Introduction

The emergence of new, functional or complex collective behaviors in social systems has fascinated many scientists. One of the primary questions in this field is how cooperation or coordination patterns originate based on elementary individual interactions. While one could think that these are a result of intelligent human actions, it turns out that much simpler models assuming automatic responses can reproduce the observations very well. This suggests that humans are using their intelligence primarily for more complicated tasks, but also that simple interactions can lead to intelligent patterns of motion. Of course, it is reasonable to assume that these interactions are the result of a previous learning process that has optimized the automatic response in terms of minimizing collisions and delays. This, however, seems to be sufficient to explain most observations.

Note, however, that research into pedestrian and crowd behavior is highly multi-disciplinary. It involves activities of traffic scientists, psychologists, sociologists, biologists, physicists, computer scientists, and others. Therefore, it is not surprising that there are sometimes different or even controversial views on the subject, e.g. with regard to the concept of "panic", the explanation of collective, spatio-temporal patterns of motion in pedestrian crowds, the best modeling concept, or the optimal number of parameters of a model.

In this contribution, we will start with a short history of pedestrian modeling and, then, introduce the wide-spread "social force model" of pedestrian interactions to illustrate further issues such as, for example, model calibration by video tracking data. Next, we will turn to the subject of crowd dynamics, since one typically finds the formation of large-scale spatio-temporal patterns of motion, when many

---

*This chapter reprints parts of a previous publication with kind permission of the copyright owner, Springer Publishers. It is requested to cite this work as follows: D. Helbing and A. Johansson (2010) Pedestrian, crowd and evacuation dynamics, in: *Encyclopedia of Complexity and Systems Science* (Springer, New York), Vol. 16, pp. 6476–6495.

D. Helbing (ed.), *Social Self-Organization*, Understanding Complex Systems, DOI 10.1007/978-3-642-24004-1_3, © Springer-Verlag Berlin Heidelberg 2012

pedestrians interact with each other. These patterns will be discussed in some detail before we will turn to evacuation situations and cases of extreme densities, where one can sometimes observe the breakdown of coordination. Finally, we will address possibilities to design improved pedestrian facilities, using special evolutionary algorithms.

3.2 Pedestrian Dynamics

3.2.1 Short History of Pedestrian Modeling

Pedestrians have been empirically studied for more than four decades [1–3]. The evaluation methods initially applied were based on direct observation, photographs, and time-lapse films. For a long time, the main goal of these studies was to develop a *level-of-service concept* [4], *design elements* of pedestrian facilities [5–8], or *planning guidelines* [9, 10]. The latter have usually the form of *regression relations*, which are, however, not very well suited for the prediction of pedestrian flows in pedestrian zones and buildings with an exceptional architecture, or in challenging evacuation situations. Therefore, a number of simulation models have been proposed, e.g. *queueing models* [11], *transition matrix models* [12], and *stochastic models* [13], which are partly related to each other. In addition, there are models for the *route choice behavior* of pedestrians [14, 15].

None of these concepts adequately takes into account the self-organization effects occurring in pedestrian crowds. These are the subject of recent experimental studies [8,16–20]. Most pedestrian models, however, were formulated before. A first modeling approach that appears to be suited to reproduce spatio-temporal patterns of motion was proposed by Henderson [21], who conjectured that pedestrian crowds behave similar to gases or fluids (see also [22]). This could be partially confirmed, but a realistic gas-kinetic or fluid-dynamic theory for pedestrians must contain corrections due to their particular interactions (i.e. avoidance and deceleration maneuvers) which, of course, do not obey momentum and energy conservation. Although such a theory can be actually formulated [23,24], for practical applications a direct simulation of *individual* pedestrian motion is favourable, since this is more flexible. As a consequence, pedestrian research mainly focusses on *agent-based models* of pedestrian crowds, which also allow one to consider local coordination problems. The "social force model" [25, 26] is maybe the most well-known of these models, but we also like to mention *cellular automata* of pedestrian dynamics [27–33] and *AI-based models* [34, 35].

3.2.2 The Social Force Concept

In the following, we shall shortly introduce the social force concept, which reproduces most empirical observations in a simple and natural way. Human behavior often seems to be "chaotic", irregular, and unpredictable. So, why and under what conditions can we model it by means of forces? First of all, we need to be confronted with a phenomenon of motion in some (quasi-)continuous space, which may be also an abstract behavioral space such as an opinion scale [36]. Moreover, it is favourable to have a system where the fluctuations due to unknown influences are not large compared to the systematic, deterministic part of motion. This is usually the case in pedestrian traffic, where people are confronted with standard situations and react "automatically" rather than taking complicated decisions, e.g. if they have to evade others.

This "automatic" behavior can be interpreted as the result of a *learning process* based on trial and error [37], which can be simulated with *evolutionary algorithms* [38]. For example, pedestrians have a preferred side of walking, since an asymmetrical avoidance behavior turns out to be profitable [25, 37]. The related *formation of a behavioral convention* can be described by means of *evolutionary game theory* [25, 39].

Another requirement is the vectorial additivity of the separate force terms reflecting different environmental influences. This is probably an approximation, but there is some experimental evidence for it. Based on quantitative measurements for animals and test persons subject to separately or simultaneously applied stimuli of different nature and strength, one could show that the behavior in conflict situations can be described by a superposition of forces [40, 41]. This fits well into a concept by Lewin [42], according to which behavioral changes are guided by so-called *social fields* or *social forces*, which has later on been put into mathematical terms [25, 43]. In some cases, social forces, which determine the amount and direction of systematic behavioral changes, can be expressed as gradients of dynamically varying potentials, which reflect the social or behavioral fields resulting from the interactions of individuals. Such a social force concept was applied to opinion formation and migration [43], and it was particularly successful in the description of collective pedestrian behavior [8, 25, 26, 37].

For reliable simulations of pedestrian crowds, we do not need to know whether a certain pedestrian, say, turns to the right at the next intersection. It is sufficient to have a good estimate what percentage of pedestrians turns to the right. This can be either empirically measured or estimated by means of route choice models [14]. In some sense, the uncertainty about the individual behaviors is averaged out at the macroscopic level of description. Nevertheless, we will use the more flexible microscopic simulation approach based on the social force concept. According to this, the temporal change of the location $r_\alpha(t)$ of pedestrian α obeys the equation

$$\frac{dr_\alpha(t)}{dt} = v_\alpha(t). \tag{3.1}$$

Moreover, if $f_\alpha(t)$ denotes the sum of social forces influencing pedestrian α and if $\xi_\alpha(t)$ are individual fluctuations reflecting unsystematic behavioral variations, the velocity changes are given by the *acceleration equation*

$$\frac{dv_\alpha}{dt} = f_\alpha(t) + \xi_\alpha(t). \tag{3.2}$$

A particular advantage of this approach is that we can take into account the flexible usage of space by pedestrians, requiring a continuous treatment of motion. It turns out that this point is essential to reproduce the empirical observations in a natural and robust way, i.e. without having to adjust the model to each single situation and measurement site. Furthermore, it is interesting to note that, if the fluctuation term is neglected, the social force model can be interpreted as a particular *differential game*, i.e. its dynamics can be derived from the minimization of a special utility function [44].

3.2.3 Specification of the Social Force Model

The social force model for pedestrians assumes that each individual α is trying to move in a desired direction e_α^0 with a desired speed v_α^0, and that it adapts the actual velocity v_α to the desired one, $v_\alpha^0 = v_\alpha^0 e_\alpha^0$, within a certain relaxation time τ_α. The systematic part $f_\alpha(t)$ of the acceleration force of pedestrian α is then given by

$$f_\alpha(t) = \frac{1}{\tau_\alpha}(v_\alpha^0 e_\alpha^0 - v_\alpha) + \sum_{\beta(\neq\alpha)} f_{\alpha\beta}(t) + \sum_i f_{\alpha i}(t), \tag{3.3}$$

where the terms $f_{\alpha\beta}(t)$ and $f_{\alpha i}(t)$ denote the repulsive forces describing attempts to keep a certain safety distance to other pedestrians β and obstacles i. In very crowded situations, additional physical contact forces come into play (see Sect. 3.4.3). Further forces may be added to reflect attraction effects between members of a group or other influences. For details see [37].

First, we will assume a simplified interaction force of the form

$$f_{\alpha\beta}(t) = f\big(d_{\alpha\beta}(t)\big), \tag{3.4}$$

where $d_{\alpha\beta} = r_\alpha - r_\beta$ is the distance vector pointing from pedestrian β to α. Angular-dependent shielding effects may be furthermore taken into account by a prefactor describing the anisotropic reaction to situations in front of as compared to behind a pedestrian [26, 45], see Sect. 3.2.4. However, we will start with a *circular specification* of the distance-dependent interaction force,

$$f(d_{\alpha\beta}) = A_\alpha e^{-d_{\alpha\beta}/B_\alpha} \frac{d_{\alpha\beta}}{\|d_{\alpha\beta}\|}, \tag{3.5}$$

where $d_{\alpha\beta} = \|d_{\alpha\beta}\|$ is the distance. The parameter A_α reflects the *interaction strength*, and B_α corresponds to the *interaction range*. While the dependence on α explicitly allows for a dependence of these parameters on the single individual, we will assume a homogeneous population, i.e. $A_\alpha = A$ and $B_\alpha = B$ in the following. Otherwise, it would be hard to collect enough data for parameter calibration.

Elliptical specification: Note that it is possible to express (3.5) as gradient of an exponentially decaying potential $V_{\alpha\beta}$. This circumstance can be used to formulate a generalized, elliptical interaction force via the potential

$$V_{\alpha\beta}(b_{\alpha\beta}) = AB\,\mathrm{e}^{-b_{\alpha\beta}/B}, \tag{3.6}$$

where the variable $b_{\alpha\beta}$ denotes the semi-minor axis $b_{\alpha\beta}$ of the elliptical equipotential lines. This has been specified according to

$$2b_{\alpha\beta} = \sqrt{(\|d_{\alpha\beta}\| + \|d_{\alpha\beta} - (v_\beta - v_\alpha)\Delta t\|)^2 - \|(v_\beta - v_\alpha)\Delta t\|^2}, \tag{3.7}$$

so that both pedestrians α and β are treated symmetrically. The repulsive force is related to the above potential via

$$f_{\alpha\beta}(d_{\alpha\beta}) = -\nabla_{d_{\alpha\beta}} V_{\alpha\beta}(b_{\alpha\beta}) = -\frac{dV_{\alpha\beta}(b_{\alpha\beta})}{db_{\alpha\beta}} \nabla_{d_{\alpha\beta}} b_{\alpha\beta}(d_{\alpha\beta}), \tag{3.8}$$

where $\nabla_{d_{\alpha\beta}}$ represents the gradient with respect to $d_{\alpha\beta}$. Considering the chain rule, $\|z\| = \sqrt{z^2}$, and $\nabla_z \|z\| = z/\sqrt{z^2} = z/\|z\|$, this leads to the explicit formula

$$f_{\alpha\beta}(d_{\alpha\beta}) = A\mathrm{e}^{-b_{\alpha\beta}/B} \cdot \frac{\|d_{\alpha\beta}\| + \|d_{\alpha\beta} - y_{\alpha\beta}\|}{2b_{\alpha\beta}} \cdot \frac{1}{2}\left(\frac{d_{\alpha\beta}}{\|d_{\alpha\beta}\|} + \frac{d_{\alpha\beta} - y_{\alpha\beta}}{\|d_{\alpha\beta} - y_{\alpha\beta}\|}\right) \tag{3.9}$$

with $y_{\alpha\beta} = (v_\beta - v_\alpha)\Delta t$. We used $\Delta t = 0.5$ s. For $\Delta t = 0$, we regain the expression of (3.5).

The elliptical specification has two major advantages compared to the circular one: First, the interactions depend not only on the distance, but also on the relative velocity. Second, the repulsive force is not strictly directed from pedestrian β to pedestrian α, but has a lateral component. As a consequence, this leads to less confrontative, smoother ("sliding") evading maneuvers. Note that further velocity-dependent specifications of pedestrian interaction forces have been proposed [7,26], but we will restrict to the above specifications, as these are sufficient to demonstrate the method of evolutionary model calibration. For suggested improvements regarding the specification of social forces see, for example, [46,47].

3.2.4 Angular Dependence

In reality, of course, pedestrian interactions are not isotropic, but dependent on the angle $\varphi_{\alpha\beta}$ of the encounter, which is given by the formula

$$\cos(\varphi_{\alpha\beta}) = \frac{v_\alpha}{\|v_\alpha\|} \cdot \frac{-d_{\alpha\beta}}{\|d_{\alpha\beta}\|}. \tag{3.10}$$

Generally, pedestrians show little response to pedestrians behind them. This can be reflected by an angular-dependent prefactor $w(\varphi_{\alpha\beta})$ of the interaction force [45]. Empirical results are represented in Fig. 3.2 (right). Reasonable results are obtained for the following specification of the prefactor:

$$w\big(\varphi_{\alpha\beta}(t)\big) = \left(\lambda_\alpha + (1-\lambda_\alpha)\frac{1+\cos(\varphi_{\alpha\beta})}{2}\right), \tag{3.11}$$

where λ_α with $0 \leq \lambda_\alpha \leq 1$ is a parameter which grows with the strength of interactions from behind. An evolutionary parameter optimization gives values $\lambda \approx 0.1$ (see Sect. 3.2.5, i.e. a strong anisotropy. Other angular-dependent specifications split up the interaction force between pedestrians into a component against the direction of motion and another one perpendicular to it. Such a description allows for even smoother avoidance maneuvers.

3.2.5 *Evolutionary Calibration with Video Tracking Data*

For parameter calibration, several video recordings of pedestrian crowds in different natural environments have been used. The dimensions of the recorded areas were known, and the floor tiling or environment provided something like a "coordinate system". The heads were automatically determined by searching for round moving structures, and the accuracy of tracking was improved by comparing actual with linearly extrapolated positions (so it would not happen so easily that the algorithm interchanged or "lost" closeby pedestrians). The trajectories of the heads were then projected on two-dimensional space in a way correcting for distortion by the camera perspective. A representative plot of the resulting trajectories is shown in Fig. 3.1. Note that trajectory data have been obtained with infra-red sensors [48] or video cameras [49, 50] for several years now, but algorithms that can simultaneously handle more than one thousand pedestrians have become available

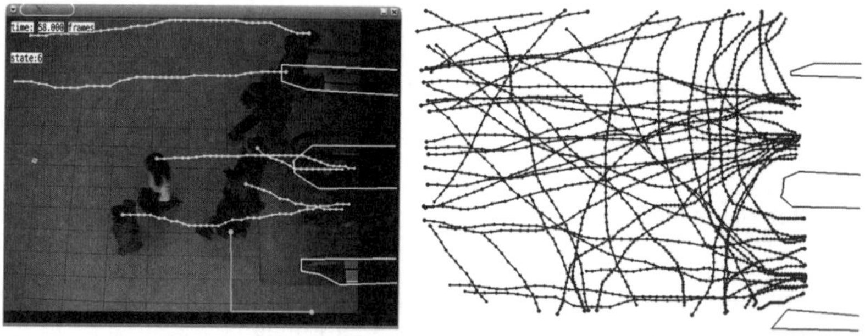

Fig. 3.1 Video tracking used to extract the trajectories of pedestrians from video recordings close to two escalators (after [45]). *Left*: Illustration of the tracking of pedestrian heads. *Right*: Resulting trajectories after being transformed onto the two-dimensional plane

only recently [51]. For model calibration, it is recommended to use a hybrid method fusing empirical trajectory data and microscopic simulation data of pedestrian movement in space. In corresponding algorithms, a virtual pedestrian is assigned to each tracked pedestrian in the simulation domain. One then starts a simulation for a time period T (e.g. 1.5 s), in which one pedestrian α is moved according to a simulation of the social force model, while the others are moved exactly according to the trajectories extracted from the videos. This procedure is performed for all pedestrians α and for several different starting times t, using a fixed parameter set for the social force model.

Each simulation run is performed according to the following scheme:

1. Define a starting point and calculate the state (position r_α, velocity v_α, and acceleration $a_\alpha = dv_\alpha/dt$) for each pedestrian α.
2. Assign a desired speed v_α^0 to each pedestrian, e.g. the maximum speed during the pedestrian tracking time. This is sufficiently accurate, if the overall pedestrian density is not too high and the desired speed is constant in time.
3. Assign a desired goal point for each pedestrian, e.g. the end point of the trajectory.
4. Given the tracked motion of the surrounding pedestrians β, simulate the trajectory of pedestrian α over a time period T based on the social force model, starting at the actual location $r_\alpha(t)$.

After each simulation run, one determines the relative distance error

$$\frac{\|r_\alpha^{\text{simulated}}(t+T) - r_\alpha^{\text{tracked}}(t+T)\|}{\|r_\alpha^{\text{tracked}}(t+T) - r_\alpha^{\text{tracked}}(t)\|}. \tag{3.12}$$

After averaging the relative distance errors over the pedestrians α and starting times t, 1 minus the result can be taken as measure of the goodness of fit (the "fitness") of the parameter set used in the pedestrian simulation. Hence, the best possible value of the "fitness" is 1, but any deviation from the real pedestrian trajectories implies lower values.

One result of such a parameter optimization is that, for each video, there is a broad range of parameter combinations of A and B which perform almost equally well [45]. This allows one to apply additional goal functions in the parameter optimization, e.g. to determine among the best performing parameter values such parameter combinations, which perform well for *several* video recordings, using a fitness function which equally weights the fitness reached in each single video. This is how the parameter values listed in Table 3.1 were determined. It turns out that, in order to reach a good model performance, the pedestrian interaction force must be specified velocity dependent, as in the elliptical model.

Note that our evolutionary fitting method can be also used to determine interaction laws without prespecified interaction functions. For example, one can obtain the distance dependence of pedestrian interactions without a pre-specified function. For this, one adjusts the values of the force at given distances $d_k = kd_1$ (with $k \in \{1, 2, 3, \ldots\}$) in an evolutionary way. To get some smoothness, linear interpolation

Table 3.1 Interaction strength A and interaction range B resulting from our evolutionary parameter calibration for the circular and elliptical specification of the interaction forces between pedestrians (see main text), with an assumed angular dependence according to (3.11). A comparison with the extrapolation scenario, which assumes constant speeds, allows one to judge the improvement in the goodness of fit ("fitness") by the specified interaction force. The calibration was based on three different video recordings, one for low crowd density, one for medium, and one for high density (see [45] for details). The parameter values are specified as mean value ± standard deviation. The best fitness value obtained with the elliptical specification for the video with the lowest crowd density was as high as 0.9

Model	A [m/s^2]	B [m]	λ	Fitness
Extrapolation	0	–	–	0.34
Circular	0.42 ± 0.26	1.65 ± 1.01	0.12 ± 0.07	0.40
Elliptical	0.04 ± 0.01	3.22 ± 0.67	0.06 ± 0.04	0.61

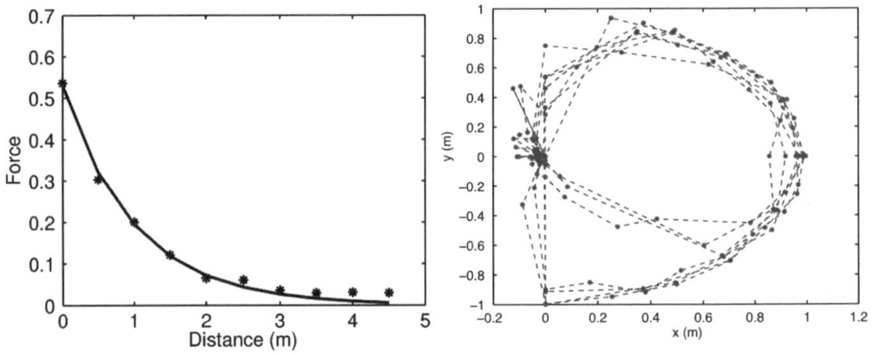

Fig. 3.2 Results of an evolutionary fitting of pedestrian interactions. *Left*: Empirically determined distance dependence of the interaction force between pedestrians. An exponential decay fits the empirical data quite well. The dashed fit curve corresponds to (3.5) with the parameters $A = 0.53$ and $B = 1.0$. *Right*: Angular dependence of the influence of other pedestrians. The direction along the positive x axis corresponds to the walking direction of pedestrians, y to the perpendicular direction (After [45])

is applied. The resulting fit curve is presented in Fig. 3.2 (left). It turns out that the empirical dependence of the force with distance can be well fitted by an exponential decay.

3.3 Crowd Dynamics

3.3.1 Analogies with Gases, Fluids, and Granular Media

When the density is low, pedestrians can move freely, and the observed crowd dynamics can be partially compared with the behavior of gases. At medium and high densities, however, the motion of pedestrian crowds shows some striking analogies with the motion of fluids:

1. Footprints of pedestrians in snow look similar to streamlines of fluids [15].
2. At borderlines between opposite directions of walking one can observe "viscous fingering" [52, 53].
3. The emergence of pedestrian streams through standing crowds [7, 37, 54] appears analogous to the formation of river beds [55, 56].

At high densities, however, the observations have rather analogies with driven granular flows. This will be elaborated in more detail in Sects. 3.4.3 and 3.4.4. In summary, one could say that fluid-dynamic analogies work reasonably well in normal situations, while granular aspects dominate at extreme densities. Nevertheless, the analogy is limited, since the self-driven motion and the violation of momentum conservation imply special properties of pedestrian flows. For example, one usually does not observe eddies.

3.3.2 Self-organization of Pedestrian Crowds

Despite its simplifications, the social force model of pedestrian dynamics describes a lot of observed phenomena quite realistically. Especially, it allows one to explain various self-organized spatio-temporal patterns that are not externally planned, prescribed, or organized, e.g. by traffic signs, laws, or behavioral conventions [7, 8, 37]. Instead, the spatio-temporal patterns discussed below emerge due to the non-linear interactions of pedestrians even without assuming strategical considerations, communication, or imitative behavior of pedestrians. Despite this, we may still interpret the forming cooperation patterns as phenomena that establish social order on short time scales. It is actually surprising that strangers coordinate each other within seconds, if they have grown up in a similar environment. People from different countries, however, are sometimes irritated about local walking habits, which indicates that learning effects and cultural backgrounds still play a role in social interactions as simple as random pedestrian encounters. Rather than on particular features, however, in the following we will focus on the common, internationally reproducible observations.

3.3.2.1 Lane Formation

In pedestrian flows one can often observe that oppositely moving pedestrians are forming lanes of uniform walking direction (see Fig. 3.3) [8, 20, 25, 26]. This phenomenon even occurs when there is not a large distance to separate each other, e.g. on zebra crossings. However, the width of lanes increases (and their number decreases), if the interaction continues over longer distances (and if perturbations, e.g. by flows entering or leaving on the sides, are low; otherwise the phenomenon of lane formation may break down [57]).

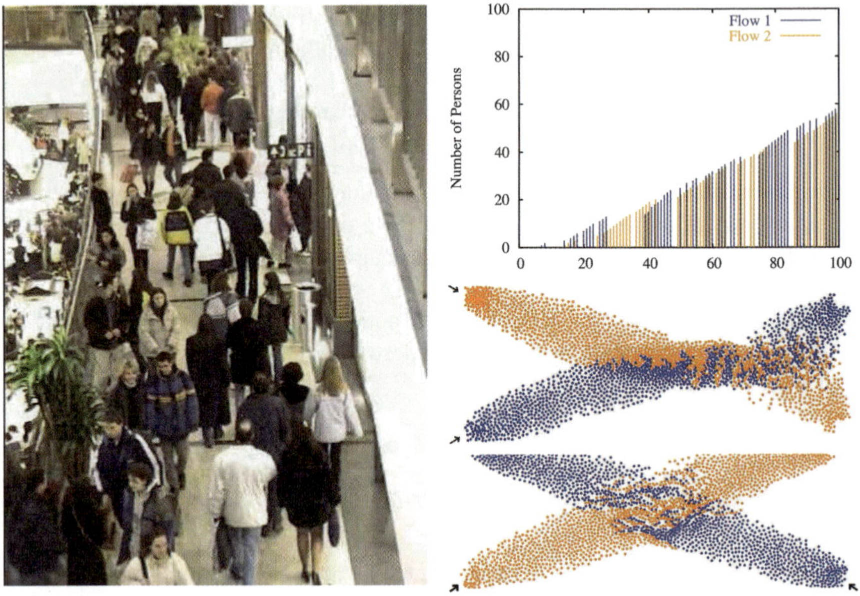

Fig. 3.3 Self-organization of pedestrian crowds. *Left*: Photograph of lanes formed in a shopping center. Computer simulations reproduce the self-organization of such lanes very well. *Top right*: Evaluation of the cumulative number of pedestrians passing a bottleneck from different sides. One can clearly see that the narrowing is often passed by groups of people in an oscillatory way rather than one by one. *Bottom right*: Multi-agent simulation of two crossing pedestrian streams, showing the phenomenon of stripe formation. This self-organized pattern allows pedestrians to pass the other stream without having to stop, namely by moving sidewards in a forwardly moving stripe (After [8])

Lane formation may be viewed as *segregation phenomenon* [58, 59]. Although there is a weak preference for one side (with the corresponding behavioral convention depending on the country), the observations can only be well reproduced when repulsive pedestrian interactions are taken into account. The most relevant factor for the lane formation phenomenon is the higher relative velocity of pedestrians walking in opposite directions. Compared to people following each other, oppositely moving pedestrians have more frequent interactions until they have segregated into separate lanes by stepping aside whenever another pedestrian is encountered. The most long-lived patterns of motion are the ones which change the least. It is obvious that such patterns correspond to lanes, as they minimize the frequency and strength of avoidance maneuvers. Interestingly enough, as computer simulations show, lane formation occurs also when there is no preference for any side.

Lanes minimize frictional effects, accelerations, energy consumption, and delays in oppositely moving crowds. Therefore, one could say that they are a pattern reflecting "collective intelligence". In fact, it is not possible for a single pedestrian to reach such a collective pattern of motion. Lane formation is a self-organized collaborative pattern of motion originating from simple pedestrian interactions. Particularly in cases of no side preference, the system behavior cannot be understood by adding up the behavior of the single individuals. This is a typical feature of complex, self-organizing systems and, in fact, a wide-spread characteristics of social systems. It is worth noting, however, that it does not require a conscious behavior to reach forms of social organization like the segregation of oppositely moving pedestrians into lanes. This organization occurs automatically, although most people are not even aware of the existence of this phenomenon.

3.3.2.2 Oscillatory Flows at Bottlenecks

At bottlenecks, bidirectional flows of moderate density are often characterized by oscillatory changes in the flow direction (see Fig. 3.3) [8, 26]. For example, one can sometimes observe this at entrances of museums during crowded art exhibitions or at entrances of staff canteens during lunch time. While these oscillatory flows may be interpreted as an effect of friendly behavior ("you go first, please"), computer simulations of the social force model indicate that the collective behavior may again be understood by simple pedestrian interactions. That is, oscillatory flows can even occur in the absence of communication, although it may be involved in reality. The interaction-based mechanism of oscillatory flows suggests to interpret them as another self-organization phenomenon, which again reduces frictional effects and delays. That is, oscillatory flows have features of "collective intelligence".

While this may be interpreted as result of a learning effect in a large number of similar situations (a "repeated game"), our simulations suggest an even simpler, "many-particle" interpretation: Once a pedestrian is able to pass the narrowing, pedestrians with the same walking direction can easily follow. Hence, the number and "pressure" of waiting, "pushy" pedestrians on one side of the bottleneck becomes less than on the other side. This eventually increases their chance to occupy the passage. Finally, the "pressure difference" is large enough to stop the flow and turn the passing direction at the bottleneck. This reverses the situation, and eventually the flow direction changes again, giving rise to oscillatory flows.

At bottlenecks, further interesting observations can be made: Hoogendoorn and Daamen [60] report the formation of layers in unidirectional bottleneck flows. Due to the partial overlap of neighboring layers, there is a zipper effect. Moreover, Kretz et al. [61] have observed that the specific flow through a narrow bottleneck decreases with a growing width of the bottleneck, as long as it can be passed by one person at a time only. This is due to mutual obstructions, if two people are trying to enter the bottleneck simultaneously. If the opening is large enough to be entered by several people in parallel, the specific flow stays constant with increasing width. Space is then used in a flexible way.

3.3.2.3 Stripe Formation in Intersecting Flows

In intersection areas, the flow of people often appears to be irregular or "chaotic". In fact, it can be shown that there are several possible collective patterns of motion, among them rotary and oscillating flows. However, these patterns continuously compete with each other, and a temporarily dominating pattern is destroyed by another one after a short time. Obviously, there has not evolved any social convention that would establish and stabilize an ordered and efficient flow at intersections.

Self-organized patterns of motion, however, are found in situations where pedestrian flows cross each other only in two directions. In such situations, the phenomenon of stripe formation is observed [62]. Stripe formation allows two flows to penetrate each other without requiring the pedestrians to stop. For an illustration see Fig. 3.3. Like lanes, stripes are a segregation phenomenon, but not a stationary one. Instead, the stripes are density waves moving into the direction of the sum of the directional vectors of both intersecting flows. Naturally, the stripes extend sidewards into the direction which is perpendicular to their direction of motion. Therefore, the pedestrians move forward *with* the stripes and sidewards *within* the stripes. Lane formation corresponds to the particular case of stripe formation where both directions are exactly opposite. In this case, no intersection takes place, and the stripes do not move systematically. As in lane formation, stripe formation allows to minimize obstructing interactions and to maximize the average pedestrian speeds, i.e. simple, repulsive pedestrian interactions again lead to an "intelligent" collective behavior.

3.4 Evacuation Dynamics

While the previous section has focussed on the dynamics of pedestrian crowds in normal situations, we will now turn to the description of situations in which extreme crowd densities occur. Such situations may arise at mass events, particularly in cases of urgent egress. While most evacuations run relatively smoothly and orderly, the situation may also get out of control and end up in terrible crowd disasters (see Table 3.2). In such situations, one often speaks of "panic", although, from a scientific standpoint, the use of this term is rather controversial. Here, however, we will not be interested in the question whether "panic" actually occurs or not. We will rather focus on the issue of crowd dynamics at high densities and under psychological stress.

3.4.1 Evacuation and Panic Research

Computer models have been also developed for emergency and evacuation situations [32, 63–71]. Most research into panic, however, has been of empirical nature (see, e.g. [72–74]), carried out by social psychologists and others.

With some exceptions, panic is thought to occur in cases of scarce or dwindling resources [75, 76], which are either required for survival or anxiously desired. They are usually distinguished into escape panic ("stampedes", bank or stock market panic) and acquisitive panic ("crazes", speculative manias) [77, 78], but in some cases this classification is questionable [79].

Table 3.2 Incomplete list of major crowd disasters since 1970 after J. F. Dickie in [89], http://www.crowddynamics.com/Main/Crowddisasters.html, http://SportsIllustrated.CNN.com/soccer/world/news/2000/07/09/stadium_disasters_ap/, and other internet sources, excluding fires, bomb attacks, and train or plane accidents. The number of injured people was usually a multiple of the fatalities

Date	Place	Venue	Deaths	Reason
1971	Ibrox, UK	Stadium	66	Collapse of barriers
1974	Cairo, Egypt	Stadium	48	Crowds break barriers
1982	Moscow, USSR	Stadium	340	Re-entering fans after last minute goal
1988	Katmandu, Nepal	Stadium	93	Stampede due to hailstorm
1989	Hillsborough, Sheffield, UK	Stadium	96	Fans trying to force their way into the stadium
1990	New York City	Bronx	87	Illegal happy land social club
1990	Mena, Saudi Arabia	Pedestrian tunnel	1,426	Overcrowding
1994	Mena, Saudi Arabia	Jamarat bridge	266	Overcrowding
1996	Guatemala City, Guatemala	Stadium	83	Fans trying to force their way into the stadium
1998	Mena, Saudi Arabia		118	Overcrowding
1999	Kerala, India	Hindu shrine	51	Collapse of parts of the shrine
1999	Minsk, Belarus	Subway station	53	Heavy rain at rock concert
2001	Ghana, West Africa	Stadium	>100	Panic triggered by tear gas
2004	Mena, Saudi Arabia	Jamarat bridge	251	Overcrowding
2005	Wai, India	Religious procession	150	Overcrowding (and fire)
2005	Bagdad, Iraque	Religious procession	>640	Rumors regarding suicide bomber
2005	Chennai, India	Disaster area	42	Rush for flood relief supplies
2006	Mena, Saudi Arabia	Jamarat bridge	363	Overcrowding
2006	Pilippines	Stadium	79	Rush for game show tickets
2006	Ibb, Yemen	Stadium	51	Rally for Yemeni president

It is often stated that panicking people are obsessed by short-term personal interests uncontrolled by social and cultural constraints [76, 77]. This is possibly a result of the reduced attention in situations of fear [76], which also causes that options like side exits are mostly ignored [72]. It is, however, mostly attributed to social contagion [73, 75–84], i.e., a transition from individual to mass psychology, in which individuals transfer control over their actions to others [78], leading to conformity [85]. This "herding behavior" is in some sense irrational, as it often leads to bad overall results like dangerous overcrowding and slower escape [72,78,79]. In this way, herding behavior can increase the fatalities or, more generally, the damage in the crisis faced.

The various socio-psychological theories for this contagion assume hypnotic effects, rapport, mutual excitation of a primordial instinct, circular reactions, social facilitation (see the summary by Brown [83]), or the emergence of normative support for selfish behavior [84]. Brown [83] and Coleman [78] add another explanation related to the prisoner's dilemma [86, 87] or common goods dilemma [88], showing that it is reasonable to make one's subsequent actions contingent upon those of others. However, the socially favourable behavior of walking orderly is unstable, which normally gives rise to rushing by everyone. These thoughtful considerations are well compatible with many aspects discussed above and with the classical experiments by Mintz [75], which showed that jamming in escape situations depends on the reward structure ("payoff matrix").

Nevertheless and despite of the frequent reports in the media and many published investigations of crowd disasters (see Table 3.2), a quantitative understanding of the observed phenomena in panic stampedes was lacking for a long time. The following sections will close this gap.

3.4.2 Situations of "Panic"

Panic stampede is one of the most tragic collective behaviors [73–75, 77, 78, 80–84], as it often leads to the death of people who are either crushed or trampled down by others. While this behavior may be comprehensible in life-threatening situations like fires in crowded buildings [72, 76], it is hard to understand in cases of a rush for good seats at a pop concert [79] or without any obvious reasons. Unfortunately, the frequency of such disasters is increasing (see Table 3.2), as growing population densities combined with easier transportation lead to greater mass events like pop concerts, sport events, and demonstrations. Nevertheless, systematic empirical studies of panic [75, 90] are rare [76, 77, 79], and there is a scarcity of quantitative theories capable of predicting crowd dynamics at extreme densities [32, 63, 64, 67, 68, 71]. The following features appear to be typical [57, 91]:

1. In situations of escape panic, individuals are getting nervous, i.e. they tend to develop blind actionism.
2. People try to move considerably faster than normal [9].
3. Individuals start pushing, and interactions among people become physical in nature.
4. Moving and, in particular, passing of a bottleneck frequently becomes incoordinated [75].
5. At exits, jams are building up [75]. Sometimes, intermittent flows or arching and clogging are observed [9].
6. The physical interactions in jammed crowds add up and can cause dangerous pressures up to 4,500 Newtons per meter [72, 89], which can bend steel barriers or tear down brick walls.
7. The strength and direction of the forces acting in large crowds can suddenly change [51], pushing people around in an uncontrollable way. This may cause people to fall.
8. Escape is slowed down by fallen or injured people turning into "obstacles".

9. People tend to show herding behavior, i.e., to do what other people do [76, 81].
10. Alternative exits are often overlooked or not efficiently used in escape situations [72, 76].

3.4.3 Force Model for Panicking Pedestrians

Additional, physical interaction forces $f^{\text{ph}}_{\alpha\beta}$ come into play when pedestrians get so close to each other that they have physical contact (i.e. $d_{\alpha\beta} < r_{\alpha\beta} = r_\alpha + r_\beta$, where r_α means the "radius" of pedestrian α) [91]. In this case, which is mainly relevant to panic situations, we assume also a "body force" $k(r_{\alpha\beta} - d_{\alpha\beta})\, n_{\alpha\beta}$ counteracting body compression and a *"sliding friction force"* $\kappa(r_{\alpha\beta} - d_{\alpha\beta})\, \Delta v^t_{\beta\alpha}\, t_{\alpha\beta}$ impeding *relative* tangential motion. Inspired by the formulas for granular interactions [92, 93], we assume

$$f^{\text{ph}}_{\alpha\beta}(t) = k\Theta(r_{\alpha\beta} - d_{\alpha\beta})n_{\alpha\beta} + \kappa\Theta(r_{\alpha\beta} - d_{\alpha\beta})\Delta v^t_{\beta\alpha}\, t_{\alpha\beta}, \tag{3.13}$$

where the function $\Theta(z)$ is equal to its argument z, if $z \geq 0$, otherwise 0. Moreover, $t_{\alpha\beta} = (-n^2_{\alpha\beta}, n^1_{\alpha\beta})$ means the tangential direction and $\Delta v^t_{\beta\alpha} = (v_\beta - v_\alpha) \cdot t_{\alpha\beta}$ the tangential velocity difference, while k and κ represent large constants. (Strictly speaking, friction effects already set in before pedestrians touch each other, because of the psychological tendency not to pass other individuals with a high relative velocity, when the distance is small.)

The interactions with the boundaries of walls and other obstacles are treated analogously to pedestrian interactions, i.e., if $d_{\alpha i}(t)$ means the distance to obstacle or boundary i, $n_{\alpha i}(t)$ denotes the direction perpendicular to it, and $t_{\alpha i}(t)$ the direction tangential to it, the corresponding interaction force with the boundary reads

$$f_{\alpha i} = \{A_\alpha \exp[(r_\alpha - d_{\alpha i})/B_\alpha] + k\Theta(r_\alpha - d_{\alpha i})\}\, n_{\alpha i} - \kappa\Theta(r_\alpha - d_{\alpha i})(v_\alpha \cdot t_{\alpha i})\, t_{\alpha i}. \tag{3.14}$$

Finally, fire fronts are reflected by repulsive social forces similar those describing walls, but they are much stronger. The physical interactions, however, are qualitatively different, as people reached by the fire front become injured and immobile ($v_\alpha = 0$).

3.4.4 Collective Phenomena in Situations of "Panic"

Inspired by the observations discussed in Sect. 3.4.2, we have simulated situations of "panic" escape in the computer, assuming the following features:

1. People are getting nervous, resulting in a higher level of fluctuations.
2. They are trying to escape from the source of panic, which can be reflected by a significantly higher desired velocity v^0_α.

3. Individuals in complex situations, who do not know what is the right thing to do, orient at the actions of their neighbours, i.e. they tend to do what other people do. We will describe this by an additional herding interaction.

We will now discuss the fundamental collective effects which fluctuations, increased desired velocities, and herding behavior can have according to simulations. Note that, in contrast to other approaches, we do not assume or imply that individuals in panic or emergency situations would behave relentless and asocial, although they sometimes do.

3.4.4.1 Herding and Ignorance of Available Exits

If people are not sure what is the best thing to do, there is a tendency to show a "herding behavior", i.e. to imitate the behavior of others. Fashion, hypes and trends are examples for this. The phenomenon is also known from stock markets, and particularly pronounced when people are anxious. Such a situation is, for example, given if people need to escape from a smoky room. There, the evacuation dynamics is very different from normal leaving (see Fig. 3.4).

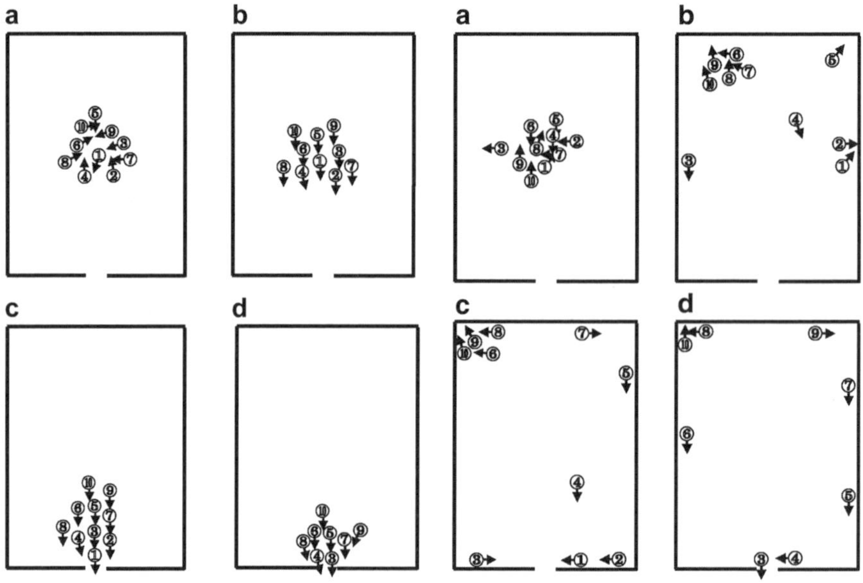

Fig. 3.4 *Left*: Normal leaving of a room, when the exit is well visible. Snapshots of a video-recorded experiment with ten people after (**a**) $t = 0$ s (initial condition), (**b**) $t = 1$ s, (**c**) $t = 3$ s, and (**d**) $t = 5$ s. The face directions are indicated by *arrows*. *Right*: Escape from a room with no visibility, e.g. due to dense smoke or a power blackout. Snapshots of an experiment with test persons, whose eyes were covered by masks, after $t = 0$ s (initial condition), $t = 5$ s, (**c**) $t = 10$ s, and (**d**) $t = 15$ s (After [18])

Under normal visibility, everybody easily finds an exit and uses more or less the shortest path. However, when the exit cannot be seen, evacuation is much less efficient and may take a long time. Most people tend to walk relatively straight into the direction in which they suspect an exit, but in most cases, they end up at a wall. Then, they usually move along it in one of the two possible directions, until they finally find an exit [18]. If they encounter others, there is a tendency to take a decision for one direction and move collectively. Also in case of acoustic signals, people may be attracted into the same direction. This can lead to over-crowded exits, while other exits are ignored. The same can happen even for normal visibility, when people are not well familiar with their environment and are not aware of the directions of the emergency exits.

Computer simulations suggest that neither individualistic nor herding behavior performs well [91]. Pure individualistic behavior means that each pedestrian finds an exit only accidentally, while pure herding behavior implies that the complete crowd is eventually moving into the same and probably congested direction, so that available emergency exits are not efficiently used. Optimal chances of survival are expected for a certain mixture of individualistic and herding behavior, where individualism allows *some* people to detect the exits and herding guarantees that successful solutions are imitated by small groups of others [91].

3.4.4.2 "Freezing by Heating"

Another effect of getting nervous has been investigated in [57]. Let us assume the individual fluctuation strength, i.e. the standard deviation of the noise term ξ_α, is given by

$$\eta_\alpha = (1 - n_\alpha)\eta_0 + n_\alpha \eta_{\max}, \tag{3.15}$$

where n_α with $0 \leq n_\alpha \leq 1$ measures the nervousness of pedestrian α. The parameter η_0 means the normal and $\eta_{\max}$ the maximum fluctuation strength. It turns out that, at sufficiently high pedestrian densities, lanes are destroyed by increasing the fluctuation strength (which is analogous to the temperature). However, instead of the expected transition from the "fluid" lane state to a disordered, "gaseous" state, a "solid" state is formed. It is characterized by a blocked, "frozen" situation so that one calls this paradoxial transition *"freezing by heating"* (see Fig. 3.5). Notably

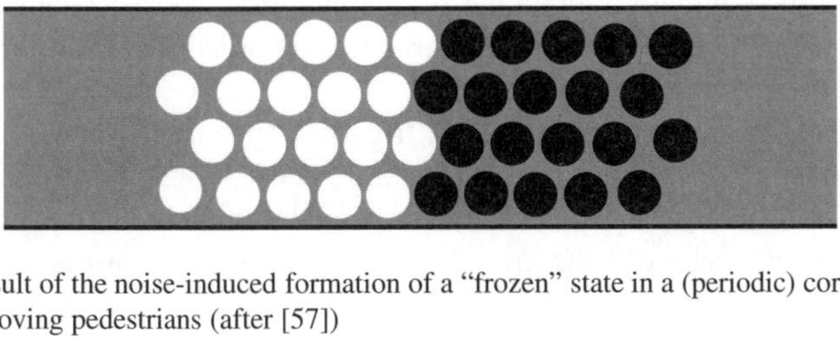

Fig. 3.5 Result of the noise-induced formation of a "frozen" state in a (periodic) corridor used by oppositely moving pedestrians (after [57])

enough, the blocked state has a *higher* degree of order, although the internal energy is *increased* [57].

The preconditions for this unusual freezing-by-heating transition are the driving term $v_\alpha^0 e_\alpha^0 / \tau_\alpha$ and the dissipative friction $-v_\alpha / \tau_\alpha$, while the sliding friction force is not required. Inhomogeneities in the channel diameter or other impurities which temporarily slow down pedestrians can further this transition at the respective places. Finally note that a transition from fluid to blocked pedestrian counter flows is also observed, when a critical density is exceeded, as impatient pedestrians enter temporary gaps in the opposite lane to overtake others [31,57]. However, in contrast to computer simulations, resulting deadlocks are usually not permanent in real crowds, as turning the bodies (shoulders) often allows pedestrians to get out of the blocked area.

3.4.4.3 Intermittent Flows, Faster-Is-Slower Effect, and "Phantom Panic"

If the overall flow towards a bottleneck is higher than the overall outflow from it, a pedestrian queue emerges [94]. In other words, a waiting crowd is formed upstream of the bottleneck. High densities can result, if people keep heading forward, as this eventually leads to higher and higher compressions. Particularly critical situations may occur if the arrival flow is much higher than the departure flow, especially if people are trying to get towards a strongly desired goal ("aquisitive panic") or away from a perceived source of danger ("escape panic") with an increased driving force $v_\alpha^0 e_\alpha^0 / \tau$. In such situations, the high density causes coordination problems, as several people compete for the same few gaps. This typically causes body interactions and frictional effects, which can slow down crowd motion or evacuation *("faster is slower effect")*.

A possible consequence of these coordination problems are intermittent flows. In such cases, the outflow from the bottleneck is not constant, but it is typically interrupted. While one possible origin of the intermittent flows are clogging and arching effects as known from granular flows through funnels or hoppers [92, 93], stop-and-go waves have also been observed in more than 10 m wide streets and in the 44 m wide entrance area to the Jamarat Bridge during the pilgrimage in January 12, 2006 [51], see Fig. 3.6. Therefore, it seems to be important that people do not move continuously, but have minimum strides [25]. That is, once a person is stopped, he or she will not move until some space opens up in front. However, increasing impatience will eventually reduce the minimum stride, so that people eventually start moving again, even if the outflow through the bottleneck is stopped. This will lead to a further compression of the crowd.

In the worst case, such behavior can trigger a "phantom panic", i.e. a crowd disaster *without* any serious reasons (e.g., in Moscow, 1982). For example, due to the "faster-is-slower effect" panic can be triggered by small pedestrian counterflows [72], which cause delays to the crowd intending to leave. Consequently, stopped pedestrians in the back, who do not see the reason for the temporary slowdown, are getting impatient and pushy. In accordance with observations [7, 25], one may model this by increasing the desired velocity, for example, by the formula

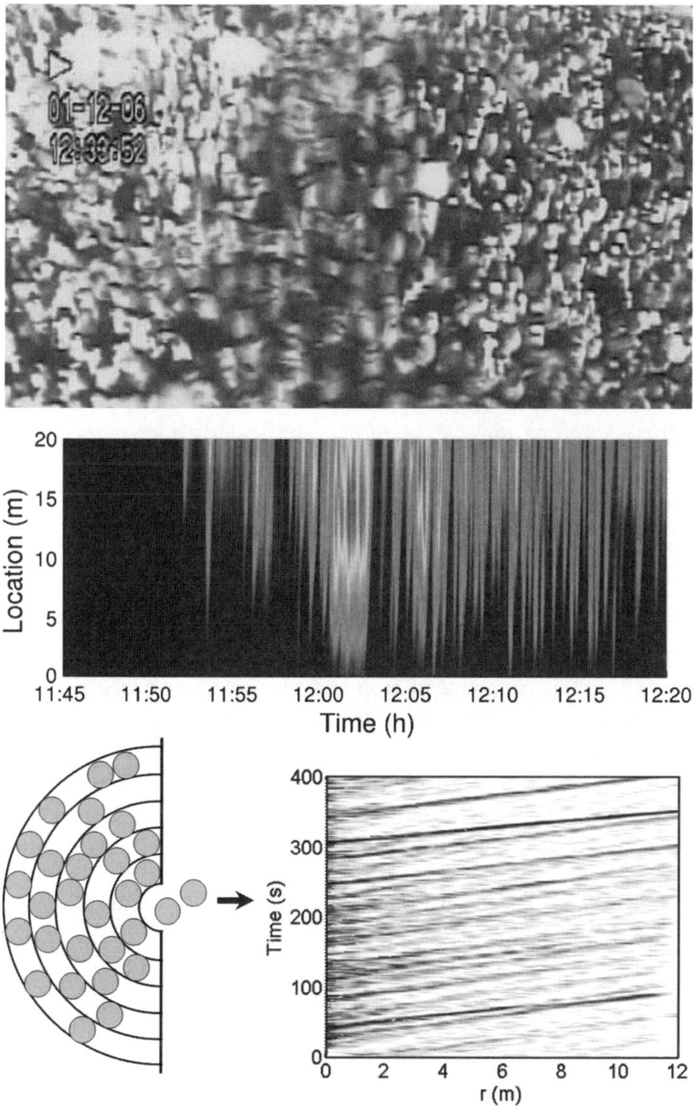

Fig. 3.6 *Top*: Long-term photograph showing stop-and-go waves in a densely packed street. While stopped people appear relatively sharp, people moving from right to left have a fuzzy appearance. Note that gaps propagate from left to right. *Middle*: Empirically observed stop-and-go waves in front of the entrance to the Jamarat Bridge on January 12, 2006 (after [51]), where pilgrims moved from left to right. *Dark areas* correspond to phases of motion, *light grey* to stop phases. The "location" coordinate represents the distance to the beginning of the narrowing, i.e. to the cross section of reduced width. *Bottom left*: Illustration of the "shell model" (see [94]), in particular of situations where several pedestrians compete for the same gap, which causes coordination problems. *Bottom right*: Simulation results of the shell model. The observed stop-and-go waves result from the alternation of forward pedestrian motion and backward gap propagation

$$v_\alpha^0(t) = [1 - n_\alpha(t)]v_\alpha^0(0) + n_\alpha(t)v_\alpha^{\max}. \tag{3.16}$$

Herein, $v_\alpha^{\max}$ is the maximum desired velocity and $v_\alpha^0(0)$ the initial one, corresponding to the expected velocity of leaving. The time-dependent parameter

$$n_\alpha(t) = 1 - \frac{\overline{v}_\alpha(t)}{v_\alpha^0(0)} \tag{3.17}$$

reflects the nervousness, where $\overline{v}_\alpha(t)$ denotes the average speed into the desired direction of motion. Altogether, long waiting times increase the desired speed v_α^0 or driving force $v_\alpha^0(t)e_\alpha^0/\tau$, which can produce high densities and inefficient motion. This further increases the waiting times, and so on, so that this tragic feedback can eventually trigger so high pressures that people are crushed or falling and trampled. It is, therefore, imperative, to have sufficiently wide exits and to prevent counterflows, when big crowds want to leave [91].

3.4.4.4 Transition to Stop-and-Go Waves

Recent empirical studies of pilgrim flows in the area of Makkah, Saudi Arabia, have shown that intermittent flows occur not only when bottlenecks are obvious. On January 12, 2006, pronounced stop-and-go waves have been even observed upstream of the 44 m wide entrance to the Jamarat Bridge [51]. While the pilgrim flows were smooth and continuous ("laminar") over many hours, at 11:53 am stop-and-go waves suddenly appeared and propagated over distances of more than 30 m (see Fig. 3.6). The sudden transition was related to a significant drop of the flow, i.e. with the onset of congestion [51]. Once the stop-and-go waves set in, they persisted over more than 20 min.

This phenomenon can be reproduced by a recent model based on two continuity equations, one for forward pedestrian motion and another one for backward gap propagation [94]. The model was derived from a "shell model" (see Fig. 3.6) and describes very well the observed alternation between backward gap propagation and forward pedestrian motion.

3.4.4.5 Transition to "Crowd Turbulence"

On the same day, around 12:19, the density reached even higher values and the video recordings showed a sudden transition from stop-and-go waves to *irregular* flows (see Fig. 3.7). These irregular flows were characterized by random, unintended displacements into all possible directions, which pushed people around. With a certain likelihood, this caused them to stumble. As the people behind were moved by the crowd as well and could not stop, fallen individuals were trampled, if they did not get back on their feet quickly enough. Tragically, the area of trampled people grew more and more in the course of time, as the fallen pilgrims became obstacles

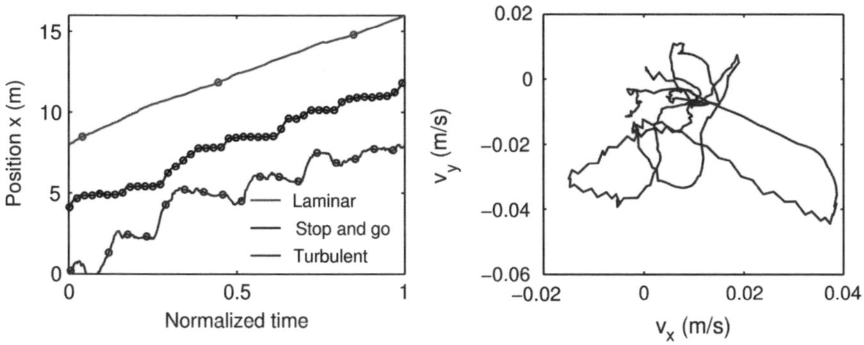

Fig. 3.7 Pedestrian dynamics at different densities. *Left*: Representative trajectories (space-time plots) of pedestrians during the laminar, stop-and-go, and turbulent flow regime. Each trajectory extends over a range of 8 m, while the time required for this stretch is normalized to 1. To indicate the different speeds, symbols are included in the curves every 5 s. While the laminar flow (*top line*) is fast and smooth, motion is temporarily interrupted in stop-and-go flow (*medium line*), and backward motion can occur in "turbulent" flows (*bottom line*). *Right*: Example of the temporal evolution of the velocity components $v_x(t)$ into the average direction of motion and $v_y(t)$ perpendicular to it in "turbulent flow", which occurs when the crowd density is extreme. One can clearly see the irregular motion into all possible directions characterizing "crowd turbulence". For details see [51]

for others [51]. The result was one of the biggest crowd disasters in the history of pilgrimage.

How can we understand this transition to irregular crowd motion? A closer look at video recordings of the crowd reveals that, at this time, people were so densely packed that they were moved involuntarily by the crowd. This is reflected by random displacements into all possible directions. To distinguish these irregular flows from laminar and stop-and-go flows and due to their visual appearance, we will refer to them as "crowd turbulence".

As in certain kinds of fluid flows, "turbulence" in crowds results from a sequence of instabilities in the flow pattern. Additionally, one finds a sharply peaked probability density function of velocity increments

$$V_x^\tau = V_x(r, t + \tau) - V_x(r, t), \tag{3.18}$$

which is typical for turbulence [95], if the time shift τ is small enough [51]. One also observes a power-law scaling of the displacements indicating self-similar behaviour [51]. As large eddies are not detected, however, the similarity with *fluid* turbulence is limited, but there is still an analogy to turbulence at currency exchange markets [95]. Instead of vortex cascades like in turbulent fluids, one rather finds a hierarchical fragmentation dynamics: At extreme densities, individual motion is replaced by mass motion, but there is a stick-slip instability which leads to "rupture" when the stress in the crowd becomes too large. That is, the mass splits up into clusters of different sizes with strong velocity correlations *inside* and distance-dependent correlations *between* the clusters.

"Crowd turbulence" has further specific features [51]. Due to the physical contacts among people in extremely dense crowds, we expect commonalities with granular media. In fact, dense driven granular media may form density waves, while moving forward [96], and can display turbulent-like states [97,98]. Moreover, under quasi-static conditions [97], force chains [99] are building up, causing strong variations in the strengths and directions of local forces. As in earthquakes [100,101] this can lead to events of sudden, uncontrollable stress release with power-law distributed displacements. Such a power-law has also been discovered by video-based crowd analysis [51].

3.4.5 *Some Warning Signs of Critical Crowd Conditions*

Turbulent waves are experienced in dozens of crowd-intensive events each year all over the world [102]. Therefore, it is necessary to understand why, where and when potentially critical situations occur. Viewing real-time video recordings is not very suited to identify critical crowd conditions: While the average density rarely exceeds values of six persons per square meter, the local densities can reach almost twice as large values [51]. It has been found, however, that even evaluating the local densities is not enough to identify the critical times and locations precisely, which also applies to an analysis of the velocity field [51]. The decisive quantity is rather the "crowd pressure", i.e. the density, multiplied with the variance of speeds. It allows one to identify critical locations and times (see Fig. 3.8).

There are even advance warning signs of critical crowd conditions: The crowd accident on January 12, 2006 started about 10 min after "turbulent" crowd motion set in, i.e. after the "pressure" exceeded a value of $0.02/s^2$ (see Fig. 3.8). Moreover, it occurred more than 30 min after stop-and-go waves set in, which can be easily detected in accelerated surveillance videos. Such advance warning signs of critical

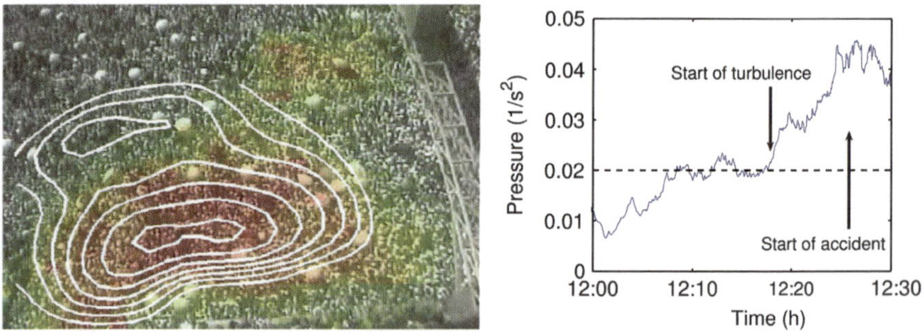

Fig. 3.8 *Left*: Snapshot of the on-line visualization of "crowd pressure". *Red colors* (see the lower ellipses) indicate areas of critical crowd conditions. In fact, the sad crowd disaster during the Muslim pilgrimage on January 12, 2006, started in this area. *Right*: The "crowd pressure" is a quantitative measure of the onset of "crowd turbulence". The crowd disaster started when the "crowd pressure" reached particularly high values. For details see [51]

crowd conditions can be evaluated on-line by an automated video analysis system. In many cases, this can help one to gain time for corrective measures like flow control, pressure-relief strategies, or the separation of crowds into blocks to stop the propagation of shockwaves [51]. Such anticipative crowd control could increase the level of safety during future mass events.

3.4.6 Evolutionary Optimization of Pedestrian Facilities

Having understood some of the main factors causing crowd disasters, it is interesting to ask how pedestrian facilities can be designed in a way that maximizes the efficiency of pedestrian flows and the level of safety. One of the major goals during mass events must be to avoid extreme densities. These often result from the onset of congestion at bottlenecks, which is a consequence of the breakdown of free flow and causes an increasing degree of compression. When a certain critical density is increased (which depends on the size distribution of people), this potentially implies high pressures in the crowd, particularly if people are impatient due to long delays or panic.

The danger of an onset of congestion can be minimized by avoiding bottlenecks. Notice, however, that jamming can also occur at widenings of escape routes [91]. This surprising fact results from disturbances due to pedestrians, who try to overtake each other and expand in the wider area because of their repulsive interactions. These squeeze into the main stream again at the end of the widening, which acts like a bottleneck and leads to jamming. The corresponding drop of efficiency E is more pronounced:

1. If the corridor is narrow.
2. If the pedestrians have different or high desired velocities.
3. If the pedestrian density in the corridor is high.

Obviously, the emerging pedestrian flows decisively depend on the geometry of the boundaries. They can be simulated on a computer already in the planning phase of pedestrian facilities. Their configuration and shape can be systematically varied, e.g. by means of evolutionary algorithms [28, 103] and evaluated on the basis of particular mathematical performance measures [7]. Apart from the *efficiency*

$$E = \frac{1}{N} \sum_{\alpha} \frac{v_\alpha \cdot e_\alpha^0}{v_\alpha^0} \tag{3.19}$$

we can, for example, define the *measure of comfort* $C = (1 - D)$ via the discomfort

$$D = \frac{1}{N} \sum_{\alpha} \frac{\overline{(v_\alpha - \overline{v_\alpha})^2}}{\overline{(v_\alpha)^2}} = \frac{1}{N} \sum_{\alpha} \left(1 - \frac{\overline{v_\alpha}^2}{\overline{(v_\alpha)^2}} \right). \tag{3.20}$$

The latter is again between 0 and 1 and reflects the frequency and degree of sudden velocity changes, i.e. the level of discontinuity of walking due to necessary avoidance maneuvers. Hence, the optimal configuration regarding the pedestrian requirements is the one with the highest values of efficiency and comfort.

During the optimization procedure, some or all of the following can be varied:

1. The location and form of planned buildings.
2. The arrangement of walkways, entrances, exits, staircases, elevators, escalators, and corridors.
3. The shape of rooms, corridors, entrances, and exits.
4. The function and time schedule (Recreation rooms or restaurants are often continuously frequented, rooms for conferences or special events are mainly visited and left at peak periods, exhibition rooms or rooms for festivities require additional space for people standing around, and some areas are claimed by queues or through traffic.)

In contrast to early evolutionary optimization methods, recent approaches allow to change not only the dimensions of the different elements of pedestrian facilities, but also to vary their topology. The procedure of such algorithms is illustrated in Fig. 3.9. Highly performing designs are illustrated in Fig. 3.10. It turns out that, for an emergency evacuation route, it is favorable if the crowd does not move completely straight towards a bottleneck. For example, a zigzag design of the evacuation route can reduce the pressure on the crowd upstream of a bottleneck (see Fig. 3.11). The proposed evolutionary optimization procedure can, of course, not only be applied to the design of new pedestrian facilities, but also to a reduction of existing bottlenecks, when suitable modifications are implemented.

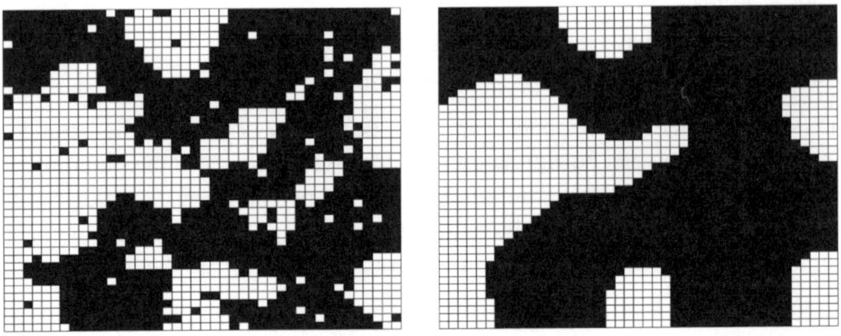

Fig. 3.9 The evolutionary optimization based on Boolean grids uses a two-stage algorithm (see [104] for details). *Left*: In the "randomization stage", obstacles are distributed over the grid with some randomness, thereby allowing for the generation and testing of new topologies (architectures). *Right*: In the "agglomeration stage", small nearby obstacles are clustered to form larger objects with smooth boundaries. After several iterations, the best performing designs are reasonably shaped. See Fig. 3.10 for examples of possible bottleneck designs

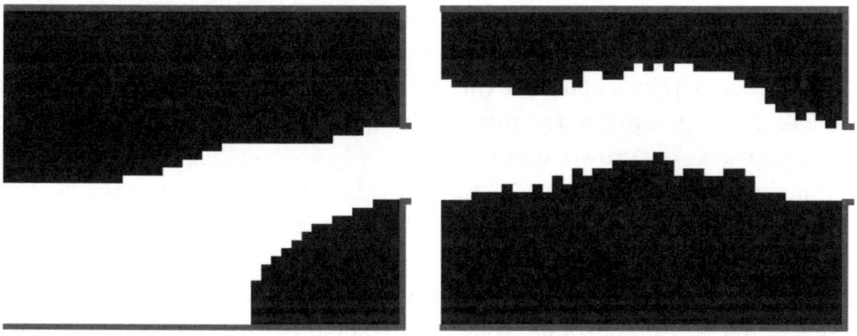

Fig. 3.10 Two examples of improved designs for cases with a bottleneck along the escape route of a large crowd, obtained with an evolutionary algorithm based on Boolean grids (after [104]). People were assumed to move from left to right only. *Left*: Funnel-shaped escape route. *Right*: Zig-zag design

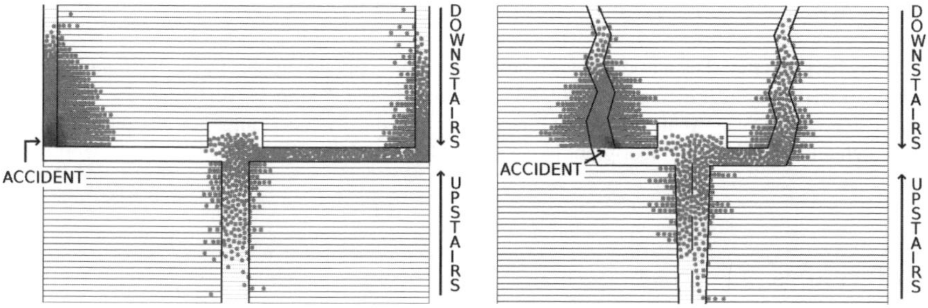

Fig. 3.11 *Left*: Conventional design of a stadium exit in an emergency scenario, where we assume that some pedestrians have fallen at the end of the downwards staircase to the left. The *dark* areas indicate high pressures, since pedestrians are impatient and pushing from behind. *Right*: In the improved design, the increasing diameter of corridors can reduce waiting times and impatience (even with the same number of seats), thereby accelerating evacuation. Moreover, the zigzag design of the downwards staircases changes the pushing direction in the crowd. Computer simulations indicate that the zig-zag design can reduce the average pressure in the crowd at the location of the incident by a factor of two (After [8])

3.5 Future Directions

In this contribution, we have presented a multi-agent approach to pedestrian and crowd dynamics. Despite the great effort required, pedestrian interactions can be well quantified by video tracking. Compared to other social interactions they turn out to be quite simple. Nevertheless, they cause a surprisingly large variety of self-organized patterns and short-lived social phenomena, where coordination or cooperation emerges spontaneously. For this reason, they are interesting to study, particularly as one can expect new insights into coordination mechanisms of social beings beyond the scope of classical game theory. Examples for observed self-organization phenomena in normal situations are lane formation, stripe formation, oscillations and intermittent clogging effects at bottlenecks, and the evolution of behavioral conventions (such as the preference of the right-hand side in continental Europe). Under extreme conditions (high densities or panic), however, coordination may break down, giving rise to "freezing-by-heating" or "faster-is-slower effects", stop-and-go waves or "crowd turbulence".

Similar observations as in pedestrian crowds are made in other social systems and settings. Therefore, we expect that realistic models of pedestrian dynamics will also promote the understanding of opinion formation and other kinds of collective behaviors. The hope is that, based on the discovered elementary mechanisms of emergence and self-organization, one can eventually also obtain a better understanding of the constituting principles of more complex social systems. At least the same underlying factors are found in many social systems: non-linear interactions of individuals, time-dependence, heterogeneity, stochasticity, competition for scarce resources (here: space and time), decision-making, and learning. Future work will certainly also address issues of perception, anticipation, and communication.

Acknowledgements The authors are grateful for partial financial support by the German Research Foundation (research projects He 2789/7-1, 8-1) and by the "Cooperative Center for Communication Networks Data Analysis", a NAP project sponsored by the Hungarian National Office of Research and Technology under grant No. KCKHA005.

References

Primary Literature

1. B.D. Hankin, R.A. Wright, Oper. Res. Q. **9**, 81–88 (1958)
2. S.J. Older, Traffic. Eng. Contr. **10**, 160–163 (1968)
3. U. Weidmann, Transporttechnik der Fußgänger, (Institut für Verkehrsplanung, Transporttechnik, Straßen- und Eisenbahnbau, ETH Zürich, 1993)
4. J.J. Fruin, Designing for pedestrians: A level-of-service concept, in *Highway Research Record, Number 355: Pedestrians* (Highway Research Board, Washington, D.C., 1971), pp. 1–15
5. J. Pauls, Fire Technol. **20**, 27–47 (1984)
6. W.H. Whyte, City. Rediscovering the Center (Doubleday, New York, 1988)
7. D. Helbing, Verkehrsdynamik (Springer, Berlin, 1997)
8. D. Helbing, L. Buzna, A. Johansson, T. Werner, Transport. Sci. **39**(1), 1–24 (2005)
9. W.M. Predtetschenski, A.I. Milinski, *Personenströme in Gebäuden – Berechnungsmethoden für die Projektierung* – (Rudolf Müller, Köln-Braunsfeld, 1971)
10. Transportation Research Board, *Highway Capacity Manual*, Special Report 209 (Transportation Research Board, Washington, D.C., 1985)
11. S.J. Yuhaski Jr., J.M. Macgregor Smith, Queueing Syst. **4**, 319–338 (1989)
12. D. Garbrecht, Traffic Q. **27**, 89–109 (1973)
13. N. Ashford, M. O'Leary, P.D. McGinity, Traffic. Eng. Contr. **17**, 207–210 (1976)
14. A. Borgers, H. Timmermans, Socio-Econ. Plann. Sci. **20**, 25–31 (1986)
15. D. Helbing, *Stochastische Methoden, nichtlineare Dynamik und quantitative Modelle sozialer Prozesse*, Ph.D. thesis (University of Stuttgart, 1992, published by Shaker, Aachen, 1993)
16. D. Helbing, M. Isobe, T. Nagatani, K. Takimoto, Phys. Rev. E **67**, 067101 (2003)

17. W. Daamen, S.P. Hoogendoorn, in *Proceedings of the 82nd Annual Meeting at the Transportation Research Board* (CDROM, Washington D.C., 2003)
18. M. Isobe, D. Helbing, T. Nagatani, Phys. Rev. E **69**, 066132 (2004)
19. A. Seyfried, B. Steffen, W. Klingsch, M. Boltes, J. Stat. Mech. P10002 (2005)
20. T. Kretz, A. Grünebohm, M. Kaufman, F. Mazur, M. Schreckenberg, J. Stat. Mech. P10001 (2006)
21. L.F. Henderson, Transport. Res. **8**, 509–515 (1974)
22. R.L. Hughes, Transport. Res. B **36**, 507–535 (2002)
23. D. Helbing, Complex Syst. **6**, 391–415 (1992)
24. S.P. Hoogendoorn, P.H.L. Bovy, Transport. Res. Record. **1710**, 28–36 (2000)
25. D. Helbing, Behav. Sci. **36**, 298–310 (1991)
26. D. Helbing, P. Molnár, Phys. Rev. E **51**, 4282–4286 (1995)
27. P.G. Gipps, B. Marksjö, Math. Comp. Simul. **27**, 95–105 (1985)
28. K. Bolay, *Nichtlineare Phänomene in einem fluid-dynamischen Verkehrsmodell* (Master's thesis, University of Stuttgart, 1998)
29. V.J. Blue, J.L. Adler, Transport. Res. Record. **1644**, 29–36 (1998)
30. M. Fukui, Y. Ishibashi, J. Phys. Soc. Jpn. **68**, 2861–2863 (1999)
31. M. Muramatsu, T. Irie, T. Nagatani, Physica A **267**, 487–498 (1999)
32. H. Klüpfel, M. Meyer-König, J. Wahle, M. Schreckenberg, in *Theory and Practical Issues on Cellular Automata*, ed. by S. Bandini, T. Worsch (Springer, London, 2000)
33. C. Burstedde, K. Klauck, A. Schadschneider, J. Zittartz, Physica A **295**, 507–525 (2001)
34. S. Gopal, T.R. Smith, in *Spatial Choices and Processes*, ed. by M.M. Fischer, P. Nijkamp, Y.Y. Papageorgiou (North-Holland, Amsterdam, 1990), pp. 169–200
35. C.W. Reynolds, in From Animals to Animats 3: Proceedings of the Third International Conference on Simulation of Adaptive Behavior, ed. by D. Cliff, P. Husbands, J.-A. Meyer, S. Wilson (MIT Press, Cambridge, Massachusetts, 1994), pp. 402–410
36. D. Helbing, Behav. Sci. **37**, 190–214 (1992)
37. D. Helbing, P. Molnár, I. Farkas, K. Bolay, Environ. Plann. B **28**, 361–383 (2001)
38. J. Klockgether, H.-P. Schwefel, in Proceedings of the Eleventh Symposium on Engineering Aspects of Magnetohydrodynamics, ed. by D.G. Elliott (California Institute of Technology, Pasadena, CA, 1970), pp. 141–148
39. D. Helbing, in Economic Evolution and Demographic Change. Formal Models in Social Sciences, ed. by G. Haag, U. Mueller, K.G. Troitzsch (Springer, Berlin, 1992), pp. 330–348
40. N.E. Miller, in Personality and the behavior disorders, ed. by J.McV. Hunt, Vol. 1 (Ronald, New York, 1944)
41. N.E. Miller, in *Psychology: A Study of Science*, ed. by S. Koch, Vol. 2 (McGraw Hill, New York, 1959)
42. K. Lewin, *Field Theory in Social Science* (Harper & Brothers, New York, 1951)
43. D. Helbing, J. Math. Sociol. **19**(3), 189–219 (1994)
44. S. Hoogendoorn, P.H.L. Bovy, Optim. Contr. Appl. Meth. **24**(3), 153–172 (2003)
45. A. Johansson, D. Helbing, P.K. Shukla, Specification of the social force pedestrian model by evolutionary adjustment to videotracking data. Advances in Complex Systems (ACS), **10**(2) 271–288 (2007)
46. T.I. Lakoba, D.J. Kaup, N.M. Finkelstein, Simulation **81**(5), 339–352 (2005)
47. A. Seyfried, B. Steffen, T. Lippert, Physica A **368**, 232–238 (2006)
48. J. Kerridge, T. Chamberlain, in Pedestrian and Evacuation Dynamics '05, ed. by N. Waldau, P. Gattermann, H. Knoflacher, M. Schreckenberg (Springer, Berlin, 2005)
49. S.P. Hoogendoorn, W. Daamen, P.H.L. Bovy, in Proceedings of the 82nd Annual Meeting at the Transportation Research Board (CDROM, Mira Digital Publishing, Washington D.C., 2003)
50. K. Teknomo, Microscopic pedestrian flow characteristics: Development of an image processing data collection and simulation model (PhD thesis, Tohoku University Japan, Sendai, 2002)
51. D. Helbing, A. Johansson, H.Z. Al-Abideen, Phys. Rev. E **75**, 046109 (2007)
52. L.P. Kadanoff, J. Stat. Phys. **39**, 267–283 (1985)

53. H.E. Stanley, N. Ostrowsky (eds.), *On Growth and Form* (Martinus Nijhoff, Boston, 1986)
54. T. Arns, Video films of pedestrian crowds (Stuttgart, 1993)
55. H.-H. Stølum, Nature **271**, 1710–1713 (1996)
56. I. Rodríguez-Iturbe, A. Rinaldo, *Fractal River Basins: Chance and Self-Organization* (Cambridge University, Cambridge, England, 1997)
57. D. Helbing, I. Farkas, T. Vicsek, Phys. Rev. Lett. **84**, 1240–1243 (2000)
58. T. Schelling, J. Math. Sociol. **1**, 143–186 (1971)
59. D. Helbing, T. Platkowski, Int. J. Chaos Theor. Appl. **5**(4), 47–62 (2000)
60. S.P. Hoogendoorn, W. Daamen, Transpn. Sci. **39**(2), 147–159 (2005)
61. T. Kretz, A. Grünebohm, M. Schreckenberg, J. Stat. Mech. P10014 (2006)
62. K. Ando, H. Oto, T. Aoki, Railway Res. Rev. **45**(8), 8–13 (1988)
63. K.H. Drager, G. Løvås, J. Wiklund, H. Soma, D. Duong, A. Violas, V. Lanèrès, in the *Proceedings of the 1992 Emergency Management and Engineering Conference* (Society for Computer Simulation, Orlando, Florida, 1992), pp. 101–108
64. M. Ebihara, A. Ohtsuki, H. Iwaki, Microcomput. Civ. Eng. **7**, 63–71 (1992)
65. N. Ketchell, S. Cole, D.M. Webber, C.A. Marriott, P.J. Stephens, I.R. Brearley, J. Fraser, J. Doheny, J. Smart, in *Engineering for Crowd Safety*, ed. by R.A. Smith, J.F. Dickie (Elsevier, Amsterdam, 1993), pp. 361–370
66. S. Okazaki, S. Matsushita, in *Engineering for Crowd Safety*, ed. by R.A. Smith, J.F. Dickie (Elsevier, Amsterdam, 1993), pp. 271–280
67. G.K. Still, New computer system can predict human behaviour response to building fires. Fire **84**, 40–41 (1993)
68. G.K. Still, *Crowd Dynamics* (Ph.D. thesis, University of Warwick, 2000)
69. P.A. Thompson, E.W. Marchant, *Modelling techniques for evacuation*, in *Engineering for Crowd Safety*, ed. by R.A. Smith, J.F. Dickie (Elsevier, Amsterdam, 1993), pp. 259–269
70. G.G. Løvås, *On the importance of building evacuation system components*, IEEE Trans. Eng. Manag. **45**, 181–191 (1998)
71. H.W. Hamacher, S.A. Tjandra, in *Pedestrian and Evacuation Dynamics*, ed. by M. Schreckenberg, S.D. Sharma (Springer, Berlin, 2001), pp. 227–266
72. D. Elliott, D. Smith, *Football stadia disasters in the United Kingdom: Learning from tragedy?*, Industrial & Environmental Crisis Quarterly **7**(3), 205–229 (1993)
73. B.D. Jacobs, P. 't Hart, in *Hazard Management and Emergency Planning*, Chap. 10, ed. by D.J. Parker, J.W. Handmer (James & James Science, London, 1992)
74. D. Canter (ed.), *Fires and Human Behaviour* (David Fulton, London, 1990)
75. A. Mintz, J. Abnorm. Norm. Soc. Psychol. **46**, 150–159 (1951)
76. J.P. Keating, Fire J., 57–61+147 (May/1982)
77. D.L. Miller, *Introduction to Collective Behavior*, Fig. 3.3 and Chap. 9 (Wadsworth, Belmont, CA, 1985)
78. J.S. Coleman, *Foundations of Social Theory*, Chaps. 9 and 33 (Belkamp, Cambridge, MA, 1990)
79. N.R. Johnson, *Panic at "The Who Concert Stampede": An empirical assessment*, Soc. Prob. **34**(4), 362–373 (1987)
80. G. LeBon, *The Crowd* (Viking, New York, 1960 [1895])
81. E. Quarantelli, Sociol. Soc. Res. **41**, 187–194 (1957)
82. N.J. Smelser, *Theory of Collective Behavior*, (The Free Press, New York, 1963)
83. R. Brown, *Social Psychology* (The Free Press, New York, 1965)
84. R.H. Turner, L.M. Killian, *Collective Behavior*, 3rd edn. (Prentice Hall, Englewood Cliffs, NJ, 1987)
85. J.L. Bryan, Fire J., 27–30+86–90 (Nov./1985)
86. R. Axelrod, W.D. Hamilton, Science **211**, 1390–1396 (1981)
87. R. Axelrod, D. Dion, Science **242**, 1385–1390 (1988)
88. N.S. Glance, B.A. Huberman, Sci. Am. **270**, 76–81 (1994)
89. R.A. Smith, J.F. Dickie (eds.), *Engineering for Crowd Safety* (Elsevier, Amsterdam, 1993)
90. H.H. Kelley, J.C. Condry Jr., A.E. Dahlke, A.H. Hill, J. Exp. Soc. Psychol. **1**, 20–54 (1965)

91. D. Helbing, I. Farkas, T. Vicsek, Nature **407**, 487–490 (2000)
92. G.H. Ristow, H.J. Herrmann, Phys. Rev. E **50**, R5–R8 (1994)
93. D.E. Wolf, P. Grassberger (eds.), *Friction, Arching, Contact Dynamics* (World Scientific, Singapore, 1997)
94. D. Helbing, A. Johansson, J. Mathiesen, M.H. Jensen, A. Hansen Phys. Rev. Lett. **97**, 168001 (2006)
95. S. Ghashghaie, W. Breymann, J. Peinke, P. Talkner, Y. Dodge, Nature **381**, 767–770 (1996)
96. G. Peng, H.J. Herrmann, Phys. Rev. E **49**, R1796–R1799 (1994)
97. F. Radjai, S. Roux, Phys. Rev. Lett. **89**, 064302 (2002)
98. K.R. Sreenivasan, Nature **344**, 192–193 (1990)
99. M.E. Cates, J.P. Wittmer, J.-P. Bouchaud, P. Claudin, Phys. Rev. Lett. **81**, 1841–1844 (1998)
100. P. Bak, K. Christensen, L. Danon, T. Scanlon, Phys. Rev. Lett. **88**, 178501 (2002)
101. P.A. Johnson, X. Jia, Nature **437**, 871–874 (2005)
102. J.J. Fruin, in *Engineering for Crowd Safety*, ed. by R.A. Smith, J.F. Dickie (Elsevier, Amsterdam, 1993), pp. 99–108
103. T. Baeck, *Evolutionary Algorithms in Theory and Practice* (Oxford University Press, New York, 1996)
104. A. Johansson, D. Helbing, in *Pedestrian and Evacuation Dynamics 2005*, ed. by N. Waldau, P. Gattermann, H. Knoflacher, M. Schreckenberg (Springer-Verlag, Berlin, 2007), pp. 267–272

Chapter 4
Opinion Formation*

4.1 Introduction

Many biological systems exhibit collective patterns, which emerge through simple interactions of large numbers of individuals. A typical example is agglomeration phenomena. Such clustering dynamics have been found in systems as different as bacterial colonies [1], gregarious animals like cockroaches [2], fish schools [3], flocks of birds [4], and animal groups [5]. Similar phenomena are observed in ecosystems [6] and human populations, as examples ranging from the formation of pedestrian groups [7] to the formation of urban agglomerations demonstrate [8, 9].

Recently, numerous studies on the structure of human interaction networks [10–12] demonstrated that clustering is *not* restricted to physical or geographical space. For instance, clustering has been extensively studied in networks of email communication [13], phone calls [12], scientific collaboration [14] and sexual contacts [15]. It is much less understood, however, how and what conditions clustering patterns emerge in behavioral or opinion space. Empirical studies suggest that opinions differ globally [16, 17], while they cluster locally within geographical regions [18], socio-demographic groups [19], or Internet communities [20]. In addition, research on dynamics in work teams demonstrates that even groups of very small size often show high opinion diversity and can even suffer from opinion polarization [21, 22].

Opinion clustering is defined as the co-existence of distinct subgroups (clusters) of individuals with similar opinions, while opinions in different subgroups are relatively large. The gaps in our theoretical understanding of opinion clustering are pressing since both local consensus and global diversity are precarious. On the one hand, cultural diversity may get lost in a world where people are increasingly exposed to influences from mass media, Internet communication, interregional

*This chapter reprints a previously published paper and should be cited as follows: *M. Mäs, A. Flache*, and D. Helbing (2010) Individualization as driving force of clustering phenomena in humans. *PLoS Comput. Biol.* **6**(10), e1000959.

D. Helbing (ed.), *Social Self-Organization*, Understanding Complex Systems,
DOI 10.1007/978-3-642-24004-1_4, © Springer-Verlag Berlin Heidelberg 2012

migration, and mass tourism, which may promote a universal monoculture [23, 24], as the extinction of languages suggests [25]. On the other hand, increasing individualization threatens to disintegrate the social structures in which individuals are embedded, with the possible consequence of the loss of societal consensus [26, 27]. This is illustrated by the recent debate on the decline of social capital binding individuals into local communities [28].

Early formal models of social influence imply that monoculture is unavoidable, unless a subset of the population is perfectly cut off from outside influences [29]. Social isolation, however, appears questionable as explanation of pluralism. In modern societies, distances in social networks are quite short on the whole, and only relatively few random links are required to dramatically reduce network distance [10].

Aiming to explain pluralism, researchers have incorporated the empirically well-supported observation of "homophily", i.e. the tendency of "birds of a feather to flock together" [30, 31], into formal models of social influence [32]. These models typically assume "bounded confidence" (BC) in the sense that only those individuals interact, whose opinions do not differ more than a given threshold level [33, 34]. As Fig. 4.1a illustrates, BC generates opinion clustering, a result that generalizes to model variants with categorical rather than continuous opinions [32, 35]. However, clustering in the BC-model is sensitive to "interaction noise": A small random chance that agents may interact even when their opinions are not similar, causes monoculture again (see Fig. 4.1b).

To avoid this convergence of opinions, it was suggested that individuals would separate themselves from negatively evaluated others [19, 36, 37]. However, recent empirical results do not support such "negative influence" [38]. Scientists also tried to avoid convergence by "opinion noise", i.e. random influences, which lead to arbitrary opinion changes with a small probability. Assuming uniformly distributed opinion noise [39] leads to sudden, large, and unmotivated opinion changes of individuals, while theories of social integration [26, 27, 40, 41] and empirical studies of individualization [42, 43] show a tendency of incremental opinion changes rather than arbitrary opinion jumps. Incremental opinion changes, however, tend to promote monoculture, even in models with categorical rather than continuous opinions [44]. Figure 4.1 demonstrates that adding a "white noise" term $(N(0, \theta))$ to an agent's current opinion in the BC model fails to explain opinion clustering. Weak opinion noise ($\theta = 5$) triggers convergence cascades that inevitably end in monoculture. Stronger noise restores opinion diversity, but not clustering. Instead, diversity is based on frequent individual deviations from a predominant opinion cluster (for $\theta = 18$). However, additional clusters can not form and persist, because opinion noise needs to be strong to separate enough agents from the majority cluster – so strong that randomly emerging smaller clusters cannot stabilize.

In conclusion, the formation of persistent opinion clusters is such a difficult puzzle that all attempts to explain them had to make assumptions that are difficult to justify by empirical evidence. The solution proposed in the following, in contrast, aims to reconcile model assumptions with sociological and psychological research. The key innovation is to integrate another decisive feature into the model, namely

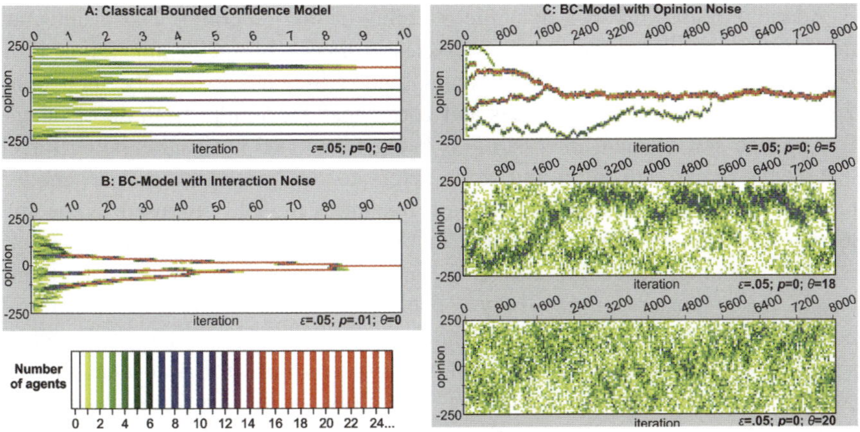

Fig. 4.1 *Opinion dynamics produced by the bounded confidence (BC) model [33] with and without noise*. Populations consist of 100 agents. Opinions vary between −250 and 250. Initial opinions are uniformly distributed. For visualization, the opinion scale is divided into 50 bins of equal size. Color coding indicates the relative frequency of agents in each bin. (**a**) Dynamics of the *BC-model without noise* [33] over ten iterations. At each simulation event, one agent's opinion is replaced by the average opinion of those other agents who hold opinions $o_j(t)$ within the focal agent's confidence interval $(o_i(t) - \epsilon \leq o_j(t) \leq o_i(t) + \epsilon)$. For $\epsilon = 0.05$, one finds several homogeneous clusters, which stabilize when the distance between all clusters exceeds the confidence threshold ϵ. (**b**) Computer simulation of the same BC-model, but considering *interaction noise*. Agents that would otherwise not have been influential, now influence the focal agent's opinion with a probability of $p = 0.01$. This small noise is sufficient to eventually generate monoculture. (**c**) Simulation of the BC-model with opinion noise. After each opinion update, a random value drawn from a normal distribution with an average of zero and a standard deviation of θ (abbreviated by $N(0, \theta)$) is added to the opinion. For weak opinion noise ($\theta = 5$), one cluster is formed, which carries out a random walk on the opinion scale. When the opinion noise is significantly increased ($\theta = 18$), there is still one big cluster, but many separated agents exist as well (cf. Fig. 4.4). With even stronger opinion noise ($\theta = 20$), the opinion distribution becomes completely random

the "striving for uniqueness" [42, 43]. While individuals are influenced by their social environment, they also show a desire to increase their uniqueness when too many other members of society hold similar opinions. We incorporate this assumption as a white noise term in the model. However, in contrast to existing models we assume that noise strength is not constant but *adaptive*. To be precise, we assume that the impact of noise on the opinion of an individual is the stronger the less unique the individual's opinion is compared to the other members of the population. Consumer behavior regarding fashions illustrates the adaptiveness of opinion noise: When new clothing styles are adopted by some people, they often tend to be imitated by others with similar spirit and taste (the "peer group"). However, when imitation turns the new style into a norm, people will seek to increase their uniqueness. This will sooner or later lead some individuals to invent new ways to dress differently from the new norm.

Adaptive noise creates a dynamic interplay of the integrating and disintegrating forces highlighted by Durkheim's classic theory of social integration [26]. Durkheim argued that integrating forces bind individuals to society, motivating them to conform and adopt values and norms that are similar to those of others. But he also saw societal integration as being threatened by disintegrating forces that foster individualization and drive actors to differentiate from one another [27, 40, 41]. The "Durkheimian opinion dynamics model" proposed in the following can explain pluralistic clustering for the case of continuously varying opinions, although it incorporates all the features that have previously been found to *undermine* clustering: (1) a fully connected influence network, (2) absence of bounded confidence, (3) no negative influence, and (4) white opinion noise. From a methodological viewpoint, our model builds on concepts from statistical physics, namely the phenomenon of "nucleation" [45], illustrated by the formation of water droplets in supersaturated vapor. However, by assuming adaptive noise, we move beyond conventional nucleation models. The model also resembles elements of Interacting Particle Systems [46] like the voter model and the anti-voter model [47–50] which have been used to study dynamics of discrete opinions ("pro" and "contra"). However, we focus here on *continuous* opinions like the degree to which individuals are in favor of or against a political party.

Computational simulation experiments reveal that, despite the continuity of opinions in our model, it generates pluralism as an intermediate phase between monoculture and individualism. When the integrating forces are too strong, the model dynamics inevitably implies monoculture, even when the individual opinions are initially distributed at random. When the disintegrating forces prevail, the result is what Durkheim called "anomie", a state of extreme individualism without a social structure, even if there is perfect consensus in the beginning. Interestingly, there is no sharp transition between these two phases, when the relative strength of both forces is changed. Instead, we observe an additional, intermediate regime, where opinion clustering occurs, which is independent of the initial condition. In this regime, adaptive noise entails robust pluralism that is stabilized by the adaptiveness of cluster size. When clusters are small, individualization tendencies are too weak to prohibit a fusion of clusters. However, when clusters grow large, individualization increases in strength, which triggers a splitting into smaller clusters ("fission"). In this way, our model solves the cluster formation problem of earlier models. While in BC models, white noise causes either monoculture or fragmentation (Fig. 4.1c), in the Durkheimian opinion dynamics model proposed here, it *enables* clustering. Therefore, rather than *endangering* cluster formation, noise supports it. In the following, we describe the model and identify conditions under which pluralism can flourish.

4.2 Model

The model has been elaborated as an agent-based model [51] addressing the opinion dynamics of interacting individuals. The simulated population consists of N agents i, representing individuals, each characterized by an opinion $o_i(t)$ at time t.

The numerical value for the opinion varies between a given minimum and maximum value on a metric scale. We use the term "opinion" here, for consistency with the literature on social influence models. However, $o_i(t)$ may also reflect behaviors, beliefs, norms, customs or any other cardinal cultural attribute that individuals consider relevant and that is changed by social influence. The dynamics is modeled as a sequence of events. Every time $t' = k/N$ (with $k \in \{1, \ldots, N\}$), the computer randomly picks an agent i and changes the opinion $o_i(t)$ by the amount

$$\Delta o_i = \frac{\sum\limits_{\substack{j=1 \\ j \neq i}}^{N} \big(o_j(t) - o_i(t)\big) w_{ij}(t)}{\sum\limits_{\substack{j=1 \\ j \neq i}}^{N} w_{ij}(t)} + \xi_i(t). \tag{4.1}$$

The first term on the rhs of (4.1) models the integrating forces of Durkheim's theory. Technically, agents tend to adopt the weighted average of the opinions $o_j(t)$ of all other members j of the population. Implementing homophily, the social influence $w_{ij}(t)$ that agent j has on agent i is the stronger, the smaller their opinion distance $d_{ij}(t) = |o_j(t) - o_i(t)|$ is. Formally, we assume

$$w_{ij}(t) = \mathrm{e}^{-d_{ij}(t)/A} = \mathrm{e}^{-|o_j(t) - o_i(t)|/A}. \tag{4.2}$$

The parameter A represents the range of social influence of agents. For small positive values of A, agents are very confident in their current opinion and are mainly influenced by individuals who hold very similar opinions, while markedly distinct opinions have little impact. The higher A is, however, the more are agents influenced by individuals with considerably different opinions and the stronger are the integrating forces in our Durkheimian theory.

The *dis*integrating forces on the opinion of agent i are modeled by a noise term $\xi_i(t)$. Specifically, the computer adds a normally distributed random value $\xi_i(t)$ ("white noise") to the first term on the rhs of (4.1). While we assume that the mean value of the random variable $\xi_i(t)$ is zero, the standard deviation has been specified as

$$\theta_i(t) = s \sum_{j=1}^{N} \mathrm{e}^{-d_{ij}(t)}. \tag{4.3}$$

The larger the standard deviation, the stronger are the individualization tendencies of an agent. Following Durkheim's theory, (4.3) implements noise in an *adaptive* way: Accordingly, an agent's striving for individualization is weak, if there are only a few others with similar opinions. Under such conditions, there is no need to increase distinctiveness. However, if many others hold a similar opinion, then individuals are more motivated to differ from others.

By including the focal agent i in the sum of (4.3), we assume that there is always some degree of opinion noise, even when agent i holds a perfectly unique opinion. These fluctuations may have a variety of reasons, such as misjudgments, trial-and-error behavior, or the influence of exogenous factors on the individual opinion. Furthermore, this assumption reflects Durkheim's notion that the seeking for uniqueness is a fundamental feature of human personality, which can not be suppressed completely [26, 52].

We use the parameter s of (4.3) to vary the strength of the disintegrating forces in society. The higher the value of s, the higher is the standard deviation of the distribution, from which $\xi_i(t)$ is drawn, and the stronger are the disintegrating forces. Finally, to keep the opinions of the agents within the bounds of the opinion scale, we set the value of $\xi_i(t)$ to zero, if the bounds of the opinion space would be left otherwise.

4.3 Results

We have studied the Durkheimian opinion dynamics model with extensive computer simulations, focusing on relatively small populations ($N = 100$), because in this case it is reasonable to assume that all members may interact with each other. For bigger populations one would have to take into account the topology of the social interaction network as well. Such networks would most likely consist of segregated components ("communities"), which are not or only loosely connected with each other [12–15]. Existing social influence models can explain how under such conditions each community develops its own shared opinion (see Fig. 4.1a). However, according to these models opinion clustering is only stable when there is no interaction between communities [29, 33], an assumption that appears *not* to be empirically correct in an increasingly connected world. Therefore, we focus on a setting for which the lack of connectedness is guaranteed to be excluded as explanation of clustering and study model dynamics in relatively small and complete interaction networks.

To illustrate the model dynamics, Fig. 4.2 shows three typical simulation runs for different strengths s of disintegrating forces, while the strength $A = 2$ of the integrating force is kept constant. In each run, all agents start with an opinion in the middle of the opinion scale ($o_i(0) = 0$), i.e. conformity. This is an initial condition for which the classical BC-model does not produce diversity. Figure 4.2a shows typical opinion trajectories for a population in which the integrating forces are much stronger than the disintegrating forces. Consequently, the population develops collective consensus, i.e. the variation of opinions remains small, even though not all agents hold exactly the same opinion. Triggered by the random influences $\xi_i(t)$, the average opinion performs a characteristic random walk.

When the disintegrating force prevails, the pattern is strikingly different. Figure 4.2b shows that for large noise strengths s, the initial consensus breaks up quickly, and the agents' opinions are soon scattered across the entire opinion

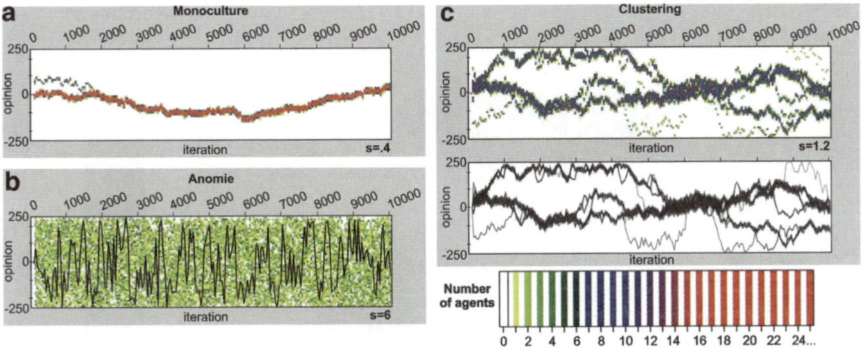

Fig. 4.2 *Opinion trajectories of three representative simulation runs with 100 agents generated by the Durkheimian model.* In all three runs, the opinions are restricted to values between -250 and 250, and all agents hold the same opinion initially ($o_i(0) = 0$ for all i). In all runs, we assume the same social influence range $A = 2$, but vary the strength s of the disintegrating force. (**a**) Monoculture, resulting in the case of a weak disintegrating force ($s = 0.4$). Agents do not hold perfectly identical opinions, but the variance is low. We studied dynamics over 10.000 iterations. (**b**) Anomie (i.e. extreme individualism), generated by a very strong disintegrating force ($s = 6$). Agents spread over the complete opinion scale. The *black line* represents the time-dependent opinion of a single, randomly picked agent, showing significant opinion changes over time, which is in contrast to the collective opinion formation dynamics found in the monocultural and pluralistic cases (**a**) and (**b**). (**c**) For a moderate disintegrating force ($s = 1.2$), the population quickly disintegrates into clusters. As long as these clusters are small, they are metastable. However, clusters perform random walks and can merge (e.g. around iteration 5,500). As the disintegrating force grows with the size of a cluster, big clusters eventually split up into subclusters (e.g. around iteration 7,000). The additional graph, in which each agent's opinion trajectory is represented by a *solid black line*, is an alternative visualization of the simulation run with $s = 1.2$. It shows that the composition of clusters persists over long time periods

space. Simulation scenarios A and B are characteristic for what Durkheim referred to as states of social cohesion and of anomie. Interestingly, however, pluralism arises as a third state in which several opinion clusters form and coexist. Figure 4.2c shows a typical simulation run, where the adaptive noise maintains pluralism despite the antagonistic impacts of integrating and disintegrating forces – in fact *because* of this. In the related region of the parameter space, disintegrating forces prevent global consensus, but the integrating forces are strong enough to also prevent the population from extreme individualization. This is in pronounced contrast to what we found for the BC-model with strong noise (Fig. 4.1c). Instead, we obtain a number of coexisting, metastable clusters of a characteristic, parameter-dependent size. Each cluster consists of a relatively small number of agents, which keeps the disintegrating forces in the cluster weak and allows clusters to persist. (Remember that the tendency of individualization according to (4.3) increases, when many individuals hold similar opinions.) However, due to opinion drift, distinct clusters may eventually merge. When this happens, the emergent cluster becomes unstable and will eventually split up into smaller clusters, because disintegrating forces increase in strength as a cluster grows.

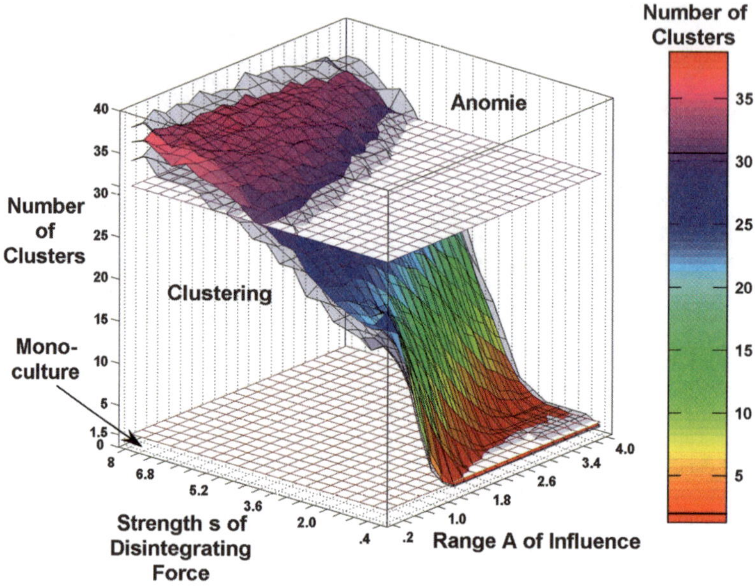

Fig. 4.3 *Conditions of clustering, monoculture and anomie.* The figure shows the dependence of the average number of clusters in the Durkheimian model on the strength s of the disintegrating force and the range A of social influence. To generate it, we conducted computer simulations with $N = 100$ agents, starting with initial consensus ($o_i(0) = 0$ for all i). We restricted opinions to values between -250 and 250. We varied the strength s of the disintegrating force between $s = 0.4$ and $s = 8$ in steps of 0.4. A varied between $A = 0.2$ and $A = 4$ in steps of 0.2. For each parameter combination, we conducted 100 independent replications and assessed the average number of clusters formed after 250,000 iterations (see z-axis and the color scale). The two transparent (*gray*) surfaces depict the inter-quartile range, which indicates a small variance in the number of clusters (and also typical cluster sizes) for each parameter combination. The *horizontal grids* indicate the borders of the three phases, as defined by us. An average cluster size below 1.5 indicates monoculture. Values between 1.5 and 31 reflect clustering. Finally, values above 31 correspond to opinion distributions that cannot be distinguished from random ones and represent a state of anomie

Strikingly, the state of diversity, in which several opinion clusters can coexist, is not restricted to a narrow set of conditions under which integrating and disintegrating forces are balanced exactly. Figure 4.3 demonstrates that opinion clusters exist in a significant area of the parameter space, i.e. the clustering state establishes another phase, which is to be distinguished from monoculture and from anomie.

To generate Fig. 4.3, we conducted a simulation experiment in which we varied the influence range A and the strength s of the disintegrating force. For each parameter combination, we ran 100 replications and measured the average number of clusters that were present after 250,000 iterations. To count the number of clusters in a population, we ordered the N agents according to their opinion. A cluster was defined as a set of agents in adjacent positions such that each set member was separated from the adjacent set members by a maximum of 5 scale points (= opinion range/N). Figure 4.3 shows that, for large social influence ranges A and small noise strengths s, the average number of clusters is below 1.5, reflecting monoculture in the population. In the other extreme, i.e. for a small influence range A and large noise strengths s, the resulting distribution contains more than 31 clusters, a number of clusters that cannot be distinguished from purely random distributions. Following Durkheim, we have classified such cases as anomie, i.e. as the state of extreme individualism. Between these two phases, there are numerous parameter combinations, for which the number of clusters is higher than 1.5 and clearly smaller than in the anomie phase. This constitutes the clustering phase. Figure 4.3 also shows that, for each parameter combination, there is a small variance in the number of clusters, which is due to a statistical equilibrium of occasional fusion and fission processes of opinion clusters (see Fig. 4.2c).

The same results were found, when starting the computer simulations with a uniform opinion distribution. This demonstrates that the simulations were run long enough (250,000 iterations) to obtain reliable results. It also suggests that clustering is an attractor in the sense that the model generates clustering independent of the initial distribution of opinions. In addition, we performed additional statistical tests with the simulation outcomes to make sure that the existence of clusters in our model indeed indicates pluralism and not fragmentation, a state in which a population consists of one big cluster and a number of isolated agents (see Fig. 4.4). To illustrate, Fig. 4.4a plots the size of the biggest cluster in the population versus the number of clusters (see the blue areas). For comparison, the yellow area depicts the corresponding distribution for randomly fragmented opinion distributions. The figure shows that the distributions hardly overlap and that the Durkheimian model generates clustering rather than fragmentation. In clear contrast, Fig. 4.4b reveals that the opinion distributions generated by the noisy BC-model are fragmented and *not* clustered.

4.4 Discussion

The phenomenon of self-organized clustering phenomena in biological and social systems is widespread and important. With the advent of mathematical and computer models for such phenomena, there has been an increasing interest to study them also in human populations. The work presented here focuses on resolving the long-standing puzzle of opinion clustering.

The emergence and persistence of pluralism is a striking phenomenon in a world in which social networks are highly connected and social influence is an ever present force that reduces differences between those who interact. We have developed a formal theory of social influence that, besides anomie and monoculture, shows a third, pluralistic phase characterized by opinion clustering. It occurs, when all individuals interact with each other and noise prevents the convergence to a single opinion, despite homophily.

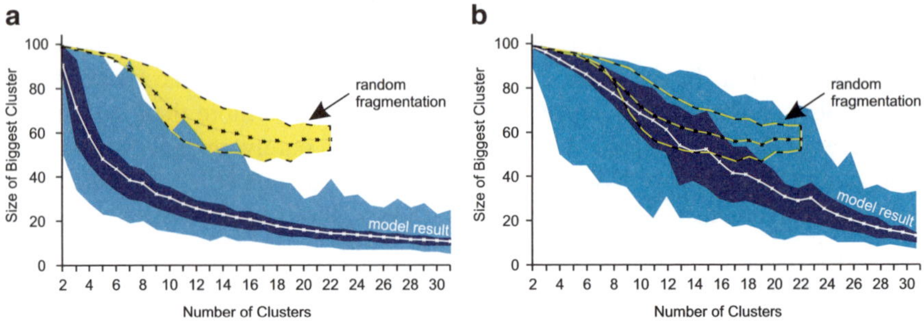

Fig. 4.4 *Comparison of the (**a**) Durkheimian model and (**b**) the noisy BC-model.* Figures plot the size of the biggest cluster versus the number of clusters and compare it to the case of random fragmentation in all simulation runs that resulted in more than one and less than 32 clusters. Figure 4.4a is based on the simulation experiment with the Durkheimian model underlying Fig. 4.3. Figure 4.4b is based on an experiment with the BC-model [33] where we varied the bounded-confidence level ϵ between 0.01 and 0.15 in steps of 0.02 and the noise level θ between 5 and 50 in steps of 5. We conducted 100 replications per parameter combination and measured the number of clusters and the size of the biggest cluster after 250,000 iterations. *White solid lines* represent the average size of the biggest cluster. The *dark blue area* shows the respective interquartile range and the *light blue area* the complete value range. For comparison, we generated *randomly fragmented opinion distributions* of $N = 100$ agents where n agents hold random opinions ($N(0, 50)$) and the remaining $N - n$ agents hold opinion $o_i = 0$ and form one big cluster. We varied the value of n between 0 and 100 in steps of 1 and generated 1,000 distributions per condition. The average size of the biggest cluster of the resulting distributions is shown by the thin *yellow-black line*. (The curve stops at 22, since this is the highest number of clusters generated.) The *bold yellow-black lines* represent the related interquartile range. We find that the value range of the Durkheimian model (*blue area*) hardly overlaps with the interquartile range of the fragmented distributions (*yellow area*). This demonstrates that the Durkheimian model shows clustering rather than fragmentation. In contrast, Fig. 4.4b illustrates that the distributions of the noisy BC-model and the results for random fragmentation overlap

Our model does not assume negative influence, and it behaves markedly different from bounded confidence models, in which white opinion noise produces fragmentation rather than clustering. Furthermore, our model does not rely on the problematic assumption of classical influence models that agents are forevermore cut-off from influence by members of distinct clusters. In order to demonstrate this, we studied model predictions in a setting where all members of the population interact with each other. However, empirical research shows that opinion clustering tends to coincide with clustered network structures [20] and spatial separation [18]. It would therefore be natural to generalize the model in a way that it also considers the structure of real social networks. Such a model is obtained by replacing the values $w_{ij}(t)$ by $w_{ij}(t)a_{ij}$, where a_{ij} are the entries of the adjacency matrix (i.e. $a_{ij} = 1$, if individuals i and j interact, otherwise $a_{ij} = 0$). Then, the resulting opinion clusters are expected to have a broad range of different sizes, similar to what is observed for the sizes of social groups.

Our model highlights the functional role that "noise" (randomness, fluctuations, or other sources of variability) plays for the organization of social systems. It furthermore shows that the combination of two mechanisms (deterministic integrating forces and stochastic disintegrating forces) can give rise to new phenomena. We also believe that our results are meaningful for the analysis of the social integration of our societies. According to Durkheim's theory of the development of societies [26], traditional human societies are characterized by "mechanical solidarity". In these societies, individuals are strongly integrated in very homogeneous communities which exert strong influence on the behavior and opinions of individuals. According to Durkheim, however, these regulating social structures dissolve as societies turn modern. In addition, Durkheim [26] and contemporary social thinkers [27] argue that in modern and globalized societies individuals are increasingly exposed to disintegrating forces, which foster individualization [26]. As a consequence, the social forces which let individuals follow societal norms may lose their power to limit individual variation. Durkheim feared that the high diversity could disintegrate societies as they modernize [26]. That is, extreme individualization in modern societies may obstruct the social structures that traditionally provided social support and guidance to individuals.

Today, modern societies are highly diverse, but at the same time they are far from a state of disintegration as foreseen by Durkheim. He argued that this is possible if societies develop what he called "organic solidarity". In this state societies are highly diverse but at the same time the division of labor creates a dense web of dependencies which integrate individuals into society and generate sufficient moral and social binding [26]. Strikingly, our formal model of Durkheim's theory revealed another possibility which does not require additional integrating structures like the division of labor: Besides monoculture and anomie, there is a third, pluralistic clustering phase, in which individualization prevents overall consensus, but at the same time, social influence can still prevent extreme individualism. The interplay between integrating and disintegrating forces leads to a plurality of opinions, while metastable subgroups occur, within which individuals find a local consensus. Individuals may identify with such subgroups and develop long-lasting social relationships with similar others. Therefore, they are not isolated and not without support or guidance, in contrast to the state of disintegration that Durkheim was worried about.

We have seen, however, that pluralism and cultural diversity require an approximate balance between integrating and disintegrating forces. If this balance is disturbed, societies may drift towards anomie or monoculture. It is, therefore, interesting to ask how the current tendency of globalization will influence society and cultural dynamics. The Internet, interregional migration, and global tourism, for example, make it easy to get in contact with members of distant and different cultures. Previous models [24, 35] suggest that this could affect cultural diversity in favor of a monoculture. However, if the individual striving for uniqueness is sufficiently strong, formation of diverse groups (a large variety of international social communities) should be able to persist even in a globalizing world. In view of the alternative futures, characterized by monoculture or pluralism, further theoretical, empirical, and experimental research should be performed to expand our knowledge of the mechanisms that will determine the future of pluralistic societies.

Acknowledgements We thank Tobias Stark, Heiko Rauhut, Jacob G. Foster and Michael Macy as well as the members of the Norms and Networks cluster at the Department of Sociology at the University of Groningen and the members of the Cooperative Relations and Social Networks Seminar at the Department of Sociology at Utrecht University for their constructive comments.

Author Contributions Conceived and designed the experiments: MM AF DH. Performed the experiments: MM. Analyzed the data: MM. Wrote the paper: MM AF DH.

References

1. E. Ben-Jacob, O. Schochet, A. Tenenbaum, I. Cohen, A. Czirok, T. Vicsek, Generic Modeling of Cooperative Growth-Patterns in Bacterial Colonies. Nature **368**, 46–49 (1994)
2. R. Jeanson, C. Rivault, J.L. Deneubourg, S. Blanco, R. Fournier, C. Jost, G. Theraulaz, Self-organized aggregation in cockroaches. Anim. Behav. **69**, 169–180 (2005)
3. J. Gautrais, C. Jost, G. Theraulaz, Key behavioural factors in a self-organised fish school model. Ann. Zool. Fenn. **45**, 415–428 (2008)
4. M. Ballerini, N. Calbibbo, R. Candeleir, A. Cavagna, E. Cisbani, et al., Interaction ruling animal collective behavior depends on topological rather than metric distance: Evidence from a field study. Proc. Natl. Acad. Sci. USA **105**, 1232–1237 (2008)
5. I.D. Couzin, J. Krause, N.R. Franks, S.A. Levin, Effective leadership and decision-making in animal groups on the move. Nature **433**, 513–516 (2005)
6. Y. Iwasa, V. Andreasen, S.A. Levin, Aggregation in model ecosystems. 1. Perfect Aggregation. Ecol. Model **37**, 287–302 (1987)
7. M. Moussaid, N. Perozo, S. Garnier, D. Helbing, G. Theraulaz, The walking behaviour of pedestrian social groups and its impact on crowd dynamics. PLoS One **5**(4), e10047 (2010)
8. H.A. Makse, S. Havlin, H.E. Stanley, Modeling Urban-Growth Patterns. Nature **377**, 608–612 (1995)
9. M. Batty, The size, scale, and shape of cities. Science **319**, 769–771 (2008)
10. D.J. Watts, S.H. Strogatz, Collective dynamics of 'small-world' networks. Nature **393**, 440–442 (1998)
11. A.L. Barabasi, R. Albert, Emergence of scaling in random networks. Science **286**, 509–512 (1999)
12. G. Palla, A.L. Barabasi, T. Vicsek, Quantifying social group evolution. Nature **446**, 664–667 (2007)
13. D. Liben-Nowell, J. Kleinberg, Tracing information flow on a global scale using Internet chain-letter data. Proc. Natl. Acad. Sci. USA **105**, 4633–4638 (2008)
14. M.E.J. Newman, Coauthorship networks and patterns of scientific collaboration. Proc. Natl. Acad. Sci. USA **101**, 5200–5205 (2004)
15. F. Liljeros, C.R. Edling, L.A.N. Amaral, H.E. Stanley, Y. Aberg, The web of human sexual contacts. Nature **411**, 907–908 (2001)
16. M.P. Fiorina, S.J. Abrams, Political polarization in the American public. Annu. Rev. Polit. Sci. **11**, 563–588 (2008)
17. P. DiMaggio, J. Evans, B. Bryson, Have Americans' social attitudes become more polarized? Am. J. Sociol. **102**, 690–755 (1996)
18. E.L. Glaeser, B.A. Ward, Myths and realities of American political geography. J. Econ. Perspect. **20**, 119–144 (2006)
19. N.P. Mark, Culture and competition: Homophily and distancing explanations for cultural niches. Am. Sociol. Rev. **68**, 319–345 (2003)
20. D. Lazer, et al., Computational Social Science. Science **323**, 721–723 (2009)

21. F.J. Milliken, L.L. Martins, Searching for Common Threads: Understanding the Multiple Effects of Diversity in Organizational Groups. Acad. Manag. J. **21**, 402–433 (1996)
22. P.C. Early, E. Mosakowski, Creating hybrid team cultures: an empirical test of transnational team functioning. Acad. Manag. J. **43**, 26–49 (2000)
23. T.L. Friedman, *The World is Flat. A brief history of the twenty-first century* (Farrar, Straus and Giroux, New York, 2005)
24. M.J. Greig, The end of geography? Globalization, communication and culture in the international system. J. Conflict Res. **46**, 225–243 (2002)
25. W.J. Sutherland, Parallel extinction risk and global distribution of languages and species. Nature **423**, 276–279 (2003)
26. E. Durkheim, *The Division of Labor in Society* (The Free Press, New York, 1997 [1893])
27. U. Beck, in *Reflexive Modernization. Politics, Tradition and Aestherics in the Modern Social Order*, ed. by U. Beck, A. Giddens, S. Lash (Polity Press, Cambridge, 1994)
28. M. McPherson, L. Smith-Lovin, M.E. Brashears, Social isolation in America: Changes in core discussion networks over two decades. Am. Sociol. Rev. **71**, 353–375; also consult the discussion on this article in Am. Sociol. Rev. **74**, 4 (2006)
29. R.P. Abelson, in *Contributions to Mathematical Psychology*, ed. by N. Frederiksen, H. Gulliksen (Rinehart Winston, New York, 1964), pp. 142–160
30. M. McPherson, L. Smith-Lovin, J.M. Cook, Birds of a feather: Homophily in social networks. Annu. Rev. Sociol. **27**, 415–444 (2001)
31. S. Aral, L. Muchnik, A. Sundararajan, Distinguishing influenced-based contagion from homophily-driven diffusion in dynamic networks. Proc. Natl. Acad. Sci. USA **106**, 21544–21549 (2009)
32. A. Nowak, J. Szamrej, B. Latané, From private attitude to public opinion: a dynamic theory of social impact. Psychol. Rev. **97**, 362–376 (1990)
33. R. Hegselmann, U. Krause, Opinion dynamics and bounded confidence models. Analysis, and Simulation J. Artif. Soc. S **5** (2002)
34. G. Deffuant, S. Huet, F. Amblard, An individual-based model of innnovation diffusion mixing social value and individual benefit. Am. J. Soc. **110**, 1041–1069 (2005)
35. R. Axelrod, The dissemination of culture - A model with local convergence and global polarization. J. Conflict Res. **41**, 203–226 (1997)
36. M.W. Macy, J. Kitts, A. Flache, S. Benard, in *Dynamic Social Network Modelling and Analysis*, ed. by R. Breiger, K. Carley, P. Pattison (The National Academies Press, Washington, DC, 2003), pp. 162–173
37. A. Flache, M. Mäs, How to get the timing right? A computational model of how demographic faultlines undermine team performance and how the right timing of contacts can solve the problem. Comput. Math. Organ. Theory **14**, 23–51 (2008)
38. Z. Krizan, R.S. Baron, Group polarization and choice-dilemmas: How important is self-categorization? Eur. J. Soc. Psychol. **37**, 191–201 (2007)
39. M. Pineda, R. Toral, E. Hernandez-Garcia, Noisy continuous-opinion dynamics. J. Stat. Mech. P08001 (2009)
40. M.J. Hornsey, J. Jetten, The individual within the group: Balancing the need to belong with the need to be different. Pers. Soc. Psychol. Rev. **8**, 248–264 (2004)
41. V.L. Vignoles, X. Chryssochoou, G.M. Breakwell, The distinctiveness principle: Identity, meaning, and the bounds of cultural relativity. Pers. Soc. Psychol. Rev. **4**, 337–354 (2000)
42. R. Imhoff, H.P. Erb, What motivates nonconformity? Uniqueness seeking blocks majority influence. Pers. Soc. Psychol. B **35**, 309–320 (2009)
43. C.R. Snyder, H.L. Fromkin, *Uniqueness. The Human Pursuit of Difference* (Plenum Press, New York and London, 1980)
44. K. Klemm, V.M. Eguiluz, R. Toral, M.S. Miguel, Global culture: A noise-induced transition in finite systems. Phys. Rev. E **67**, 045101(R) (2003)
45. H.E. Stanley, *Introduction to Phase Transitions and Critical Phenomena* (Oxford University Press, Oxford and New York, 1971)

46. W. Weidlich, The statistical description of polarization phenomena in society. Br. J. Math. Stat. Psychol. **24**, 251–266 (1971)
47. T.M. Liggett, *Interacting Particle Systems* (Springer, New York, 1985)
48. K. Sznajd-Weron, J. Sznajd, Opinion Evolution in Closed Community. Int. J. Mod. Phys. C **11**, 1157–1165 (2000)
49. S. Galam, Heterogeneous beliefs, segregation, and extremism in the making of public opinions. Phys. Rev. E **71**, 46123 (2005)
50. L. Behera, F. Schweitzer, On spatial consensus formation: Is the Sznajd model different from a voter model? International Journal of Modern Physics C **14**(10), 1331–1354 (2003)
51. E. Bonabeau, Agent-based modeling: Methods and techniques for simulating human systems. Proc. Natl. Acad. Sci. USA **99**, 7280–7287 (2002)
52. E. Durkheim, *Suicide. A study in Sociology* (The Free Press, New York, 1997 [1897])

Chapter 5
Spatial Self-organization Through Success-Driven Mobility*

5.1 Introduction

Although the biological, social, and economic world are full of self-organization phenomena, many people believe that the dynamics behind them is too complex to be modelled in a mathematical way. Reasons for this are the huge number of interacting variables, most of which cannot be quantified, plus the assumed freedom of decision-making or large fluctuations within biological and socio-economic systems. However, in many situations, the living entities making up these systems decide for some (more or less) optimal behavior, which can make the latter describable or predictable to a certain extend [1–10]. This is even more the case for the behavior shown under certain constraints like, for example, in pedestrian or vehicle dynamics [11–13]. While pedestrians or vehicles can move freely at small traffic densities, at large densities the interactions with the others and with the boundaries of the street confines them to a small spectrum of moving behaviors. Consequently, empirical traffic dynamics can be reproduced by simulation models surprisingly well [11–17].

In this connection, it is also interesting to mention some insights gained in statistical physics and complex systems theory: Non-linearly interacting variables do not change independently of each other, and in many cases there is a separation of the time scales on which they evolve. This often allows to "forget" about the vast number of rapidly changing variables, which are usually determined by a small number of "order parameters" and treatable as fluctuations [18, 19]. In the above mentioned examples of traffic dynamics, the order parameters are the traffic density and the average velocity of pedestrians or vehicles.

---

*This chapter reprints a previous publication with kind permission of the copyright owner, the Technique & Technologies Press Ltd. It is requested to cite this work as follows: D. Helbing and *T. Platkowski*, Self-organization in space and induced by fluctuations. *International Journal of Chaos Theory and Applications* 5(4), 47–62 (2000).

D. Helbing (ed.), *Social Self-Organization*, Understanding Complex Systems,
DOI 10.1007/978-3-642-24004-1_5, © Springer-Verlag Berlin Heidelberg 2012

Another discovery is that, by proper transformations or scaling, many different models can by mapped onto each other, i.e. they behave basically the same [13, 18–20]. That is, a certain class of models displays the same kinds of states, shows the same kinds of transitions among them, and can be described by the same "phase diagram", displaying the respective states as a function of some "control parameters" [18, 19]. We call such a class of models a "universality class", since any of these models shows the same kind of "universal" behavior, i.e., the same phenomena. Consequently, one usually tries to find the simplest model having the properties of the universality class. While physicists like to call it a "minimal model", "prototype model", or "toy model", mathematicians named the corresponding mathematical equations "normal forms" [18, 19, 21, 22].

Universal behavior is the reason of the great success of systems theory [23–25] in comparing phenomena in seemingly completely different systems, like physical, biological, or social ones. However, since these systems are composed of different entities and their corresponding interactions can be considerably different, it is not always easy to identify the variables and parameters behind their dynamics. It can be helpful to take up game-theoretical ideas, here, quantifying interactions in terms of payoffs [2–7, 26, 27]. This can be applied to positive (profitable, constructive, cooperative, symbiotic) or negative (competitive, destructive) interactions in socio-economic or biological systems, but to attractive and repulsive interactions in physical systems as well [28].

In the following, we will investigate a simple model for interactive motion in space allowing to describe (1) various self-organized agglomeration phenomena, like settlement formation, and segregation phenomena, like ghetto formation, emerging from different kinds of interactions and (2) fluctuation-induced ordering or self-organization phenomena.

Noise-related phenomena can be quite surprising and have, therefore, recently attracted the interest of many researchers. For example, we mention stochastic resonance [29], noise-driven motion [30, 31], and "freezing by heating" [32].

The issue of order through fluctuations has already a considerable history. Prigogine has discussed it in the context of structural instability with respect to the appearance of a new species [33, 34], but this is not related to the approach considered in the following.

Moreover, since both, the initial conditions and the interaction strengths in our model are assumed independent of the position in space, the fluctuation-induced self-organization discussed later on must be distinguished from so-called "noise-induced transitions" as well, where we have a space-dependent diffusion coefficient which can induce a transition [35].

Although our model is related to diffusive processes, it is also different from reaction-diffusion systems that can show fluctuation-induced self-organization phenomena known as Turing patterns [36–41], which are usually periodic in space. The noise-induced self-organization that we find seems to have (1) no typical length scale and (2) no attractor, since our model is translation-invariant. This, however, is not yet a final conclusion and still subject to investigations.

We also point out that, in the case of spatial invariance, self-organization directly implies spontaneous symmetry-breaking, and we expect a pronounced history-dependence of the resulting state. Nevertheless, when averaging over a large ensemble of simulation runs with different random seeds, we again expect a homogeneous distribution, since this is the only result compatible with translation invariance.

Finally, we mention that our results do not fit into the concept of noise-induced transitions from a metastable disordered state (local optimum) to a stable ordered state (global optimum), which are, for example, found for undercooled fluids, metallic glasses, or some granular systems [42–44].

5.2 Discrete Model of Interactive Motion in Space

Describing motion in space has the advantage that the essential variables like positions, densities, and velocities are well measurable, which allows to calibrate, test, and verify or falsify the model. Although we will focus on motion in "real" space like the motion of pedestrians or bacteria, our model may also be applied to changes of positions in abstract spaces, e.g. to opinion changes on an opinion scale [7,46]. There exist, of course, already plenty of models for motion in space, and we can mention only a few [1,7,11–21,26,28,30–32,35–41,43,45–53]. Most of them are, however, rather specific for certain systems, e.g., for fluids or for migration behavior.

For simplicity, we will restrict the following considerations to a one-dimensional space, but a generalization to higher dimensions is straightforward. The space is divided into I equal cells i which can be occupied by the entities. We will apply periodic boundary conditions, i.e. the space can be imagined as a circle. In our model, we group the N entities α in the system into homogeneously behaving subpopulations a. If $n_i^a(t)$ denotes the number of entities of subpopulation a in cell i at time t, we have the relations

$$\sum_i n_i^a(t) = N_a, \qquad \sum_a N_a = N. \tag{5.1}$$

We will assume that the numbers N_a of entities belonging to the subpopulations a do not change. It is, however, easy to take additional birth and death processes and/or transitions of individuals from one subpopulation to another into account [1].

In order not to introduce any bias, we start our simulations with a completely uniform distribution of the entities in each subpopulation over the I cells of the system, i.e., $n_i^a(0) = n_{\text{hom}}^a = N_a/I$, for which we choose a natural number. At times $t \in \{1, 2, 3, ...\}$, we apply the following update steps, using a random sequential update (although a parallel update is possible as well, which is more efficient [54, 55], but normally less realistic [56] due to the assumed synchronous updating):

1st step: For updating of the state of entity α, given it is a member of subpopulation a and located in cell i, determine the so-called (expected) "success" according to the formula

$$S_a(i,t) = \sum_b P_{ab}\, n_i^b(t) + \xi_\alpha(t)\,. \tag{5.2}$$

Here, P_{ab} is the "payoff" in interactions of an entity of subpopulation a with an entity of subpopulation b. The payoff P_{ab} is positive for attractive, profitable, constructive, or symbiotic interactions, while it is negative for repulsive, competitive, or destructive interactions. Notice that P_{ab} is assumed to be independent of the position (i.e., translation invariant), while the total payoff $\sum_b P_{ab}\, n_i^b(t)$ due to interactions depends on the distribution of entities over the system. The latter is an essential point for the possibility of fluctuation-induced self-organization. We also point out that, in formula (5.2), pair interactions are restricted to the cell in which the individual is located. Therefore, we do not assume spin-like or Ising-like interactions, in contrast to other quantitative models proposed for the description of social behavior [9, 10].

The quantities $\xi_\alpha(t)$ are random variables allowing to consider individual variations of the success, which may be "real" or due to uncertainty in the evaluation or estimation of success. In our simulation program, they are uniformly distributed in the interval $[0, D_a]$, where D_a is the fluctuation strength (not to be mixed up with a diffusion constant). However, other specifications of the noise term are possible as well.

2nd step: Determine the (expected) successes $S_a(i \pm 1, t)$ for the nearest neighbors $(i \pm 1)$ as well.

3rd step: Keep entity α in its previous cell i, if $S_a(i,t) \geq \max\{S(i-1,t), S(i+1,t)\}$. Otherwise, move to cell $(i-1)$, if $S(i-1,t) > S(i+1,t)$, and move to cell $(i+1)$, if $S(i-1,t) < S(i+1,t)$. In the remaining case $S(i-1,t) = S(i+1,t)$, jump randomly to cell $(i-1)$ or $(i+1)$ with probability 1/2.

If there is a maximum density $\rho_{\max} = N_{\max}/I$ of entities, overcrowding can be avoided by introducing a saturation factor

$$c(j,t) = 1 - \frac{N_j(t)}{N_{\max}}\,, \quad N_j(t) = \sum_a n_j^a(t)\,, \tag{5.3}$$

and performing the update steps with the generalized success

$$S'_a(j,t) = c(j,t) S_a(j,t) \tag{5.4}$$

instead of $S_a(j,t)$, where $j \in \{i-1, i, i+1\}$. The model can be also easily extended to include long distance interactions, jumps to more remote cells, etc. (cf. Sect. 5).

5.3 Simulation Results

We consider two subpopulations $a \in \{1,2\}$ and $N_1 = N_2 = 100$ entities in each subpopulation, which are distributed over $I = 20$ cells. The payoff matrix (P_{ab}) will be represented by the vector $\boldsymbol{P} = (P_{11}, P_{12}, P_{21}, P_{22})$, where we will restrict ourselves to $|P_{ab}| \in \{1,2\}$ for didactical reasons. For symmetric interactions between subpopulations, we have $P_{ab} = P_{ba}$, while for asymmetric interactions, there is $P_{ab} \neq P_{ba}$, if $a \neq b$. For brevity, the interactions within the same population will be called self-interactions, those between different populations cross-interactions.

To characterize the level of self-organization in each subpopulation a, we can, for example, use the overall successes

$$S_a(t) = \frac{1}{I^2} \sum_i \sum_b n_i^a(t)\, P_{ab}\, n_i^b(t)\,, \tag{5.5}$$

the variances

$$V_a(t) = \frac{1}{I^2} \sum_i [n_i^a(t) - n_{\text{hom}}^a]^2\,, \tag{5.6}$$

or the alternation strengths

$$A_a(t) = \frac{1}{I^2} \sum_i [n_i^a(t) - n_{i-1}^a(t)]^2\,. \tag{5.7}$$

5.3.1 *Symmetric Interactions*

By analogy with a more complicated model [28] it is expected that the global overall success $S(t) = \sum_a S_a(t)$ is an increasing function in time, if the fluctuation strengths D_a are zero. However, what happens at finite noise amplitudes D_a is not exactly known. One would usually expect that finite noise tends to obstruct or suppress self-organization, which will be investigated in the following.

We start with the payoff matrix $\boldsymbol{P} = (2, -1, -1, 2)$ corresponding to positive (or attractive) self-interactions and negative (or repulsive) cross-interactions. That is, entities of the same subpopulation like each other, while entities of different subpopulations dislike each other. The result will naturally be segregation ("ghetto formation") [1, 57], if the noise amplitude is small. However, segregation is suppressed by large fluctuations, as expected (see Fig. 5.1).

However, for medium noise amplitudes D_a, we find a much more pronounced self-organization (segregation) than for small ones (compare Fig. 5.2 with Fig. 5.1). The effect is systematic insofar as the degree of segregation (and, hence, the overall success) increases with increasing noise amplitude, until segregation breaks down above a certain critical noise level.

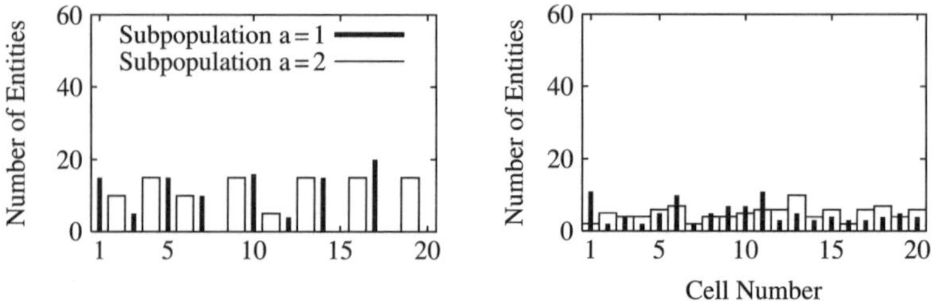

Fig. 5.1 Resulting distribution of entities at $t = 4,000$ for the payoff matrix $\boldsymbol{P} = (2, -1, -1, 2)$ at small fluctuation strength $D_a = 0.1$ (*left*) and large fluctuations strength $D_a = 5$ (*right*)

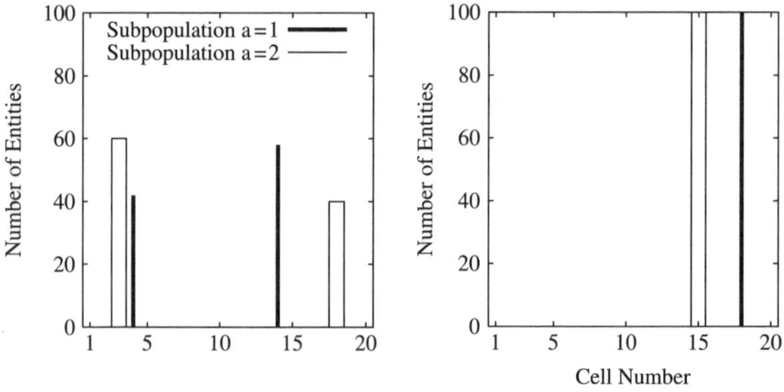

Fig. 5.2 As Fig. 5.1, but with medium fluctuation strength $D_a = 2$ (*left*) and $D_a = 3$ (*right*)

Let us investigate some other cases: For the structurally similar payoff matrix $(1, -2, -2, 1)$, we find segregation as well, which is not surprising. In contrast, we find agglomeration for the payoff matrices $(1, 2, 2, 1)$ and $(2, 1, 1, 2)$. This agrees with intuition, since all entities like each other in these cases, which makes them move to the same places, like in the formation of settlements [1], the development of trail systems [17, 52, 53], or the build up of slime molds [34, 50]. More interesting is the case corresponding to the payoff matrix $(-1, 2, 2, -1)$, where the cross-interactions are positive (attractive), while the self-interactions are negative (repulsive). One may think that this causes the entities of the same subpopulation to spread homogeneously over the system, and in all cells would result an equal number of entities of both subpopulations, which is compatible with mutual attraction. However, this homogeneous distribution turns out to be unstable with respect to fluctuations. Instead, we find agglomeration! This result is more intuitive if we imagine one subpopulation to represent women and the other one men (without taking this example too serious). While the interaction between women and men is normally strongly attractive, the interactions among men or among women may be considered to be weakly competitive. As we all know, the result is a tendency of young men and women to move into cities. Corresponding simulation results

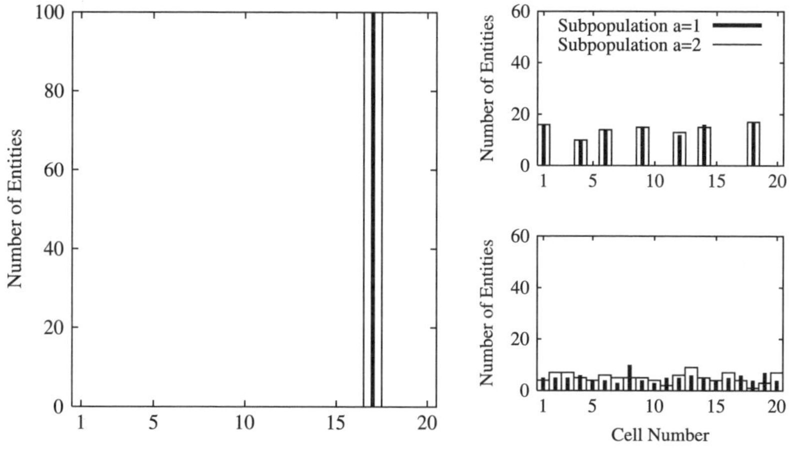

Fig. 5.3 As Fig. 5.1, but for the payoff matrix $\boldsymbol{P} = (-1, 2, 2, -1)$ and $D_a = 0.05$ (*left*), $D_a = 1.5$ (*middle*), and $D_a = 5$ (*right*)

for different noise strengths are depicted in Fig. 5.3. Again, we find that the self-organized pattern is destroyed by strong fluctuations in favour of a more or less homogeneous distribution, while medium noise strengths further self-organization.

For the payoff matrices $(-2, 1, 1, -2)$ and $(-2, -1, -1, -2)$, i.e. cases of strong negative self-interactions, we find a more or less homogeneous distribution of entities in both subpopulations, irrespective of the noise amplitude. In contrast, the payoff matrix $(-1, -2, -2, -1)$ corresponding to negative self-interactions but even stronger negative cross-interactions, leads to another self-organized pattern. We may describe it as the formation of lanes, as it is observed in pedestrian counterflows [12, 28] or in sheared granular media with different kinds of grains [47]. While both subpopulations tend to separate from each other, at the same time they tend to spread over all the available space (see Fig. 5.4), in contrast to the situation depicted in Figs. 5.1 and 5.2. Astonishingly enough, a medium level of noise again supports self-organized ordering, since it helps the subpopulations to separate from each other.

We finally mention that a finite saturation level suppresses self-organization in a surprisingly strong way, as is shown in Fig. 5.5. Instead of pronounced segregation, we will find a result similar to lane formation, and even strong agglomeration will be replaced by an almost homogeneous distribution.

Noise-Induced Ordering

A possible interpretation for noise-induced ordering would be that fluctuations allow the system to leave local minima (corresponding to partial agglomeration or segregation only). This could trigger a transition to a more stable state with

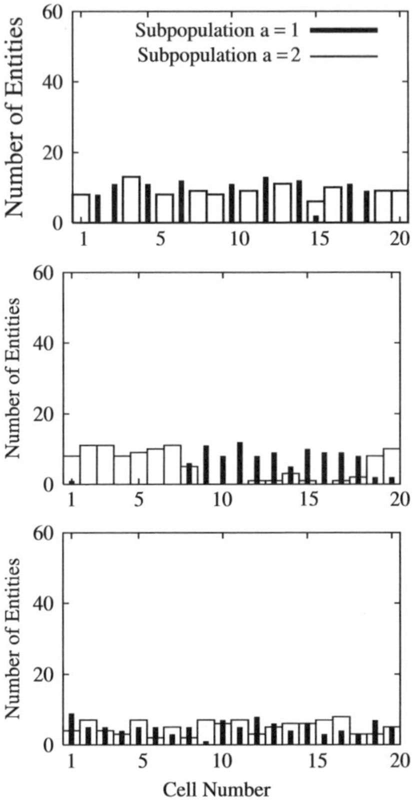

Fig. 5.4 As Fig. 5.1, but for the payoff matrix $\boldsymbol{P} = (-1, -2, -2, -1)$ and $D_a = 0.05$ (*top*), $D_a = 0.5$ (*middle*), and $D_a = 5$ (*bottom*)

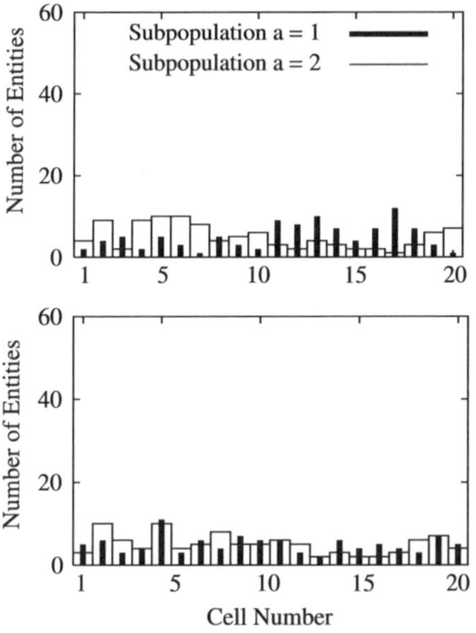

Fig. 5.5 Resulting distribution of entities at $t = 4,000$ with saturation level $N_{\max} = 50$. *Top*: $\boldsymbol{P} = (2, -1, -1, 2)$ and $D_a = 3$. *Bottom*: $\boldsymbol{P} = (-1, 2, 2, -1)$ and $D_a = 1.5$

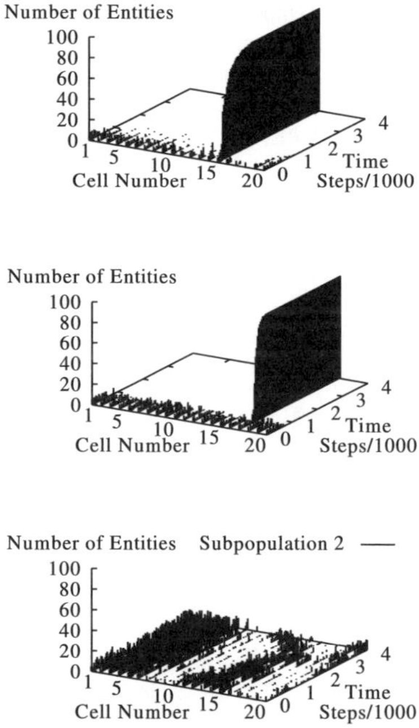

Fig. 5.6 Temporal evolution of the distribution of entitities within subpopulation $a = 2$ for $\boldsymbol{P} = (2, -1, -1, 2)$ and $D_a = 3$ (*top*), $\boldsymbol{P} = (-1, 2, 2, -1)$ and $D_a = 1.5$ (*middle*), and $\boldsymbol{P} = (-1, -2, -2, -1)$ and $D_a = 0.5$ (*bottom*)

more pronounced ordering. However, although this interpretation is consistent with a related example discussed in [28], the idea of a step-wise coarsening process is not supported by the temporal evolution of the distribution of entities (see Fig. 5.6) and the time-dependence of the overall success within the subpopulations (see Fig. 5.7). This idea is anyway not applicable to segregation, since, in the one-dimensional case, the repulsive clusters of different subpopulations cannot simply pass each other in order to join others of the same subpopulation.

According to Figs. 5.6 and 5.7, segregation and agglomeration rather take place in three phases: First, there is a certain time interval, during which the distribution of entities remains more or less homogeneous. Second, there is a short period of rapid self-organization. Third, there is a continuing period, during which the distribution and overall success do not change anymore. The latter is a consequence of the short-range interactions within our model, which are limited to the nearest neighbors. Therefore, the segregation or aggregation process practically stops, after separate peaks have evolved. This is not the case for lane formation, where the entities redistribute, but all cells remain occupied, so that we have ongoing interactions. This is reflected in the non-stationarity of the lanes and by the oscillations of the overall success.

We suggest the following interpretation for the three phases mentioned above: During the first time interval, which is characterized by a quasi-continuous distribution of entities over space, a long-range pre-ordering process takes place. After this "phase of preparation", order develops in the second phase similar to crystallization, and it persists in the third phase. The role of fluctuations seems

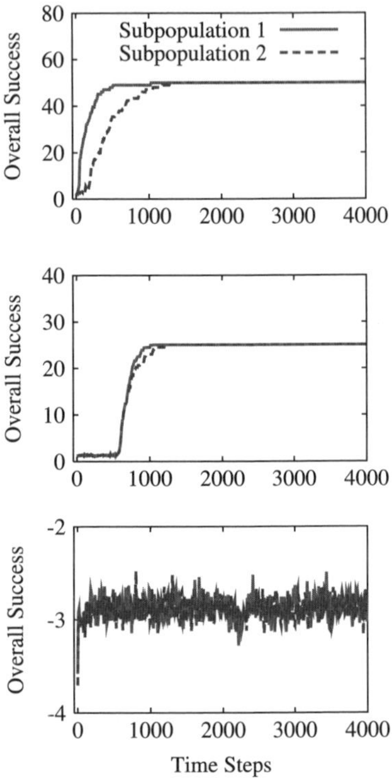

Fig. 5.7 Temporal evolution of the overall success within both subpopulations for $\boldsymbol{P} = (2, -1, -1, 2)$ and $D_a = 3$ (*top*), $\boldsymbol{P} = (-1, 2, 2, -1)$ and $D_a = 1.5$ (*middle*), and $\boldsymbol{P} = (-1, -2, -2, -1)$ and $D_a = 0.5$ (*bottom*)

to be the following: An increased noise level avoids a rash local self-organization by keeping up a quasi-continuous distribution of entities, which is required for a redistribution of entities over larger distances. In this way, a higher noise level increases the effective interaction range by extending the first phase, the "interaction phase". As a consequence, the resulting structures are more extended in space (but probably without a characteristic length scale, see Introduction).

It would be interesting to investigate, whether this mechanism has something to do with the recently discovered phenomenon of "freezing by heating", where a medium noise level causes a transition to a highly ordered (but energetically less stable) state, while extreme noise levels produce a disordered, homogeneous state again [32].

5.3.2 Asymmetric Interactions

Even more intriguing transitions than in the symmetric case can be found for asymmetric interactions between the subpopulations. Here, we will focus on the payoff matrix $(-1, 2, -2, 1)$, only. This example corresponds to the curious case, where individuals of subpopulation 1 weakly dislike each other, but strongly like individuals of the other subpopulation. In contrast, individuals of subpopulation 2 weakly like each other, but they strongly dislike the other subpopulation. A good example for this is hard to find. With some good will, one may imagine subpopulation 1 to represent poor people, while subpopulation 2 corresponds to rich people. What will be the outcome? In simple terms, the rich are expected to agglomerate in a few areas, if the poor are moving too nervously (see Fig. 5.8). In detail, however, the situation is quite complex, as discussed in the next paragraph.

Noise-Induced Self-organization

At small noise levels D_a, we will just find more or less homogeneous distributions of the entities. This is already different from the cases of agglomeration, segregation, and lane formation we have discussed before. Self-organization is also not found at higher noise amplitudes D_a, as long as we assume that they are the same in both subpopulations (i.e., $D_1 = D_2$). However, given that the fluctuation amplitude D_2 in subpopulation 2 is small, we find an agglomeration in subpopulation 2, if the noise level D_1 in subpopulation 1 is medium or high, so that subpopulation 1 remains homogeneously distributed. The order in subpopulation 2 breaks down, as soon as we have a relevant (but still small) noise level D_2 in subpopulation 2 (see Fig. 5.8).

Hence, we have a situation where asymmetric noise with $D_1 \neq D_2$ can facilitate self-organization in a system with completely homogeneous initial conditions and

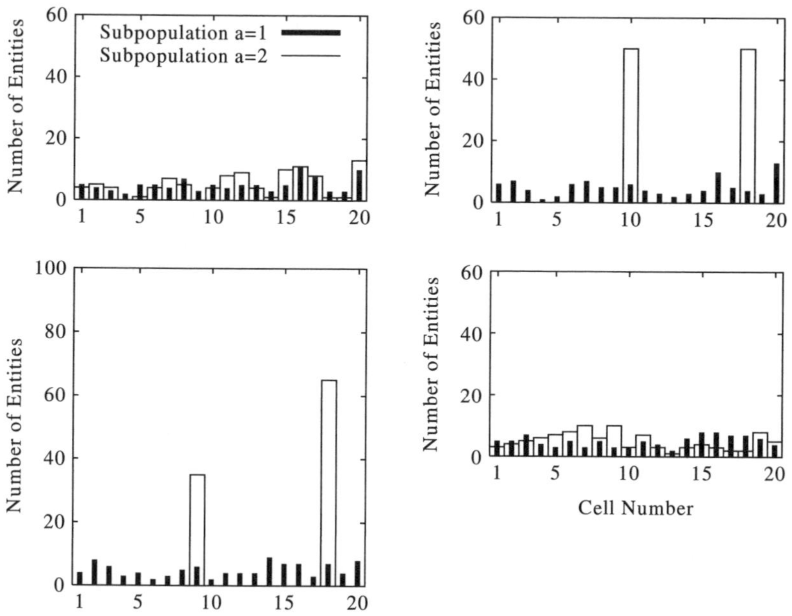

Fig. 5.8 Distributions for $\boldsymbol{P} = (-1, 2, -2, 1)$ and $D_1 = D_2 = 0.5$ (*top left*), $D_1 = 50$, $D_2 = 0.5$ (*top right*), $D_1 = 5{,}000$, $D_2 = 0.5$ (*bottom left*), $D_1 = 5{,}000$, $D_2 = 5$ (*bottom right*)

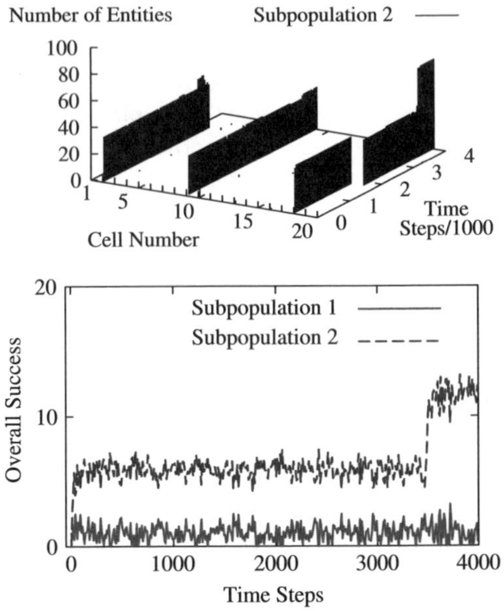

Fig. 5.9 Temporal evolution of the distribution of entitities within subpopulation $a = 2$ (*top*) and of the overall successes (*bottom*) for $\boldsymbol{P} = (-1, 2, -2, 1)$ and $D_1 = 50$, $D_2 = 0.5$

interaction laws, where we would not have ordering without any noise. We call this phenomenon noise-induced self-organization. It is to be distinguished from the noise-induced increase in the degree of ordering discussed above, where we have self-organization even without noise, if only the initial conditions are not fully homogeneous.

The role of the noise in subpopulation 1 seems to be the following: Despite of the attractive interaction with subpopulation 2, it suppresses an agglomeration in subpopulation 1, in particular at the places where subpopulation 2 agglomerates. Therefore, the repulsive interaction of subpopulation 2 with subpopulation 1 is effectively reduced. As a consequence, the attractive self-interaction within subpopulation 2 dominates, which gives rise to the observed agglomeration.

The temporal development of the distribution of entities and of the overall success in the subpopulations gives additional information (see Fig. 5.9). As in the case of lane formation, the overall success fluctuates strongly, because the subpopulations do not separate from each other, causing ongoing interactions. Hence, the resulting distribution is not stable, but changes continuously. It can, therefore, happen, that clusters of subpopulation 2 merge, which is associated with an increase of overall success in subpopulation 2 (see Fig. 5.9).

5.4 Conclusions

We have proposed a game theoretical model for self-organization in space, which is applicable to many kinds of biological, economic, and social systems with various types of profitable or competitive self- and cross-interactions between subpopulations of the system. Depending on the structure of the payoff matrix, we found several different self-organization phenomena like agglomeration, segregation, or lane formation. It turned out that medium noise strengths can increase the resulting level of order, while a high noise level leads to more or less homogeneous distributions of entities over the available space. The mechanism of noise-induced ordering in the above discussed systems with short-range interactions seems to be the following: Noise extends a "pre-ordering" phase by keeping up a quasi-continuous distribution of entities, which allows a long-range ordering. For asymmetric payoff matrices, we can even have the phenomenon of noise-induced self-organization, although we start with completely homogeneous distributions and homogeneous (translation-invariant) payoffs. However, the phenomenon requires different noise amplitudes in both subpopulations. The role of noise is to suppress agglomeration in one of the subpopulations, in this way reducing repulsive effects that would suppress agglomeration in the other subpopulation.

We point out that all the above results can be semi-quantitatively understood by means of a linear stability analysis of a related continuous version of the model [28]. This continuous version indicates that the linearly most unstable modes are the ones with the shortest wave length, so that one does not expect a characteristic length scale in the system. This is different from reaction-diffusion systems, where the most unstable mode has a finite wave length, which gives rise to the formation of periodic patterns. Nevertheless, the structures evolving in our model are spatially extended, but non-periodic. The spatial extension is increasing with the fluctuation strength, unless a critical noise amplitude is exceeded.

For a better agreement with real systems, the model can be generalized in many ways. The entities may perform a biased or unbiased random walk in space. One can allow random jumps to neigboring cells with some prescribed probability. This probability may depend on the subpopulation, and thus we can imitate different mobilities of the considered subpopulations. Evolution is slowed down by introducing a threshold, fixed or random, so that the entities change to other cells only if the differences in the relevant successes are bigger than the imposed threshold. The model can be also generalized to higher dimensions, with expected interesting patterns of self-organized structures.

In general, the random variables $\xi_\alpha(t)$ in the definition of the success functions can be allowed to have different variances for the considered cell i and the neighboring cells, with the interpretation that the uncertainty in the evaluation of the success in the considered cell is different (e.g. smaller) than that in the neighboring cells. Moreover, the uncertainties can be different for various subpopulations, which could reflect to some extent their different knowledge and behavior.

One can as well study systems with more than two subpopulations, the influence of long-range interactions, etc. The entities can also be allowed to jump to more remote cells. As an example, the following update rule could be implemented: Move entity α from cell i to the cell $(i + l)$ for which

$$S''_a(i+l,t) = d^{|l|}c(i+l,t)S_a(i+l,t) \tag{5.8}$$

is maximal ($|l| = 0, 1, \dots, l_{\max}$). If there are m cells in the range $\{(x - l_{\max}), \dots, (x + l_{\max})\}$ with the same maximal value, choose one of them randomly with probability $1/m$. According to this, when spontaneously moving to another cell, the entity prefers cells in the neighborhood with higher success. The indirect interaction behind this transition, which is based on the observation or estimation of the success in the neighborhood, is short-ranged if $l_{\max} \ll I$, otherwise long-ranged. Herein, $l_{\max}$ denotes the maximum number of cells which an entity can move within one time step. The factor containing d with $0 < d < 1$ allows to consider that it is less likely to move for large distances, if this is not motivated by a higher success. A value $d < 1$ may also reflect the fact that the observation or estimation of the success over large distances becomes more difficult and less reliable.

Acknowledgements D.H. thanks Eörs Szathmáry and Tamás Vicsek for inspiring discussions and the German Research Foundation (DFG) for financial support by a Heisenberg scholarship. T.P. is grateful to the Alexander-von-Humboldt Foundation for financial support during his stay in Stuttgart.

References

1. W. Weidlich, Physics and social science—The approach of synergetics. Phys. Reports **204**, 1–163 (1991)
2. J. von Neumann, O. Morgenstern, *Theory of Games and Economic Behavior.* (Princeton University, Princeton, 1944)
3. R. Axelrod, W.D. Hamilton, The evolution of cooperation. Science **211**, 1390–1396 (1981)
4. R. Axelrod, D. Dion, The further evolution of cooperation. Science **242**, 1385–1390 (1988)
5. J. Hofbauer, K. Sigmund, *The Theory of Evolution and Dynamical Systems.* (Cambridge University Press, Cambridge, 1988)
6. N.S. Glance, B.A. Huberman, The dynamics of social dilemmas. Sci. Am. **270**, 76–81 (1994)
7. D. Helbing, *Quantitative Sociodynamics. Stochastic Methods and Models of Social Interaction Processes.* (Kluwer Academics, Dordrecht, 1995)
8. F. Schweitzer (ed.), *Self-Organization of Complex Structures*, in *From individual to Collective Dynamics*. (Gordon and Breach, Amsterdam, 1997)
9. M. Lewenstein, A. Nowak, B. Latané, Statistical mechanics of social impact. Phys. Rev. A **45**, 763–776 (1992)
10. S. Galam, Rational group decision making. Physica A **238**, 66–80 (1997)
11. D. Helbing, *Verkehrsdynamik [Traffic Dynamics].* (Springer, Berlin, 1997)
12. D. Helbing, P. Molnár, Social force model for pedestrian dynamics. Phys. Rev. E **51**, 4282–4286 (1995)
13. D. Helbing, A. Hennecke, M. Treiber, Phase diagram of traffic states in the presence of inhomogeneities. Phys. Rev. Lett. **82**, 4360–4363 (1999)
14. D. Helbing, B.A. Huberman, Coherent moving states in highway traffic. Nature **396**, 738–740 (1998)
15. M. Treiber, D. Helbing, Macroscopic simulation of widely scattered synchronized traffic states. J. Phys. A: Math. Gen. **32**, L17-L23 (1999)

16. M. Treiber, A. Hennecke, D. Helbing, Congested traffic states in empirical observations and microscopic simulations. Phys. Rev. E **62**(2), 1805–1824 (2000)
17. D. Helbing, J. Keltsch, P. Molnár, Modelling the evolution of human trail systems. Nature **388**, 47–50 (1997)
18. H. Haken *Synergetics.* (Springer, Berlin, 1977)
19. H. Haken *Advanced Synergetics.* (Springer, Berlin, 1983)
20. D. Helbing, D. Mukamel, G.M. Schütz, Global phase diagram of a one-dimensional driven lattice gas. Phys. Rev. Lett. **82**, 10–13 (1999)
21. P. Manneville, *Dissipative Structures and Weak Turbulence.* (Academic Press, New York, 1990)
22. E.C. Zeeman (ed.), *Catastrophe Theory.* (Addison-Wesley, London, 1977)
23. L. von Bertalanffy, *General System Theory.* (Braziller, New York, 1968)
24. W. Buckley, *Sociology and Modern Systems Theory.* (Prentice-Hall, Englewood Cliffs, NJ, 1967)
25. A. Rapoport, *General System Theory. Essential Concepts and Applications.* (Abacus Press, Tunbridge Wells, Kent, 1986)
26. R. Feistel, W. Ebeling, *Evolution of Complex Systems.* (Kluwer Academic, Dordrecht, 1989)
27. D. Helbing, Stochastic and Boltzmann-like models for behavioral changes, and their relation to game theory. Physica A **193**, 241–258 (1993)
28. D. Helbing, T. Vicsek, Optimal self-organization. New J. Phys. **1**, 13.1–13.17 (1999)
29. L. Gammaitoni, P. Hänggi, P. Jung, F. Marchesoni, Stochastic resonance. Rev. Modern Phys. **70**, 223–288 (1998)
30. J. Łuczka, R. Bartussek, P. Hänggi, White noise induced transport in periodic structures. Europhys. Lett. **31**, 431–436 (1995)
31. P. Reimann, R. Bartussek, R. Häußler, P. Hänggi, Brownian motors driven by temperature oscillations. Phys. Lett. A **215**, 26–31 (1996)
32. D. Helbing, I.J. Farkas, T. Vicsek, Freezing by heating in a driven mesoscopic system. Phys. Rev. Lett. **84**, 1240–1243 (2000)
33. G. Nicolis, I. Prigogine, *Self-Organization in Nonequilibrium Systems. From Dissipative Structures to Order through Fluctuations.* (Wiley, New York, 1977)
34. I. Prigogine, Order through fluctuation: Self-organization and social system, in *Evolution and Consciousness. Human Systems in Transition*, ed. by E. Jantsch and C.H. Waddington (Addison-Wesley, Reading, MA, 1976), pp. 93–130
35. W. Horsthemke, R. Lefever, *Noise-Induced Transitions.* (Springer, Berlin, 1984)
36. A.M. Turing, The chemical basis of morphogenesis. Phil. Trans. Roy. Soc. Lond. **B237**, 37–72 (1952)
37. J.D. Murray, *Lectures on Nonlinear Differential Equation-Models in Biology.* (Clanderon Press, Oxford, 1977)
38. P.C. Fife, *Mathematical aspects of reacting and diffusing systems.* (Springer, New York, 1979)
39. E. Convay, D. Hoff, J.A. Smoller, Large time behavior of systems of nonlinear diffusion equations. SIAM J. Appl. Math. **35**, 1–16 (1978)
40. D.A. Kessler, H. Levine, Fluctuation-induced diffusive instabilities. Nature **394**, 556–558 (1998)
41. H. Zhonghuai, Y. Lingfa, X. Zuo, X. Houwen, Noise induced pattern transition and spatiotemporal stochastic resonance. Phys. Rev. Lett. **81**, 2854–2857 (1998)
42. A. Rosato, K.J. Strandburg, F. Prinz, R.H. Swendsen, Why the Brazil nuts are on top: Size segregation of particulate matter by shaking. *Physical Review Letters* **58**, 1038–1041 (1987)
43. J.A.C. Gallas, H.J. Herrmann, S. Sokołowski, Convection cells in vibrating granular media. Phys. Rev. Lett. **69**, 1371–1374 (1992)
44. P.B. Umbanhowar, F. Melo, H.L. Swinney, Localized excitations in a vertically vibrated granular layer. Nature **382**, 793–796 (1996)
45. J. Keizer, *Statistical Thermodynamics of Nonequilibrium Processes.* (Springer, New York, 1987)

46. D. Helbing, Boltzmann-like and Boltzmann-Fokker-Planck equations as a foundation of behavioral models. Physics A **196**, 546–573 (1993)
47. S.B. Santra, S. Schwarzer, H. Herrmann, Fluid-induced particle-size segregation in sheared granular assemblies. Phys. Rev. E **54**, 5066–5072 (1996)
48. Ben-Jacob, E., O. Schochet, A. Tenenbaum, I. Cohen, A. Czirók, T. Vicsek, Generic modelling of cooperative growth patterns in bacterial colonies. Nature **368**, 46–49 (1994)
49. Ben-Jacob, E., From snowflake formation to growth of bacterial colonies, Part II: Cooperative formation of complex colonial patterns. Contemp. Phys. **38**, 205–241 (1997)
50. D.A. Kessler, H. Levine, Pattern formation in *Dictyostelium* via the dynamics of cooperative biological entities. Phys. Rev. E **48**, 4801–4804 (1993)
51. T. Vicsek, A. Czirók, E. Ben-Jacob, I. Cohen, O. Shochet, Novel type of phase transition in a system of self-driven particles. Phys. Rev. Lett. **75**, 1226–1229 (1995)
52. F. Schweitzer, K. Lao, F. Family, Active random walkers simulate trunk trail formation by ants. BioSystems, **41**, 153–166 (1997)
53. E.M. Rauch, M.M. Millonas, D.R. Chialvo, Pattern formation and functionality in swarm models. Phys. Lett. A, **207**, 185–193 (1995)
54. S. Wolfram, Cellular automata as models of complexity. Nature **311**, 419–424 (1984)
55. D. Stauffer, Computer simulations of cellular automata. J. Phys. A: Math. Gen. **24**, 909–927 (1991)
56. B.A. Huberman, N.S. Glance, Evolutionary games and computer simulations. Proc. Nat. Acad. Sci. USA **90**, 7716–7718 (1993)
57. T.C. Schelling, Dynamic models of segregation. J. Math. Sociol. **1**, 143–186 (1971)

Chapter 6
Cooperation in Social Dilemmas*

6.1 Introduction

Game theory goes back to von Neumann [1], one of the superminds of quantum mechanics. Originally intended to describe interactions in economics, sociology, and biology [1–4], it has recently become a quickly growing research area in physics, where methods from non-linear dynamics and pattern formation [5–11], agent-based or particle-like models [11–13], network theory [14–18] and statistical physics [19–21] are applied. There are even quantum theoretical contributions [22].

When two entities characterized by the states, "strategies", or "behaviors" i and j interact with each other, game theory formalizes the result by payoffs P_{ij}, and the structure of the payoff matrix (P_{ij}) determines the kind of the game. The dynamics of a system of such entities is often delineated by the so-called replicator equations

$$\frac{dp(i,t)}{dt} = p(i,t)\left[\sum_j P_{ij}\, p(j,t) - \sum_{j,l} p(l,t) P_{lj}\, p(j,t)\right] \tag{6.1}$$

[3, 4]. $p(i,t)$ represents the relative frequency of behavior i in the system, which increases when the expected "success" $F_i = \sum_j P_{ij}\, p(j,t)$ exceeds the average one, $\sum_i F_i\, p(i,t)$.

Many collective phenomena in physics such as agglomeration or segregation phenomena can be studied in a game-theoretical way [11–13]. Applications also include the theory of evolution [23, 24] and the study of ecosystems [25–27]. Another exciting research field is the study of mechanisms supporting the cooperation between selfish individuals [1–4] in situations like the "prisoner's dilemma"

*This chapter reprints a previous publication with kind permission of the Americal Physical Society. It is requested to cite this work as follows: D. Helbing and S. *Lozano* (2010) Phase transitions to cooperation in the prisoner's dilemma. *Physical Review E* **81**(5), 057102. DOI: 10.1103/PhysRevE.81.057102.

D. Helbing (ed.), *Social Self-Organization*, Understanding Complex Systems, DOI 10.1007/978-3-642-24004-1_6, © Springer-Verlag Berlin Heidelberg 2012

or public goods game, where they would usually defect (free-ride or cheat). Contributing to public goods and sharing them constitute ubiquitous situations, where cooperation is crucial, for example, in order to maintain a sustainable use of natural resources or a well-functioning health or social security system.

In the following, we will give an overview of the stationary solutions of the replicator equations (6.1) and their stability properties. Based on this, we will discuss several "routes to cooperation", which transform the prisoner's dilemma into other games via different sequences of continuous or discontinuous phase transitions. These routes will then be connected to different biological or social mechanisms accomplishing such phase transitions [28]. Finally, we will introduce the concept of "equilibrium creation" and distinguish it from routes to cooperation based on "equilibrium selection" or "equilibrium displacement". A new cooperation-promoting mechanism based on adaptive group pressure will exemplify it.

6.2 Stability Properties of Different Games

Studying games with two strategies i only, the replicator equations (6.1) simplify, and we remain with

$$\frac{dp(t)}{dt} = p(t)[1-p(t)]\{\lambda_1[1-p(t)] - \lambda_2\, p(t)\}\,, \tag{6.2}$$

where $p(t) = p(1,t)$ represents the fraction of cooperators and $1 - p(t) = p(2,t)$ the fraction of defectors. $\lambda_1 = P_{12} - P_{22}$ and $\lambda_2 = P_{21} - P_{11}$ are the eigenvalues of the two stationary solutions $p = p_1 = 0$ and $p = p_2 = 1$. If $0 < \lambda_1/(\lambda_1+\lambda_2) < 1$, there is a third stationary solution $p = p_3 = \lambda_1/(\lambda_1 + \lambda_2)$ with eigenvalue $\lambda_3 = -(1 - p_3)\lambda_1$. For the sake of our discussion, we imagine an additional fluctuation term $\xi(t)$ on the right-hand-side of (6.2), reflecting small perturbations of the strategy distribution.

Four different cases can be classified [3, 4]: (1) If $\lambda_1 < 0$ and $\lambda_2 > 0$, the stationary solution p_1 corresponding to defection by everybody is stable, while the stationary solution p_2 corresponding to cooperation by everyone is unstable. That is, any small perturbation will drive the system away from full cooperation towards full defection. This situation applies to the *prisoner's dilemma (PD)* defined by payoffs with $P_{21} > P_{11} > P_{22} > P_{12}$. According to this, strategy $i = 1$ ("cooperation") is risky, as it can yield the lowest payoff P_{12}, while strategy $i = 2$ ("defection") is tempting, since it can give the highest payoff P_{21}. (2) If $\lambda_1 > 0$ and $\lambda_2 < 0$, the stationary solution p_1 is unstable, while p_2 is stable. This means that the system will end up with cooperation by everybody. Such a situation occurs for the so-called *harmony game (HG)* with $P_{11} > P_{21} > P_{12} > P_{22}$, as mutual cooperation gives the highest payoff P_{11}. (3) If $\lambda_1 > 0$ and $\lambda_2 > 0$, the stationary solutions p_1 and p_2 are unstable, but there exists a third stationary solution p_3, which turns out to be stable. As a consequence, the system is driven towards a situation, where a fraction

p_3 of cooperators is expected to coexist with a fraction $(1 - p_3)$ of defectors. Such a situation occurs for the *snowdrift game (SD)* (also known as hawk-dove or chicken game). This game is characterized by $P_{21} > P_{11} > P_{12} > P_{22}$ and assumes that unilateral defection is tempting, as it yields the highest payoff P_{21}, but also risky, as mutual defection gives the lowest payoff P_{22}. (4) If $\lambda_1 < 0$ and $\lambda_2 < 0$, the stationary solutions p_1 and p_2 are both stable, while the stationary solution p_3 is unstable. As a consequence, full cooperation is possible, but not guaranteed. In fact, the final state of the system depends on the initial condition $p(0)$ (the "history"): If $p(0) < p_3$, the system is expected to end up in the stationary solution p_1, i.e. with full defection. If $p(0) > p_3$, the system is expected to move towards $p_2 = 1$, corresponding to cooperation by everybody. The history-dependence implies that the system is multistable (here: bistable), as it has several (locally) stable solutions. This case is found for the *stag hunt game (SH)* (also called assurance). This game is characterized by $P_{11} > P_{21} > P_{22} > P_{12}$, i.e. cooperation is rewarding, as it gives the highest payoff P_{11} in case of mutual cooperation, but it is also risky, as it yields the lowest payoff P_{12}, if the interaction partner is uncooperative.

6.3 Phase Transitions and Routes to Cooperation

When facing a prisoner's dilemma, it is of vital interest to transform the payoffs in such a way that cooperation between individuals is supported. Starting with the payoffs P^0_{ij} of a prisoner's dilemma, one can reach different payoffs P_{ij}, for example, by introducing strategy-dependent taxes $T_{ij} = P^0_{ij} - P_{ij} > 0$. When increasing the taxes T_{ij} from 0 to T^0_{ij}, the eigenvalues will change from $\lambda^0_1 = P^0_{12} - P^0_{22}$ and $\lambda^0_2 = P^0_{21} - P^0_{11}$ to $\lambda_1 = \lambda^0_1 + T_{22} - T_{12}$ and $\lambda_2 = \lambda^0_2 + T_{11} - T_{21}$. In this way, one can create a variety of routes to cooperation, which are characterized by different kinds of phase transitions. We define *route 1* [PD→HG] by a direct transition from a prisoner's dilemma to a harmony game. It is characterized by a discontinuous transition from a system, in which defection by everybody is stable, to a system, in which cooperation by everybody is stable (see Fig. 6.1a). *Route 2* [PD→SH] is defined by a direct transition from the prisoner's dilemma to a stag hunt game. After the moment t_*, where λ_2 changes from positive to negative values, the system behavior becomes history-dependent: When the fluctuations $\xi(t)$ for $t > t_*$ exceed the critical threshold $p_3(t) = \lambda_1/[\lambda_1 + \lambda_2(t)]$, the system will experience a sudden transition to cooperation by everybody. Otherwise one will find defection by everyone, as in the prisoner's dilemma (see Fig. 6.1b). In order to make sure that the perturbations $\xi(t)$ will eventually exceed $p_3(t)$ and trigger cooperation, the value of λ_2 must be reduced to sufficiently large negative values. It is also possible to have a continuous rather than sudden transition to cooperation: We define *route 3* [PD→SD] by a transition from a prisoner's dilemma to a snowdrift game. As λ_1 is changed from negative to positive values, a fraction $p_3(t) = \lambda_1(t)/[\lambda_1(t) + \lambda_2]$ of cooperators is expected to result (see Fig. 6.1c). When increasing λ_1, this fraction

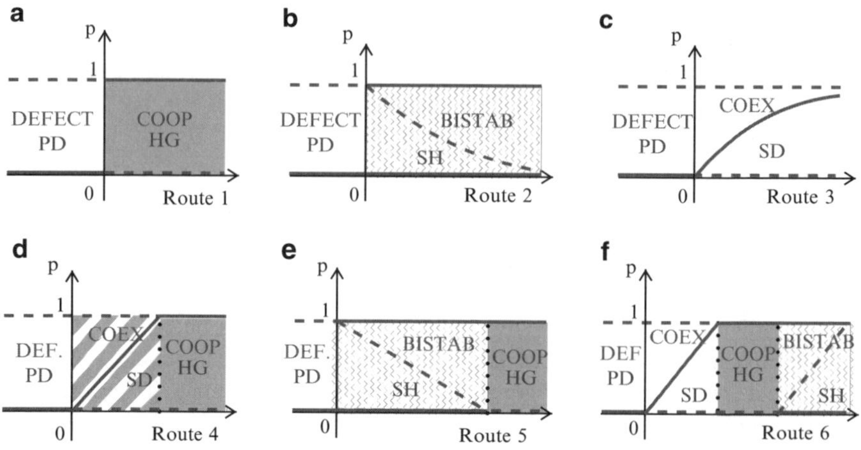

Fig. 6.1 Schematic illustration of the phase transitions defining the different routes to cooperation. The order parameter is the stationary frequency of cooperators, while the control parameters are the parameters r, w, k, m, or q in Nowak's cooperation-enhancing rules [28] (see main text) or, more generally, (non-)linear combination of the model parameters b and c. *Solid lines* represent *stable* stationary proportions of cooperators, *dashed lines unstable* fix points. *Diagonal lines* show the additional stationary solution p_3, where $0 \leq p_3 \leq 1$ (p = proportion of cooperators; DEFECT = defection is stable, i.e. everybody defects; COOP = cooperation is stable, i.e. everybody cooperates; COEX = mixture of defectors with a proportion p_3 of cooperators; BISTAB = cooperation is stable if $p_3 < p(0)$, where $p(0)$ means the initial proportion of cooperators, otherwise everybody defects)

rises continuously. One may also implement more complicated transitions. *Route 4*, for example, establishes the transition sequence PD→SD→HG (see Fig. 6.1d), while we define *route 5* by the transition PD→SH→HG (see Fig. 6.1e). One may also implement the transition PD→SD→HG→SH (*route 6*, see Fig. 6.1f), establishing a path-dependence, which can guarantee cooperation by everybody in the end. (When using *route 2*, the system remains in a defective state, if the perturbations do not exceed the critical value p_3.)

6.4 Relationship with Cooperation-Supporting Mechanisms

We will now discuss the relationship of the above introduced routes to cooperation with biological and social mechanisms ("rules") promoting the evolution of cooperation. Martin A. Nowak performs his analysis of five such rules with the reasonable specifications $T = b > 0$, $R = b - c > 0$, $S = -c < 0$, and $P = 0$ in the limit of weak selection [28]. Cooperation is assumed to require a contribution $c > 0$ and to produce a benefit $b > c$ for the interaction partner, while defection generates no payoff ($P = 0$). As most mechanisms leave λ_1 or $\lambda = (\lambda_1 + \lambda_2)/2$ unchanged, we will now focus on the payoff-dependent parameters λ_1 and λ (rather than λ_1 and λ_2). The basic prisoner's dilemma is characterized by $\lambda_1^0 = -c$ and $\lambda^0 = 0$.

According to the Supporting Online Material of [28], *kin selection* (genetic relatedness) tranforms the payoffs into $P_{11} = P_{11}^0 + r(b - c)$, $P_{12} = P_{12}^0 + br$, $P_{21} = P_{21}^0 - cr$, and $P_{22} = P_{22}^0$. Therefore, it leaves λ unchanged and increases λ_1 by $T_{22} - T_{12} = br$, where r represents the degree of genetic relatedness. *Direct reciprocity* (repeated interaction) does not change λ_1, but it reduces λ by $-\frac{1}{2}(b - c)[1/(1 - w) - 1] < 0$, where w is the probability of a future interaction. *Network reciprocity* (clustering of individuals playing the same strategy) leaves λ unchanged and increases λ_1 by $H(k)$, where $H(k)$ is a function of the number k of neighbors. Finally, *group selection* (competition between different populations) increases λ_1 by $(b - c)(m - 1)$, where m is the number of groups, while λ is not modified. However, λ_1 and λ may also change simultaneously. For example, *indirect reciprocity* (based on trust and reputation) increases λ_1 by cq and reduces λ by $-\frac{1}{2}(b - c)q < 0$, where q quantifies social acquaintanceship.

Summarizing this, kin selection, network reciprocity, and group selection preserve $\lambda = 0$ and increase the value of λ_1 (see route 1 in Fig. 6.2). Direct reciprocity, in contrast, preserves the value of λ_1 and reduces λ (see route 2a in Fig. 6.2). Indirect reciprocity promotes the same transition (see route 2b in Fig. 6.2). Supplementary, one can analyze *costly punishment*. Using the payoff specifications made in the Supporting Information of [29], costly punishment changes λ by $-(\beta + \gamma)/2 < 0$ and λ_1 by $-\gamma$ [29], i.e. when γ is increased, the values of λ and λ_1 are simultaneously reduced (see route 2c in Fig. 6.2). Here, $\gamma > 0$ represents the punishment cost invested by a cooperator to impose a punishment fine $\beta > 0$ on a defector, which decreases the payoffs of both interaction partners. Route 3 can be generated by the formation of friendship networks [30]. Route 4 may occur by kin selection, network reciprocity, or group selection, when starting with a prisoner's dilemma with $\lambda^0 < 0$ (rather than $\lambda^0 = 0$ as assumed before). Route 5 may be generated by the same mechanisms, if $\lambda^0 > 0$. Finally, route 6 can be implemented by time-dependent taxation (see Fig. 6.2).

6.5 Further Kinds of Transitions to Cooperation

The routes to cooperation discussed so far change the eigenvalues λ_1 and λ_2, and leave the stationary solutions p_1 and p_2 unchanged. However, transitions to cooperation can also be generated by shifting the stationary solutions or creating new ones, as we will show now. For this, we generalize the replicator equation (6.2) by replacing λ_1 with $f(p)$ and λ with $g(p)$, and by adding a term $h(p)$, which can describe effects of spontaneous transitions like mutations. To guarantee $0 \leq p(t) \leq 1$, we must have $h(p) = v(p) - pw(p)$ with functions $w(p) \geq v(p) \geq 0$. The resulting equation is $dp/dt = F(p(t))$ with $F(p) = (1 - p)[f(p) - 2g(p)p]p + h(p)$, and its stationary solutions p_k are given by $F(p_k) = (1 - p_k)[f(p_k) - 2g(p_k)p_k]p_k + h(p_k) = 0$. The associated

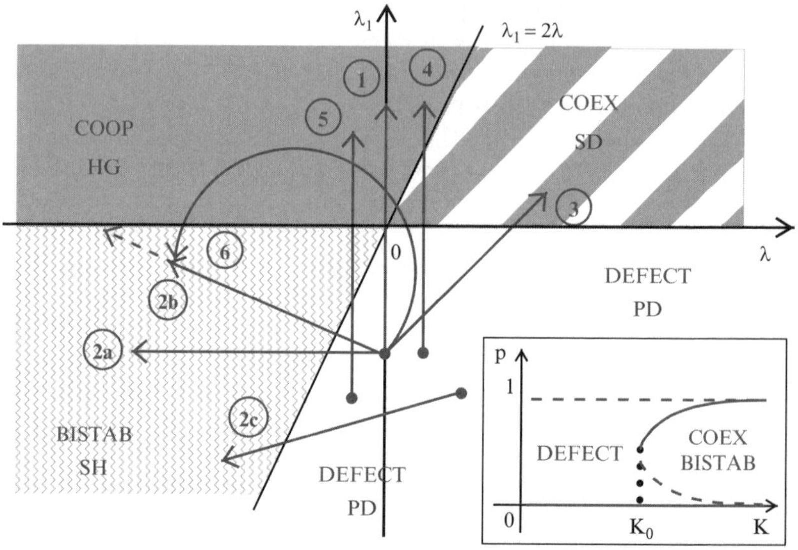

Fig. 6.2 Phase diagram of expected system behaviors, based on an analysis of the game-dynamical replicator equation (6.2) as a function of the parameters λ and λ_1. The different routes to cooperation are illustrated by *arrows*. Terms in capital letters are defined in Fig. 6.1. *Inset*: Stable stationary solutions (*solid lines*) and unstable ones (*broken lines*) as functions of the parameter K, when the reward depends on the proportion of cooperators. The bifurcation at the "tipping point" $K = K_0$ "inverts" the system behavior (see main text)

eigenvalues $\lambda_k = dF(p_k)/dp$ determining the stability of the stationary solutions p_k are

$$\lambda_k = (1-2p_k)(f_k - 2p_k g_k) + p_k(1-p_k)(f_k' - 2p_k g_k' - 2g_k) + h_k',$$

where we have used the abbreviations $f_k = f(p_k)$, $g_k = g(p_k)$, $h_k = h(p_k)$. $f_k' = f'(p_k)$, $g_k' = g'(p_k)$ and $h_k = h'(p_k)$ are the derivatives of the functions $f(p)$, $g(p)$ and $h(p)$ in the points $p = p_k$.

Classification. We can now distinguish different *kinds* of transitions from defection to cooperation: If the stationary solutions $p_1 = 0$ and $p_2 = 1$ of the prisoner's dilemma are modified, we talk about transitions to cooperation by *equilibrium displacement*. This case occurs, for example, when random mutations are not weak ($h \neq 0$). If the eigenvalues λ_1 or λ_2 of the stationary solutions $p_1 = 0$ and $p_2 = 1$ are changed, we speak of *equilibrium selection*. This case applies to all routes to cooperation discussed before. If a new stationary solution appears, we speak of *equilibrium creation*. The different cases often appear in combination with each other (see the Summary below). In the following, we will discuss an interesting case, where cooperation occurs solely through equilibrium creation, i.e. the stationary solutions p_1 and p_2 of the replicator equation for the prisoner's dilemma as well as their eigenvalues λ_1 and λ_2 remain unchanged. We illustrate this by the example of an *adaptive kind of group pressure* that rewards mutual cooperation ($T_{11} < 0$) or sanctions unilateral defection ($T_{21} > 0$). Both, rewarding and sanctioning reduces the value of λ_2, while λ_1 remains unchanged. Assuming here that the group pressure vanishes, when everybody cooperates (as it is not needed then), while it is maximum when everybody defects (to encourage cooperation) [31], we may set $f(p) = \lambda_1^0$ and $g(p) = \lambda^0 - K[1 - p(t)]$, corresponding to $\lambda_2(t) = \lambda_2^0 - 2K[1 - p(t)]$. It is obvious that we still have the two stationary solutions $p_1 = 0$ and $p_2 = 1$ with the eigenvalues $\lambda_1 = \lambda_1^0 < 0$ and $\lambda_2 = 2\lambda^0 - \lambda_1^0 > 0$ of the original prisoners dilemma with parameters λ_1^0 and λ_2^0 or λ^0. However, for large enough values of K [namely for $K > K_0 = \lambda^0 + |\lambda_1^0| + \sqrt{|\lambda_1^0|(2\lambda^0 + |\lambda_1^0|)}$], we find two additional stationary solutions

$$p_\pm = \frac{1}{2} - \frac{\lambda^0}{2K} \pm \sqrt{\left(\frac{1}{2} - \frac{\lambda^0}{2K}\right)^2 - \frac{|\lambda_1^0|}{2K}} . \tag{6.3}$$

p_- is an *unstable* stationary solution with $p_1 < p_- < p_+$ and $\lambda_- = dF(p_-)/dp > 0$, while p_+ is a *stable* stationary solution with $p_- < p_+ < p_2$ and $\lambda_+ = dF(p_+)/dp < 0$ (see inset of Fig. 6.2). Hence, the assumed dependence of the payoffs on the proportion p of cooperators generates a *bistable* situation (BISTAB), with the possibility of a coexistence of a few defectors with a large proportion p_+ of cooperators, given $K > K_0$. If $p(0) < p_-$, where $p(0)$ denotes the initial condition, defection by everybody results, while a stationary proportion p_+ of cooperators is established for $p_- < p(0) < 1$. Surprisingly, in the limit $K \to \infty$, cooperation is established for *any* initial condition $p(0) \neq 0$ (or through fluctuations).

6.6 Summary

We have discussed from a physical point of view what must happen that social or biological, payoff-changing interaction mechanisms can create cooperation in the prisoner's dilemma. The possible ways are (1) moving the stable stationary solution away from pure defection (routes 3, 4, and 6), (2) stabilizing the unstable solution (routes 1, 2, 4, 5 and 6), or (3) creating new stationary solutions, which are stable (routes 3, 4 and 6). Several of these points can be combined. If (1) is fulfilled, we speak of "equilibrium displacement", if their eigenvalues change, we called this "equilibrium selection", and if (3) is the case, we talk of "equilibrium creation". The first case can result from mutations, the second one applies to many social or biological cooperation-enhancing mechanisms [28]. We have discussed an interesting case of equilibrium creation, in which the outcome of the replicator equation is changed, although the stationary solutions of the PD and their eigenvalues remained unchanged. This can, for example, occur by adaptive group pressure [31], which introduces an adaptive feedback mechanism and thereby increases the order of non-linearity of the replicator equation. Surprisingly, already a linear dependence of the payoff values P_{ij} on the endogeneous dynamics $p(t)$ of the system is enough to destabilize defection and stabilize cooperation, thereby inverting the outcome of the prisoner's dilemma.

Acknowledgements This work was partially supported by the Future and Emerging Technologies programme FP7-COSI-ICT of the European Commission through the project QLectives (grant no.: 231200).

References

1. J. von Neumann, O. Morgenstern, *Theory of Games and Economic Behavior* (Princeton University, Princeton, 1944)
2. R. Axelrod, *The Evolution of Cooperation* (Basic, New York, 1984)
3. J. Hofbauer, K. Sigmund, *Evolutionary Games and Population Dynamics* (Cambridge University Press, Cambridge, 1998)
4. J.W. Weibull, *Evolutionary Game Theory* (MIT Press, Cambridge, MA, 1996)
5. N.F. Johnson, P.M. Hui, R. Jonson, T.S. Lo, Phys. Rev. Lett. **82**, 3360 (1999)
6. D. Challet, M. Marsili, R. Zecchina, Phys. Rev. Lett. **84**, 1824 (2000)
7. G. Szabó, C. Hauert, Phys. Rev. Lett. **89**, 118101 (2002)
8. C. Hauert, M. Doebeli, Nature **428**, 643 (2004)
9. J.C. Claussen, A. Traulsen, Phys. Rev. Lett. **100**, 058104 (2008)
10. C.P. Roca, J.A. Cuesta, A. Sánchez, Phys. Rev. Lett. **97**, 158701 (2006)
11. D. Helbing, W. Yu, PNAS **106**, 3680 (2009)
12. D. Helbing, T. Vicsek, New J. Phys. **1**, 13 (1999)
13. D. Helbing, T. Platkowski, Europhys. Lett. **60**, 227 (2002)
14. G. Szabó, G. Fath, Phys. Rep. **446**, 97 (2007)
15. J.M. Pacheco, A. Traulsen, M.A. Nowak, Phys. Rev. Lett. **97**, 258103 (2006)
16. F.C. Santos, J.M. Pacheco, Phys. Rev. Lett. **95**, 098104 (2005)
17. J. Gómez-Gardeñes, M. Campillo, L.M. Floría, Y. Moreno, Phys. Rev. Lett. **98**, 108103 (2007)
18. S. VanSegbroeck, F.C. Santos, T. Lenaerts, J.M. Pacheco, Phys. Rev. Lett. **102**, 058105 (2009)
19. J. Berg, A. Engel, Phys. Rev. Lett. **81**, 4999 (1998)
20. A. Traulsen, J.C. Claussen, C. Hauert, Phys. Rev. Lett. **95**, 238701 (2005)
21. H. Ohtsuki, M.A. Nowak, J.M. Pacheco, Phys. Rev. Lett. **98**, 108106 (2007)
22. J. Eisert, M. Wilkens, M. Lewenstein, Phys. Rev. Lett. **83**, 3077 (1999)
23. M. Eigen, P. Schuster, *The Hypercycle* (Springer, Berlin, 1979)
24. R.A. Fisher, *The Genetical Theory of Natural Selection* (Oxford University Press, Oxford, 1930)
25. M. Opper, S. Diederich, Phys. Rev. Lett. **69**, 1616 (1992)
26. V.M. de Oliveira, J.F. Fontanari, Phys. Rev. Lett. **89**, 148101 (2002)
27. J.Y. Wakano, M.A. Nowak, C. Hauert, PNAS **106**, 19 (2009)
28. M.A. Nowak, Science **314**, 1560 (2006)
29. A. Traulsen, C. Hauert, H. De Silva, M.A. Nowak, K. Sigmund, PNAS **106**(3), 709 (2009)
30. H. Ohtsuki, M.A. Nowak, J. Theor. Biol. **243**, 86–97 (2006)
31. O. Gurerk, B. Irlenbusch, B. Rockenbach, Science **312**, 108–111 (2006)

Chapter 7
Co-evolution of Social Behavior and Spatial Organization*

7.1 Introduction

While the availability of new data of human mobility has revealed relations with social communication patterns [1] and epidemic spreading [2], its significance for the cooperation among individuals is still widely unknown. This is surprising, as migration is a driving force of population dynamics as well as urban and interregional dynamics [3–5].

Below, we model cooperation in a game-theoretical way [6–8], and integrate a model of stylized relocations. This is motivated by the observation that individuals prefer better neighborhoods, e.g. a nicer urban quarter or a better work environment. To improve their situation, individuals are often willing to migrate. In our model of success-driven migration, individuals consider different alternative locations within a certain migration range, reflecting the effort they are willing or able to spend on identifying better neighborhoods. How favorable a new neighborhood is expected to be is determined by test interactions with individuals in that area ("neighborhood testing"). The related investments are often small compared to the potential gains or losses after relocating, i.e. exploring new neighborhoods is treated as "fictitious play". Finally, individuals are assumed to move to the tested neighborhood that promises to be the best.

So far, the role of migration has received relatively little attention in game theory [9–16], probably because it has been found that mobility can undermine cooperation by supporting defector invasion [11, 12]. However, this primarily applies to cases, where individuals choose their new location in a random (e.g. diffusive) way. In contrast, extending spatial games by the specific mechanism of *success-driven* migration can support the survival and spreading of cooperation. As we will show,

---

*This chapter reprints a previous publication with kind permission of the National Academy of Sciences of the USA. It is requested to cite this work as follows: D. Helbing and W. Yu, The outbreak of cooperation among success-driven individuals under noisy conditions. *Proceedings of the National Academy of Sciences USA* **106**(8), 3680–3685 (2009).

D. Helbing (ed.), *Social Self-Organization*, Understanding Complex Systems,
DOI 10.1007/978-3-642-24004-1_7, © Springer-Verlag Berlin Heidelberg 2012

it even promotes the spontaneous *outbreak* of prevalent cooperation in a world of selfish individuals with various sources of randomness ("noise"), starting with defectors only.

7.2 Model

Our study is carried out for the prisoner's dilemma game (PD). This has often been used to model selfish behavior of individuals in situations, where it is risky to cooperate and tempting to defect, but where the outcome of mutual defection is inferior to cooperation on both sides [7, 17]. Formally, the so-called "reward" R represents the payoff for mutual cooperation, while the payoff for defection on both sides is the "punishment" P. T represents the "temptation" to unilaterally defect, which results in the "sucker's payoff" S for the cooperating individual. Given the inequalities $T > R > P > S$ and $2R > T+S$, which define the classical prisoner's dilemma, it is more profitable to defect, no matter what strategy the other individual selects. Therefore, rationally behaving individuals would be expected to defect when they meet *once*. However, defection by everyone is implied as well by the game-dynamical replicator equation [10], which takes into account imitation of superior strategies, or payoff-driven birth-and-death processes. In contrast, a coexistence of cooperators and defectors is predicted for the snowdrift game (SD). While it is also used to study social cooperation, its payoffs are characterized by $T > R > S > P$ (i.e. $S > P$ rather than $P > S$).

As is well-known [17], cooperation can, for example, be supported by repeated interactions [7], by intergroup competition with or without altruistic punishment [18–20], and by network reciprocity based on the clustering of cooperators [21–23]. In the latter case, the level of cooperation in two-dimensional spatial games is further enhanced by "disordered environments" (approximately 10% unaccessible empty locations) [24], and by diffusive mobility, provided that the mobility parameter is in a suitable range [16]. However, strategy mutations, random relocations, and other sources of stochasticity ("noise") can significantly challenge the formation and survival of cooperative clusters. When no mobility or undirected, random mobility are considered, the level of cooperation in the spatial games studied by us is sensitive to noise (see Figs. 7.1d and 7.3c), as favorable correlations between cooperative neighbors are destroyed. *Success-driven* migration, in contrast, is a robust mechanism: By leaving unfavorable neighborhoods, seeking more favorable ones, and remaining in cooperative neighborhoods, it supports cooperative clusters very efficiently against the destructive effects of noise, thus preventing defector invasion in a large area of payoff parameters. We assume N individuals on a square lattice with periodic boundary conditions and $L \times L$ sites, which are either empty or occupied by one individual. Individuals are updated asynchronously, in a random sequential order. The randomly selected individual performs simultaneous interactions with the $m = 4$ direct neighbors and compares the overall payoff with that of the m neighbors. Afterwards, the strategy of the best performing neighbor

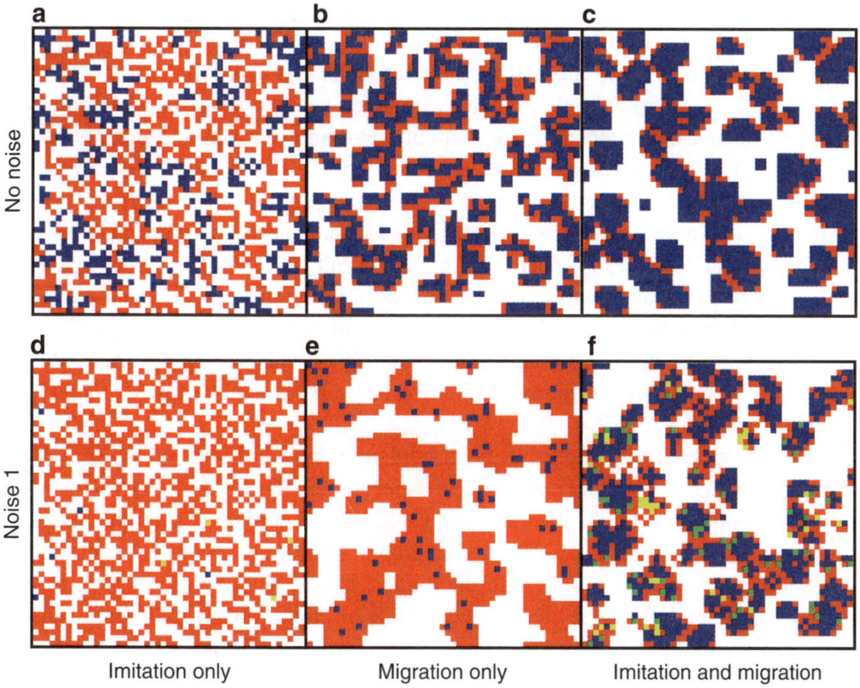

Fig. 7.1 Representative simulation results for the spatial prisoner's dilemma with payoffs $T = 1.3$, $R = 1$, $P = 0.1$, and $S = 0$ after $t = 200$ iterations. The simulations are for 49×49-grids with 50% empty sites. At time $t = 0$ we assumed 50% of the individuals to be cooperators and 50% defectors. Both strategies were homogeneously distributed over the whole grid. For reasons of comparison, all simulations were performed with identical initial conditions and random numbers (*red* = defector, *blue* = cooperator, *white* = empty site, *green* = defector who became a cooperator in the last iteration, *yellow* = cooperator who turned into a defector). Compared to simulations without noise (*top*), the strategy mutations of noise 1 with $r = q = 0.05$ do not only reduce the resulting level of cooperation, but also the outcome and pattern formation dynamics, even if the payoff values, initial conditions, and update rules are the same (*bottom*): In the imitation-only case with $M = 0$ that is displayed on the left, the initial fraction of 50% cooperators is quickly reduced due to imitation of more successful defectors. The result is a "frozen" configuration without any further strategy changes. (**a**) In the noiseless case, a certain number of cooperators can survive in small cooperative clusters. (**d**) When noise 1 is present, random strategy mutations destroy the level of cooperation almost completely, and the resulting level of defection reaches values close to 100%. The illustrations in the center show the migration-only case with mobility range $M = 5$: (**b**) When no noise is considered, small cooperative clusters are formed, and defectors are primarily located at their boundaries. (**e**) In the presence of noise 1, large clusters of defectors are formed instead, given $P > 0$. The illustrations on the right show the case, where imitation is combined with success-driven migration (here, $M = 5$): (**d**) In the noiseless case, cooperative clusters grow and eventually freeze (i.e. strategy changes or relocations do not occur any longer). (**f**) Under noisy conditions, in contrast, the cooperative clusters continue to adapt and reconfigure themselves, as the existence of yellow and green sites indicates

is copied with probability $1 - r$ ("imitation"), if the own payoff was lower. With probability r, however, the strategy is randomly "reset": *Noise 1* assumes that an individual spontaneously chooses to cooperate with probability q or to defect with probability $1 - q$ until the next strategy change. The resulting strategy mutations may be considered to reflect deficient imitation attempts or trial-and-error behavior. As a side effect, such noise leads to an independence of the finally resulting level of cooperation from the initial one at $t = 0$, and a *qualitatively different* pattern formation dynamics for the same payoff values, update rules, and initial conditions (see Fig. 7.1). Using the alternative Fermi update rule [22] would have been possible as well. However, resetting strategies rather than inverting them, combined with values q much smaller than 1/2, has here the advantage of creating particularly adverse conditions for cooperation, independently of what strategy prevails. Below, we want to learn, if predominant cooperation can survive or even emerge under such adverse conditions.

"Success-driven migration" has been implemented as follows [9, 25]: Before the imitation step, an individual explores the expected payoffs for the empty sites in the migration neighborhood of size $(2M + 1) \times (2M + 1)$ (the Moore neighborhood of range M). If the fictitious payoff is higher than in the current location, the individual is assumed to move to the site with the highest payoff and, in case of several sites with the same payoff, to the closest one (or one of them); otherwise it stays put.

7.3 Results

Computer simulations of the above model show that, in the *imitation-only* case of classical spatial games with noise 1, but *without* a migration step, the resulting fraction of cooperators in the PD tends to be very low. It basically reflects the fraction rq of cooperators due to strategy mutations. For $r = q = 0.05$, we find almost frozen configurations, in which only a small number of cooperators survive (see Fig. 7.1d). In the *migration-only* case without an imitation step, the fraction of cooperators changes only by strategy mutations. Even when the initial strategy distribution is uniform, one observes the formation of spatio-temporal patterns, but the patterns get almost frozen after some time (see Fig. 7.1e).

It is interesting that, although for the connectivity structure of our PD model neither imitation only (Fig. 7.1d) nor migration only (Fig. 7.1e) can promote cooperation under noisy conditions, their *combination* does: Computer simulations show the formation of cooperative clusters with a few defectors at their boundaries (see Fig. 7.1f). Once cooperators are organized in clusters, they tend to have more neighbors and to reach higher payoffs on average, which allows them to survive [9, 10, 25]. It will now have to be revealed, how success-driven migration causes the *formation* of clusters at all, considering the opposing noise effects. In particular, we will study, why defectors fail to invade cooperative clusters and to erode them from within, although a cooperative environment is most attractive to them.

To address these questions, Fig. 7.2 studies a "defector's paradise" with a single defector in the center of a cooperative cluster. In the noisy *imitation-only* spatial prisoner's dilemma, defection tends to spread up to the boundaries of the cluster, as cooperators imitate more successful defectors (see Figs. 7.2a–d). But if imitation is combined with *success-driven migration*, the results are in sharp contrast: Although defectors still spread initially, cooperative neighbors who are M steps away from the boundary of the cluster can now evade them. Due to this defector-triggered migration, the neighborhood reconfigures itself adaptively. For example, a large cooperative cluster may split up into several smaller ones (see Figs. 7.2e–h). Eventually, the defectors end up at the boundaries of these cooperative clusters, where they often turn into cooperators by imitation of more successful cooperators in the cluster, who tend to have more neighbors. This promotes the spreading of cooperation [9, 10, 25]. Since evasion takes time, cooperative clusters could still be destroyed when continuously challenged by defectors, as it happens under noisy conditions. Therefore, let us now study the effect of different kinds of randomness [10,26]. *Noise 1* (defined above) assumes *strategy mutations*, but leaves

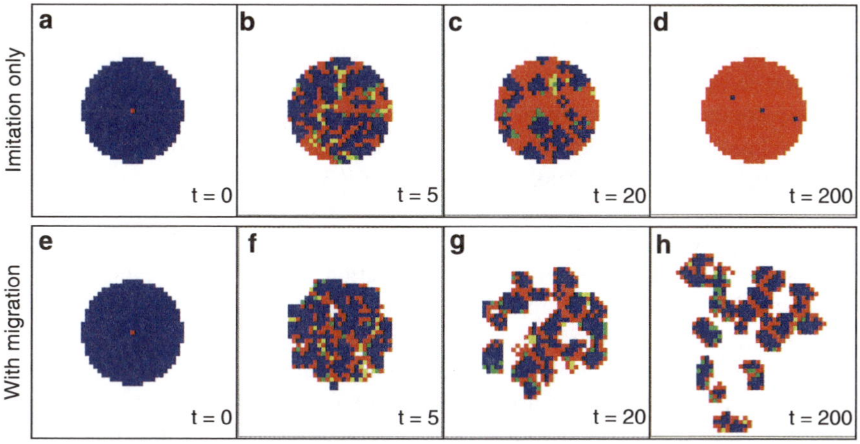

Fig. 7.2 Representative simulation results after $t = 200$ iterations in the "defector's paradise" scenario, starting with a single defector in the center of a cooperative cluster at $t = 0$. The simulations are performed on 49×49-grids with $N = 481$ individuals, corresponding to a circle of diameter 25. They are based on the spatial prisoner's dilemma with payoffs $T = 1.3$, $R = 1$, $P = 0.1$, $S = 0$ and noise parameters $r = q = 0.05$ (*red* = defector, *blue* = cooperator, *white* = empty site, *green* = defector who became a cooperator, *yellow* = cooperator who turned into a defector in the last iteration). For reasons of comparison, all simulations were carried out with identical initial conditions and random numbers. **(a–d)** In the noisy imitation-only case with $M = 0$, defection (*red*) eventually spreads all over the cluster. The few remaining cooperators (*blue*) are due to strategy mutations. **(e–h)** When we add success-driven motion, the result is very different. Migration allows cooperators to evade defectors. That triggers a splitting of the cluster, and defectors end up on the surface of the resulting smaller clusters, where most of them can be turned into cooperators. This mechanism is crucial for the unexpected survival and spreading of cooperators

the spatial distribution of individuals unchanged (see Fig. 7.3a). *Noise 2*, in contrast, assumes that individuals, who are selected with probability r, move to a randomly chosen free site without considering the expected success *(random relocations).* Such random moves may potentially be of long distance and preserve the number of cooperators, but have the potential of destroying spatial patterns (see Fig. 7.3b). *Noise 3* combines noise 1 and noise 2, assuming that individuals randomly relocate with probability r and additionally reset their strategy as in noise 1 (see Fig. 7.3c).

While cooperation in the imitation-only case is quite sensitive to noise (see Figs. 7.3a–c), the combination of imitation with success-driven motion is not (see Fig. 7.3d–f): Whenever an empty site inside a cluster of cooperators occurs, it is more likely that the free site is entered by a cooperator than by a defector, as long as cooperators prevail within the migration range M. In fact, the formation of small cooperative clusters was observed for *all* kinds of noise. That is, the combination of imitation with success-driven migration is a robust mechanism to maintain and even spread cooperation under various conditions, given there are enough cooperators in the beginning.

It is interesting, whether this mechanism is also able to facilitate a spontaneous *outbreak* of predominant cooperation in a noisy world dominated by selfishness, without a "shadow of the future" [7,27]. Our simulation scenario assumes defectors only in the beginning (see Fig. 7.4a), strategy mutations in favor of defection, and short-term payoff-maximizing behavior in the vast majority of cases. In order to study conditions under which a significant fraction of cooperators is unlikely, our simulations are performed with noise 3 and $r = q = 0.05$, as it tends to destroy spatial clusters and cooperation (see Fig. 7.3c): By relocating 5% randomly chosen individuals in each time step, noise 3 dissolves clusters into more or less separate individuals in the imitation-only case (see Figs. 7.3b+c). In the case with success-driven migration, random relocations break up large clusters into many smaller ones, which are distributed all over the space (see Figs. 7.3b+c and 7.4b). Therefore, even the clustering tendency by success-driven migration can only partially compensate for the dispersal tendency by random relocations. Furthermore, the strategy mutations involved in noise 3 tend to destroy cooperation (see Figs. 7.3a+c, where the strategies of 5% randomly chosen individuals were replaced by defection in 95% of the cases and by cooperation otherwise, to create conditions favoring defection, i.e. the dominant strategy in the prisoner's dilemma). Overall, as a result of strategy mutations (i.e. without the consideration of imitation processes), only a fraction $rq = 0.0025$ of all defectors turn into cooperators in each time step, while a fraction $r(1 - q) \approx 0.05$ of all cooperators turn into defectors (i.e. 5% in each time step). This setting is extremely unfavorable for the spreading of cooperators. In fact, defection prevails for an extremely long time (see Figs. 7.4b and 7.5a). But suddenly, when a small, supercritical cluster of cooperators has occurred by coincidence (see Fig. 7.4c), the fraction of cooperators spreads quickly (see Fig. 7.5a), and soon cooperators prevail (see Figs. 7.4d and 7.5b). Note that this spontaneous birth of predominant cooperation in a world of defectors does not occur in the noisy imitation-only case and demonstrates that success-driven migration can overcome the dispersive tendency of noises 2 and 3, if r is moderate and q has a

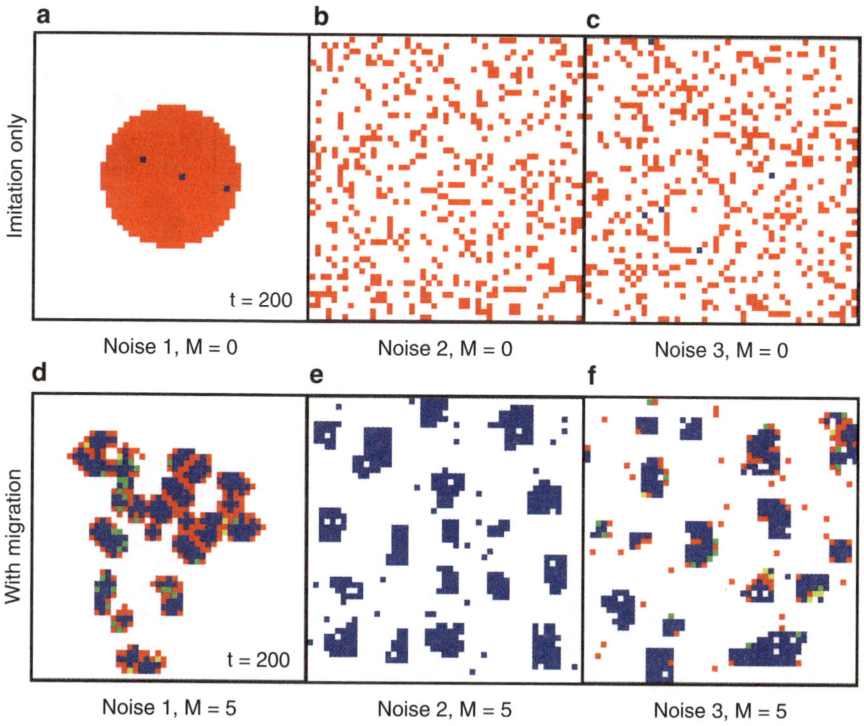

Fig. 7.3 Representative simulation results for the invasion scenario with a defector in the center of a cooperative cluster ("defector's paradise"). The chosen payoffs $T = 1.3$, $R = 1$, $P = 0.1$, and $S = 0$ correspond to a prisoner's dilemma. The simulations are for 49×49-grids with $N = 481$ individuals, corresponding to a circle of diameter 25 (*red* = defector, *blue* = cooperator, *white* = empty site, *green* = defector who became a cooperator, *yellow* = cooperator who turned into a defector in the last iteration). *Top*: Typical numerical results for the imitation-only case ($M = 0$) after $t = 200$ iterations (**a**) for noise 1 (strategy mutations) with mutation rate $r = 0.05$ and creation of cooperators with probability $q = 0.05$, (**b**) for noise 2 (random relocations) with relocation rate $r = 0.05$, and (**c**) for noise 3 (a combination of random relocations and strategy mutations) with $r = q = 0.05$. As cooperators imitate defectors with a higher overall payoff, defection spreads easily. The different kinds of noise influence the dynamics and resulting patterns considerably: While strategy mutations in (**a**) and (**c**) strongly reduce the level of cooperation, random relocations in (**b**) and (**c**) break up spatial clusters, leading to a dispersion of individuals in space. Their combination in case (**c**) essentially destroys both, clusters and cooperation. *Bottom*: Same for the case of imitation *and* success-driven migration with mobility range $M = 5$ (**d**) for noise 1 with $r = q = 0.05$, (**e**) for noise 2 with $r = 0.05$, and (**f**) for noise 3 with $r = q = 0.05$. Note that noise 1 just mutates strategies and does not support a spatial spreading, while noise 2 causes random relocations, but does not mutate strategies. This explains why the clusters in Fig. 7.3d do not spread out over the whole space and why no new defectors are created in Fig. 7.3e. However, the creation of small cooperative clusters is found in all three scenarios. Therefore, it is robust with respect to various kinds of noise, in contrast to the imitation-only case

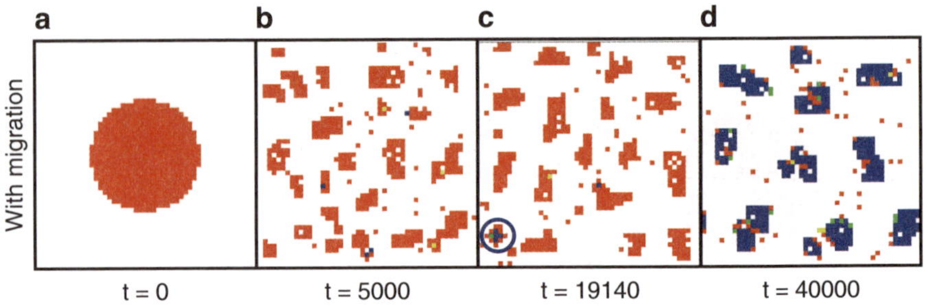

Fig. 7.4 Spontaneous outbreak of prevalent cooperation in the spatial prisoner's dilemma with payoffs $T = 1.3$, $R = 1$, $P = 0.1$, $S = 0$ in the presence of noise 3 (random relocations and strategy mutations) with $r = q = 0.05$. The simulations are for 49×49-grids (*red* = defector, *blue* = cooperator, *white* = empty site, *green* = defector who became a cooperator, *yellow* = cooperator who turned into a defector in the last iteration). (**a**) Initial cluster of defectors, which corresponds to the *final* stage of the *imitation-only* case with strategy mutations according to noise 1 (see Fig. 7.2d). (**b**) Dispersal of defectors by noise 3, which involves random relocations. A few cooperators are created randomly by strategy mutations with the very small probability $rq = 0.0025$ (0.25%). (**c**) Occurrence of a supercritical cluster of cooperators after a very long time. This cooperative "nucleus" originates by random coincidence of favorable strategy mutations in neighboring sites. (**d**) Spreading of cooperative clusters in the whole system. This spreading despite the destructive effects of noise requires an effective mechanism to form growing cooperative clusters (such as success-driven migration) and cannot be explained by random coincidence. See the supplementary video for an animation of the outbreak of cooperation for a different initial condition

finite value. That is, success-driven migration generates spatial correlations between cooperators more quickly than these noises can destroy them. This changes the outcome of spatial games essentially, as a comparison of Figs. 7.2a–d with 7.4a–d shows.

The conditions for the spreading of cooperators from a supercritical cluster ("nucleus") can be understood by configurational analysis [26, 28] (see Fig. 7.1), but the underlying argument can be both, simplified and extended: According to Fig. 7.6a, the level of cooperation changes when certain lines (or, more generally, certain hyperplanes) in the payoff-parameter space are crossed. These hyperplanes are all of the linear form

$$n_1 R + n_2 S = n_3 T + n_4 P, \tag{7.1}$$

where $n_k \in \{0, 1, 2, 3, 4\}$. The left-hand side of)7.1) represents the payoff of the most successful cooperative neighbor of a focal individual, assuming that this has n_1 cooperating and n_2 defecting neighbors, which implies $n_1 + n_2 \leq m = 4$. The right-hand side reflects the payoff of the most successful defecting neighbor, assuming that n_3 is the number of his/her cooperating neighbors and n_4 the number of defecting neighbors, which implies $n_3 + n_4 \leq m = 4$. Under these conditions, the best-performing cooperative neighbor earns a payoff of $n_1 R + n_2 S$, and the best-performing defecting neighbor earns a payoff of $n_3 T + n_4 P$. Therefore, the focal individual will imitate the cooperator, if $n_1 R + n_2 S > n_3 T + n_4 P$, but copy

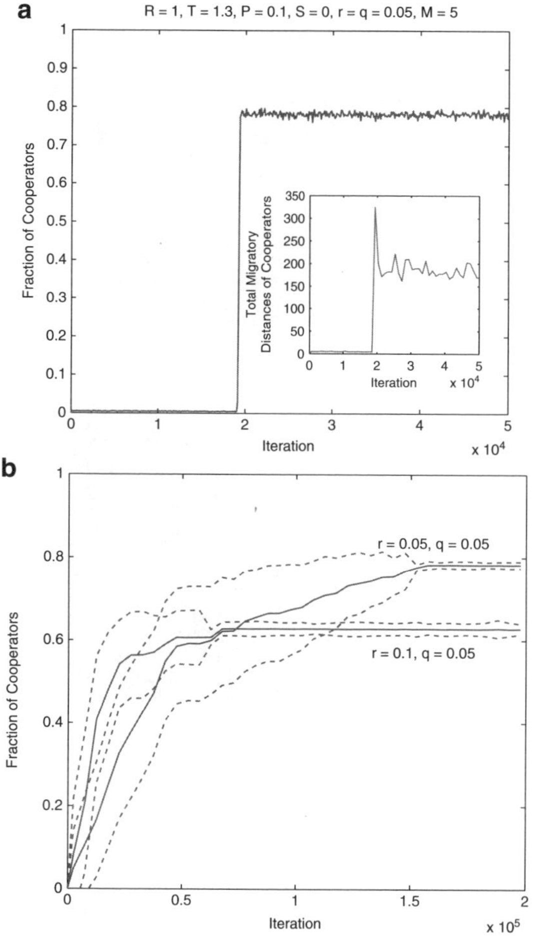

Fig. 7.5 Representative example for the outbreak of predominant cooperation in the prisoner's dilemma with payoffs $T = 1.3$, $R = 1$, $P = 0.1$, $S = 0$, in the presence of noise 3 with $r = q = 0.05$. The simulations are for 49×49-grids with a circular cluster of defectors and no cooperators in the beginning (see Fig. 7.4a). (**a**) After defection prevails for a very long time (here for almost 20,000 iterations), a sudden transition to a large majority of cooperators is observed. *Inset*: The overall distance moved by all individuals during one iteration has a peak at the time when the outbreak of cooperation is observed. Before, the rate of success-driven migration is very low, while it stabilizes at an intermediate level afterwards. This reflects a continuous evasion of cooperators from defectors and, at the same time, the continuous effort to form and maintain cooperative clusters. The graph displays the amount of success-driven migration only, while the effect of random relocations is not shown. (**b**) Evaluating 50 simulation runs, the error bars (representing three standard deviations) show a large variation of the time points when prevalent cooperation breaks out. Since this time point depends on the coincidence of random cooperation in neighboring sites, the large error bars have their natural reason in the stochasticity of this process. After a potentially very long time period, however, all systems end up with a high level of cooperation. The level of cooperation decreases with the noise strength r, as expected, but moderate values of r can even *accelerate* the transition to predominant cooperation. Using the parameter values $r = 0.1$ and $q = 0.2$, the outbreak of prevalent cooperation takes often less than 200 iterations

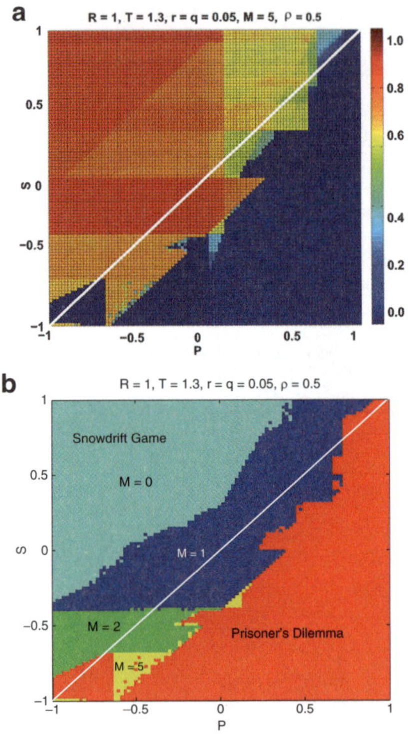

Fig. 7.6 Dependence of the fraction of cooperators for given payoff parameters $T = 1.3$ and $R = 1$ on the parameters P and S. The area above the solid diagonal line corresponds to the snowdrift game, the area below to the prisoner's dilemma. Our simulations were performed for grids with $L \times L = 99 \times 99$ sites and $N = L^2/2$ individuals, corresponding to a density $\rho = N/L^2 = 0.5$. At time $t = 0$ we assumed 50% of the individuals to be cooperators and 50% defectors. Both strategies were homogeneously distributed over the whole grid. The finally resulting fraction of cooperators was averaged at time $t = 200$ over 50 simulation runs with different random realizations. The simulations were performed with noise 3 (random relocations with strategy mutations) and $r = p = 0.05$. An enhancement in the level of cooperation (often by more than 100%) is observed mainly in the area with $P - 0.4 < S < P + 0.4$ and $P < 0.7$. Results for the noiseless case with $r = 0$ are shown in Fig. 7.5. (**a**) The fraction of cooperators is represented by color codes (see the bar to the right of the figure, where *dark orange*, for example, corresponds to 80% cooperators). It can be seen that the fraction of cooperators is approximately constant in areas limited by straight lines (mostly triangular and rectangular ones). These lines correspond to (7.1) for different specifications of n_1, n_2, n_3, and n_4 (see main text for details). (**b**) The *light blue* area reflects the parameters for which cooperators reach a majority in the imitation-only case with $M = 0$: For all payoffs P and S corresponding to a prisoner's dilemma, cooperators are clearly in the minority, as expected. However, taking into account success-driven migration changes the situation in a pronounced way: For a mobility range $M = 1$, the additional area with more than 50% cooperators is represented by *dark blue*, the further extended area of prevailing cooperation for $M = 2$ by *green* color, and for $M = 5$ in *yellow*. If $M = 5$, defectors are in the majority only for parameter combinations falling into the *red* area. This demonstrates that success-driven migration can promote predominant cooperation in considerable areas, where defection would prevail without migration. For larger interaction neighborhoods m, e.g. $m = 8$, the area of prevalent cooperation is further increased overall (not shown). Note that the irregular shape of the separating lines is no artefact of the computer simulation or initial conditions. It results by superposition of the areas defined by (7.1), see Fig. 7.6a

the strategy of the defector if $n_1R + n_2S < n_3T + n_4P$. Equation (7.1) is the line separating the area where cooperators spread (above the line) from the area of defector invasion (below it) for a certain spatial configuration of cooperators and defectors (see Fig. 7.6a). Every spatial configuration is characterized by a set of n_k-parameters. As expected, the relative occurrence frequency of each configuration depends on the migration range M (see Fig. 7.6b): Higher values of M naturally create better conditions for the spreading of cooperation, as there is a larger choice of potentially more favorable neighborhoods.

Figure 7.6b also shows that success-driven migration extends the parameter range, in which cooperators prevail, from the parameter range of the snowdrift game with $S > P$ to a considerable parameter range of the prisoner's dilemma. For this to happen, it is important that the attraction of cooperators is mutual, while the attraction of defectors to cooperators is not. More specifically, the attraction of cooperators is proportional to $2R$, while the attraction between defectors and cooperators is proportional to $T + S$. The attraction between cooperators is stronger, because the prisoner's dilemma usually assumes the inequality $2R > T + S$.

Besides the speed of finding neighbors to interact with, the time scales of configurational changes and correlations matter as well: By entering a cooperative cluster, a defector triggers an avalanche of strategy changes and relocations, which quickly destroys the cooperative neighborhood. During this process, individuals may alter their strategy many times, as they realize opportunities by cooperation or defection immediately. In contrast, if a cooperator joins a cooperative cluster, this will stabilize the cooperative neighborhood. Although cooperative clusters continuously adjust their size and shape, the average time period of their existence is longer than the average time period after which individuals change their strategy or location. This coevolution of social interactions and strategic behavior reflects features of many social environments: While the latter come about by individual actions, a suitable social context can make the average behavior of individuals more predictable, which establishes a reinforcement process. For example, due to the clustering tendency of cooperators, the likelihood of finding another cooperator in the neighborhood of a cooperator is greater than 1/2, and also the likelihood that a cooperator will cooperate in the next iteration.

7.4 Discussion

It is noteworthy that all the above features – the survival of cooperation in a large parameter area of the PD, spatio-temporal pattern formation, noise-resistance, and the outbreak of predominant cooperation – can be captured by considering a mechanism as simple as success-driven migration: Success-driven migration *destabilizes* a homogeneous strategy distribution (compare Fig. 7.1c with 7.1a and Fig. 7.1f with 7.1d). This triggers the spontaneous formation of agglomeration and segregation patterns [29], where noise or diffusion would cause dispersal in the imitation-only case. The self-organized patterns create self-reinforcing social environments characterized by behavioral correlations, and imitation promotes the further growth of supercritical cooperation clusters. While each mechanism by itself tends to produce frozen spatial structures, the combination of imitation and migration supports adaptive patterns (see Fig. 7.1f). This facilitates, for example, the regrouping of a cluster of cooperators upon invasion by a defector, which is crucial for the survival and success of cooperators (see Fig. 7.2e–h).

By further simulations we have checked that our conclusions are robust with respect to using different update rules, adding birth and death processes, or introducing a small fraction of individuals defecting unconditionally. The same applies to various kinds of "noise". Noise can even trigger cooperation in a world full of defectors, when the probability of defectors to turn spontaneously into cooperators is 20 times smaller than the probability of cooperators to turn into defectors. Compared to the implications of the game-dynamical replicator equation, this is remarkable: While the replicator equation predicts that the stationary solution with a majority of cooperators is *un*stable with respect to perturbations and the stationary solution with a majority of defectors is stable [10], success-driven migration *inverts* the situation: The state of 100% defectors becomes *unstable* to noise, while a majority of cooperators is stabilized in a considerable area of the payoff parameter space.

Our results help to explain why cooperation can be frequent even if individuals would behave selfishly in the vast majority of interactions. Although one may think that migration would weaken social ties and cooperation, there is another side of it which helps to establish cooperation in the first place, without the need to modify the payoff structure. We suggest that, besides the ability for strategic interactions and learning, the ability to *move* has played a crucial role for the evolution of large-scale cooperation and social behavior. Success-driven migration can reduce unbalanced social interactions, where cooperation is unilateral, and support local agglomeration. In fact, it has been pointed out that local agglomeration is an important precondition for the evolution of more sophisticated kinds of cooperation [30]. For example, the level of cooperation could be further improved by combining imitation and success-driven migration with other mechanisms such as costly punishment [19,20], volunteering [22], or reputation [31–33].

Acknowledgements The authors would like to thank Christoph Hauert, Heiko Rauhut, Sergi Lozano, Michael Maes, Carlos P. Roca, and Didier Sornette for their comments.

References

1. M.C. Gonzáles, C.A. Hidalgo, A.L. Barabási, Understanding individual human mobility patterns. Nature **453**, 779–782 (2008)
2. L. Hufnagel, D. Brockmann, T. Geisel, The scaling laws of human travel. Nature **439**, 462–465 (2006)
3. M. Batty, *Cities and Complexity* (MIT Press, Cambridge, MA, 2005)
4. W. Weidlich, *Sociodynamics. A Systematic Approach to Mathematical Modelling in the Social Sciences* (Harwood Academic, Amsterdam, 2000)

5. D. Pumain (ed.), *Spatial Analysis and Population Dynamics* (John Libbey Eurotext, France, 1991)
6. J.V. Neumann, O. Morgenstern, *Theory of Games and Economic Behavior* (Princeton University, Princeton, NJ, 1944)
7. R. Axelrod, *The Evolution of Cooperation* (Basic Books, New York, 1984)
8. B. Skyrms, *Evolution of The Social Contract* (Cambridge University, New York, 1996)
9. A. Flache, R. Hegselmann, Do irregular grids make a difference? Relaxing the spatial regularity assumption in cellular models of social dynamics. Artif. Soc. Soc. Simulat. **4**(4) (2001)
10. J.M. Epstein, Zones of cooperation in demographic prisoner's dilemma. Complexity **4**(2), 36–48 (1998)
11. L.A. Dugatkin, D.S. Wilson, ROVER: A strategy for exploiting cooperators in a patchy environment. Am. Naturalist **138**(3), 687–701 (1991)
12. M. Enquist, O. Leimar, The evolution of cooperation in mobile organisms. Animal Behav. **45**, 747–757 (1993)
13. J.-F. Le Galliard, R. Ferrière, U. Dieckmann, Adaptive evolution of social traits: Origin, trajectories, and correlations of altruism and mobility. Am. Naturalist **165**(2), 206–224 (2005)
14. T. Reichenbach, M. Mobilia, E. Frey, Mobility promotes and jeopardizes biodiversity in rock-paper-scissors games. Nature **448**, 1046–1049 (2007)
15. C.A. Aktipis, Know when to walk away: contingent movement and the evolution of cooperation. J. Theor. Biol. **231**, 249–260 (2004)
16. M.H. Vainstein, A.T.C. Silva, J.J. Arenzon, Does mobility decrease cooperation? J. Theor. Biol. **244**, 722–728 (2007)
17. M.A. Nowak, Five rules for the evolution of cooperation. Science **314**, 1560–1563 (2006)
18. A. Traulsen, M.A. Nowak, Evolution of cooperation by multilevel selection. Proc. Natl. Acad. Sci. (USA) **103**, 10952–10955 (2006)
19. E. Fehr, S. Gächter, Altruistic punishment in humans. Nature **415**, 137–140 (2002)
20. R. Boyd, H. Gintis, S. Bowles, P.J. Richerson, The evolution of altruistic punishment. Proc. Natl. Acad. Sci. (USA) **100**, 3531–3535 (2003)
21. M.A. Nowak, R.M. May, Evolutionary games and spatial chaos. Nature **359**, 826–829 (1992)
22. G. Szabó, C. Hauert, Phase transitions and volunteering in spatial public goods games. Phys. Rev. Lett. **89**, 118101 (2002)
23. C. Hauert, M. Doebeli, Spatial structure often inhibits the evolution of cooperation in the snowdrift game. Nature **428**, 643–646 (2004)
24. M.H. Vainstein, J.J. Arenzon, Disordered environments in spatial games. Phys. Rev. E **64**, 051905 (2001)
25. D. Helbing, W. Yu, Migration as a mechanism to promote cooperation. Adv. Complex Syst. **11**(4), 641–652 (2008)
26. H.P. Young, *Individual Strategy and Social Structure: An Evolutionary Theory of Institutions* (Princeton University, Princeton, NJ, 1998)
27. N.S. Glance, B.A. Huberman, The outbreak of cooperation. J. Math. Soc. **17**(4), 281–302 (1993)
28. C. Hauert, Fundamental clusters in spatial 2×2 games. Proc. R. Soc. Lond. B **268**, 761–769 (2000)
29. T.C. Schelling, Dynamic models of segregation. J. Math. Sociol. **1**, 143–186 (1971)
30. J.L. Deneubourg, A. Lioni, C. Detrain, Dynamics of aggregation and emergence of cooperation. Biol. Bull. **202**, 262–267 (2002)
31. M.A. Nowak, K. Sigmund, Evolution of indirect reciprocity by image scoring. Nature **393**, 573–577 (1998)
32. M. Milinski, D. Semmann, H.J. Krambeck, Reputation helps solve the "tragedy of the commons". Nature **415**, 424–426 (2002)
33. B. Rockenbach, M. Milinski, The efficient interaction of indirect reciprocity and costly punishment. Nature **444**, 718–723 (2006)

Chapter 8
Evolution of Moral Behavior*

8.1 Introduction

Public goods such as environmental resources or social benefits are particularly prone to exploitation by non-cooperative individuals ("defectors"), who try to increase their benefit at the expense of fair contributors or users, the "cooperators". This implies a tragedy of commons [1]. It was proposed that costly punishment of non-cooperative individuals can establish cooperation in public goods dilemmas [2–8], and it is effective indeed [9–11]. Nonetheless, why would cooperators choose to punish defectors at a personal cost [12–14]? One would expect that evolutionary pressure should eventually eliminate such "moralists" due to their extra costs compared to "second-order free-riders" (i.e. cooperators, who do not punish). These, however should finally be defeated by "free-riders" (defectors). To overcome this problem [15, 16], it was proposed that cooperators who punish defectors (called "moralists" by us) would survive through indirect reciprocity [17], reputation effects [18] or the possibility to abstain from the joint enterprize [19–21] by "volunteering" [22, 23]. Without such mechanisms, cooperators who punish will usually vanish. Surprisingly, however, the second-order free-rider problem is naturally resolved, without assuming additional mechanisms, if spatial or network interactions are considered. This will be shown in the following.

In order to study the conditions for the disappearance of non-punishing cooperators and defectors, we simulate the public goods game with costly punishment, considering two cooperative strategies (C, M) and two defective ones (D, I). For illustration, one may imagine that cooperators (C) correspond to countries trying to meet the CO_2 emission standards of the Kyoto protocol [24], and "moralists" (M) to cooperative countries that additionally enforce the standards by international pressure (e.g. embargoes). Defectors (D) would correspond to those countries

*This chapter reprints a previous publication, which should be cited as follows: D. Helbing, *A. Szolnoki, M. Perc, and G. Szabó*, Evolutionary establishment of moral and double moral standards through spatial interactions. *PLoS Computational Biology* **6**(4), e1000758 (2010).

D. Helbing (ed.), *Social Self-Organization*, Understanding Complex Systems,
DOI 10.1007/978-3-642-24004-1_8, © Springer-Verlag Berlin Heidelberg 2012

ignoring the Kyoto protocol, and immoralists (I) to countries failing to meet the Kyoto standards, but nevertheless imposing pressure on other countries to fulfil them. According to the classical game-theoretical prediction, all countries would finally fail to meet the emission standards, but we will show that, in a spatial setting, interactions between the four strategies C, D, M, and I can promote the spreading of moralists. Other well-known public goods problems are over-fishing, the pollution of our environment, the creation of social benefit systems, or the establishment and maintenance of cultural institutions (such as a shared language, norms, values, etc.).

Our simplified game-theoretical description of such problems assumes that cooperators (C) and moralists (M) make a contribution of 1 to the respective public good under consideration, while nothing is contributed by defectors (D) and "immoralists" (I), i.e. defectors who punish other defectors. The sum of all contributions is multiplied by a factor r reflecting *synergy effects* of cooperation, and the resulting amount is equally shared among the $k + 1$ interacting individuals. Moreover, moralists and immoralists impose a fine β/k on each defecting individual (playing D or I), which produces an additional cost γ/k per punished defector to them (see Methods for details). The division by k scales for the group size, but for simplicity, the parameter β is called the *punishment fine* and γ the *punishment cost*.

Given the same interaction partners, an immoralist never gets a higher payoff than a defector, but does equally well in a cooperative environment. Moreover, a cooperator tends to outperform a moralist, given the interaction partners are the same. However, a cooperator can do better than a defector when the punishment fine β is large enough.

It is known that punishment in the public goods game and similar games can promote cooperation above a certain critical threshold of the synergy factor r [11, 25]. Besides cooperators who punish defectors, Heckathorn considered "full cooperators" (moralists) and "hypocritical cooperators" (immoralists) [26]. For well-mixed interactions (where individuals interact with a representative rather than local strategy distribution), Eldakar and Wilson find that altruistic punishment (moralists) can spread, if second-order free-riders (non-punishing altruists) are excluded, and that selfish punishers (immoralists) can survive together with altruistic non-punishers (cooperators), provided that selfish nonpunishers (defectors) are sufficiently scarce [27].

Besides well-mixed interactions, some researchers have also investigated the effect of spatial interactions [5, 11, 28, 29], since it is known that they can support the survival or spreading of cooperators [30] (but this is not always the case [31, 32]). In this way, Brandt et al. discovered a coexistence of cooperators and defectors for certain parameter combinations [11]. Compared to these studies, our model assumes somewhat different replication and strategy updating rules. The main point, however, is that we have chosen long simulation times and scanned the parameter space more extensively, which revealed several new insights, for example, the possible coexistence of immoralists and moralists, even when a substantial number of defectors is present initially. When interpreting our results within the context of moral dynamics [33], our main discoveries for a society facing public goods games may be summarized as follows:

1. *Victory over second-order free-riders*: Over a long enough time period, moralists fully eliminate cooperators, thereby solving the "second-order free-rider problem". This becomes possible by spatial segregation of the two cooperative strategies C and M, where the presence of defectors puts moralists in a advantageous position, which eventually allows moralists to get rid of non-punishing cooperators.
2. *Who laughs last laughs best effect*: Moralists defeat cooperators even when the defective strategies I and D are eventually eliminated, but this process is very slow. That is, the system behavior changes its character significantly even after very long times. This is the essence of the "who laughs last laughs best effect". The finally winning strategy can be in a miserable situation in the beginning, and its victory may take very long.
3. *Lucifer's positive side effect*: By permanently generating a number of defectors, small mutation rates can considerably accelerate the spreading of moralists.
4. *Unholy collaboration of moralists with immoralists*: Under certain conditions, moralists can survive by profiting from immoralists. This actually provides the first explanation for the existence of defectors, who hypocritically punish other defectors, although they defect themselves. The occurrence of this strange behavior is well-known in reality and even experimentally confirmed [34, 35].

These discoveries required a combination of theoretical considerations and extensive computer simulations on multiple processors over long time horizons.

8.2 Results

For well-mixed interactions, defectors are the winners of the evolutionary competition among the four behavioral strategies C, D, M, and I [36], which implies a tragedy of the commons despite punishment efforts. The reason is that cooperators (second-order free-riders) spread at the cost of moralists, while requiring them for their own survival.

Conclusions from computer simulations are strikingly different, if the assumption of well-mixed interactions is replaced by the more realistic assumption of spatial interactions. When cooperators and defectors interact in space [5, 11, 37–44], it is known that some cooperators can survive through spatial clustering [45]. However, it is not clear how the spatiotemporal dynamics and the frequency of cooperation would change in the presence of moralists and immoralists. Would spatial interactions be able to promote the spreading of punishment and thereby eliminate second-order free-riders?

In order to explore this, we have scanned a large parameter space. Figure 8.1 shows the resulting state of the system as a function of the punishment cost γ and punishment fine β after a sufficiently long transient time. If the fine-to-cost ratio β/γ and the synergy factor r are low, defectors eliminate all other strategies.

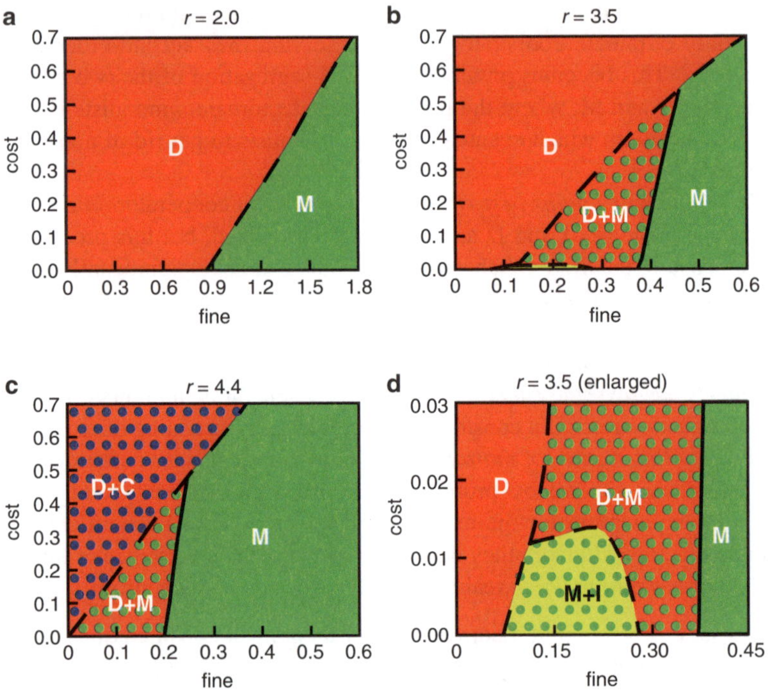

Fig. 8.1 Phase diagrams showing the remaining strategies in the spatial public goods game with cooperators (C), defectors (D), moralists (M) and immoralists (I), after a sufficiently long transient time. Initially, each of the four strategies occupies 25% of the sites of the square lattice, and their distribution is uniform in space. However, due to their evolutionary competition, two or three strategies die out after some time. The finally resulting state depends on the synergy r of cooperation, the punishment cost γ, and the punishment fine β. The displayed phase diagrams are for (**a**) $r = 2.0$, (**b**) $r = 3.5$, and (**c**) $r = 4.4$. (**d**) Enlargement of the small-cost area for $r = 3.5$. *Solid separating lines* indicate that the resulting fractions of all strategies change continuously with a modification of the model parameters β and γ, while *broken lines* correspond to discontinuous changes. All diagrams show that cooperators cannot stop the spreading of moralists, if only the fine-to-cost ratio is large enough. Furthermore, there are parameter regions where moralist can crowd out cooperators in the presence of defectors. Note that the spreading of moralists is extremely slow and follows a voter model kind of dynamics [47], if their competition with cooperators occurs in the absence of defectors. Therefore, computer simulations had to be run over extremely long times (up to 10^7 iterations for a systems size of 400×400). For similar reasons, a small level of strategy mutations (which permanently creates a small number of strategies of all kinds, in particular defectors) can largely accelerate the spreading of moralists in the M phase, while it does not significantly change the resulting fractions of the four strategies [53]. The existence of immoralists is usually not relevant for the outcome of the evolutionary dynamics. Apart from a very small parameter area, where immoralists and moralists coexist, immoralists are quickly extinct. Therefore, the 4-strategy model usually behaves like a model with the three strategies C, D, and M only. As a consequence, the phase diagrams for the latter look almost the same like the ones presented here [46]

However, for large enough fines β, cooperators and defectors are always eliminated, and moralists prevail (Fig. 8.1).

At larger r values, when the punishment costs are moderate, we find a coexistence of moralists with defectors without any cooperators. To understand why moralists can outperform cooperators despite additional punishment costs, it is important to analyze the dynamics of spatial interactions. Starting with a homogeneous strategy distribution (Fig. 8.2a), the imitation of better-performing neighbors generates small clusters of individuals with identical strategies (Fig. 8.2b). "Immoralists" die out quickly, while cooperators and moralists form separate clusters in a sea of defectors (Fig. 8.2c). The further development is determined by the interactions at the interfaces between clusters of different strategies (Figs. 8.2d–f). In the presence of defectors, the fate of moralists is not decided by a *direct* competition with cooperators, but rather by the success of both cooperative strategies against

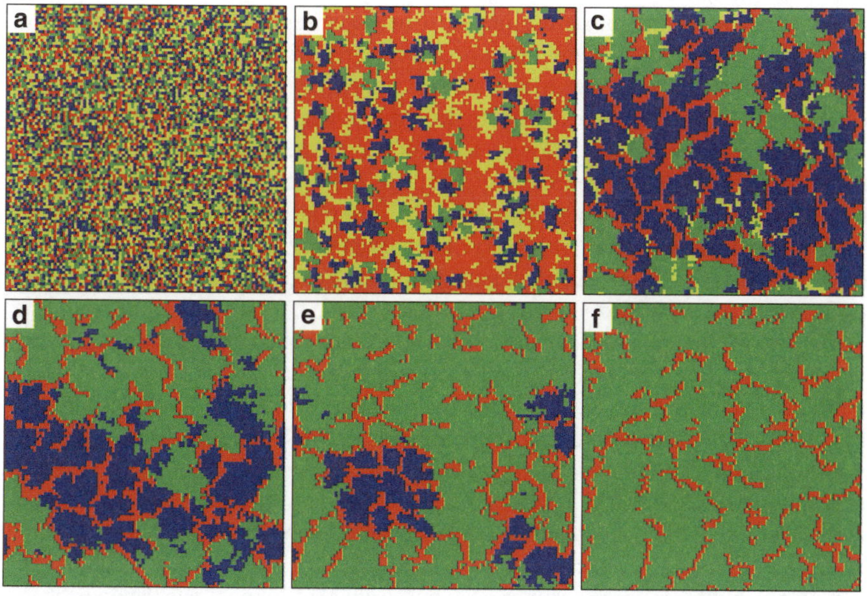

Fig. 8.2 Elimination of second-order free-riders (non-punishing cooperators) in the spatial public goods game with costly punishment for $r = 4.4$, $\beta = 0.1$, and $\gamma = 0.1$. (**a**) Initially, at time $t = 0$, cooperators (*blue*), defectors (*red*), moralists (*green*) and immoralists (*yellow*) are uniformly distributed over the spatial lattice. (**b**) After a short time period (here, at $t = 10$), defectors prevail. (**c**) After 100 iterations, immoralists have almost disappeared, and cooperators prevail, since cooperators earn high payoffs when organized in clusters. (**d**) At $t = 500$, there is a segregation of moralists and cooperators, with defectors in between. (**e**) The evolutionary battle continues between cooperators and defectors on the one hand, and defectors and moralists on the other hand (here at $t = 1,000$). (**f**) At $t = 2,000$, cooperators have been eliminated by defectors, and a small fraction of defectors survives among a large majority of moralists. Interestingly, each strategy (apart from I) has a time period during which it prevails, but only moralists can maintain their majority. While moralists perform poorly in the beginning, they are doing well in the end. We refer to this as the "who laughs last laughs best" effect

invasion attempts by defectors. If the β/γ-ratio is appropriate, moralists respond better to defectors than cooperators. Indeed, moralists can spread so successfully in the presence of defectors that areas lost by cooperators are quickly occupied by moralists (supplementary Video S1). This indirect territorial battle ultimately leads to the extinction of cooperators (Fig. 8.2f), thus resolving the second-order free-rider problem.

In conclusion, the presence of some *conventional* free-riders (defectors) supports the elimination of *second-order* free-riders. However, if the fine-to-cost ratio is high, defectors are eliminated after some time. Then, the final struggle between moralists and cooperators takes such a long time that cooperators and moralists seem to coexist in a stable way. Nevertheless, a very slow coarsening of clusters is revealed, when simulating over extremely many iterations. This process is finally won by moralists, as they are in the majority by the time the defectors disappear, while they happen to be in the minority during the first stage of the simulation (see Fig. 8.2). We call this the "who laughs last laughs best effect". Since the payoffs of cooperators and moralists are identical in the absence of other strategies, the underlying coarsening dynamics is expected to agree with the voter model [47].

Note that there is always a punishment fine β, for which moralists can outcompete all other strategies. The higher the synergy factor r, the lower the β/γ-ratio required to reach the prevalence of moralists. Yet, for larger values of r, the system behavior also becomes richer, and there are areas for small fines or high punishment costs, where clusters with different strategies can coexist (see Figs. 8.1b–d). For example, we observe the coexistence of clusters of moralists and defectors (see Fig. 8.2 and supplementary Video S1) or of cooperators and defectors (see supplementary Video S2).

Finally, for low punishment costs γ but moderate punishment fines and synergy factors r (see Fig. 8.1d), the survival of moralists may require the coexistence with "immoralists" (see Fig. 8.3 and supplementary Video S3). Such immoralists are often called "sanctimonious" or blamed for "double moral standards", as they defect themselves, while enforcing the cooperation of others (for the purpose of exploitation). This is actually the main obstacle for the spreading of immoralists, as they have to pay punishment costs, while suffering from punishment fines as well. Therefore, immoralists need small punishment costs γ to survive. As cooperators die out quickly for moderate values of r, the survival of immoralists depends on the existence of moralists they can exploit, otherwise they cannot outperform defectors. Conversely, moralists benefit from immoralists by supporting the punishment of defectors. Note, however, that this mutually profitable interaction between moralists and immoralists, which appears like an "unholy collaboration", is fragile: If β is increased, immoralists suffer from fines, and if γ is increased, punishing becomes too costly. In both cases, immoralists die out, and the coexistence of moralists and immoralists breaks down. Despite this fragility, "hypocritical" defectors, who punish other defectors, are known to occur in reality. Their existence has even been found in experiments [34, 35]. Here, we have revealed conditions for their occurrence.

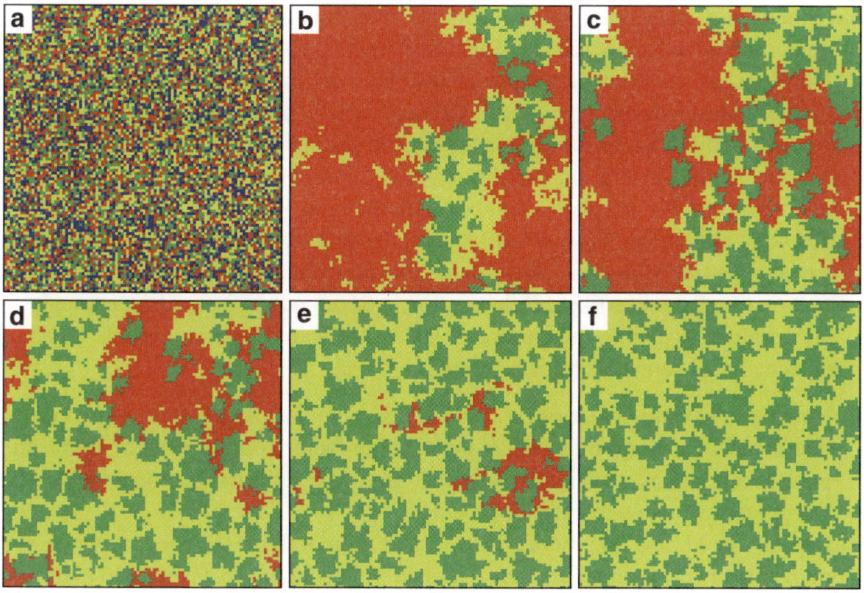

Fig. 8.3 Coexistence of moralists and immoralists for $r = 3.5$, $\beta = 0.12$, and $\gamma = 0.005$, supporting the occurrence of individuals with "double moral standards" (who punish defectors, while defecting themselves). (**a**) Initially, at time $t = 0$, cooperators (*blue*), defectors (*red*), moralists (*green*) and immoralists (*yellow*) are uniformly distributed over the spatial lattice. (**b**) After 250 iterations, cooperators have been eliminated in the competition with defectors (as the synergy effect r of cooperation is not large enough), and defectors are prevailing. (**c**–**e**) The snapshots at $t = 760$, $t = 2{,}250$, and $t = 6{,}000$ show the interdependence of moralists and immoralists, which appears like a tacit collaboration. It is visible that the two punishing strategies win the struggle with defectors by staying together. On the one hand, due to the additional punishment cost, immoralists can survive the competition with defectors only by exploiting moralists. On the other hand, immoralists support moralists in fighting defectors. (**f**) After 12,000 iterations, defectors have disappeared completely, leading to a coexistence of clusters of moralists with immoralists

8.3 Discussion

In summary, the second-order free-rider problem finds a natural and simple explanation, without requiring additional assumptions, if the local nature of most social interactions is taken into account and punishment efforts are large enough. In fact, the presence of spatial interactions can change the system behavior so dramatically that we do not find the dominance of free-riders (defectors) as in the case of well-mixed interactions, but a prevalence of moralists via a "who laughs last laughs best" effect (Fig. 8.2). Moralists can escape disadvantageous kinds of competition with cooperators by spatial segregation. However, their triumph over all the other strategies requires the temporary presence of defectors, who diminish the cooperators (second-order free-riders). Finally, moralists can take over, as they have reached a superiority over cooperators (which is further growing) and as they can outcompete defectors (conventional free-riders).

Our findings stress how crucial spatial or network interactions in social systems are. Their consideration gives rise to a rich variety of possible dynamics and a number of continuous or discontinuous transitions between qualitatively different system behaviors. Spatial interactions can even *invert* the finally expected system behavior and, thereby, explain a number of challenging puzzles of social, economic, and biological systems. This includes the higher-than-expected level of cooperation in social dilemma situations, the elimination of second-order free-riders, and the formation of what looks like a collaboration between otherwise inferior strategies.

By carefully scanning the parameter space, we found several possible kinds of coexistence between two strategies each:

- Moralists (M) and defectors (D) can coexist, when the disadvantage of cooperative behavior is not too large (i.e. the synergy factor is high enough), and if the punishment fine is sufficiently large that moralists can survive among defectors, but not large enough to get rid of them.
- Instead of M and D, moralists (M) and immoralists (I) coexist, when the punishment cost is small enough. The small punishment cost is needed to ensure that the disadvantage of punishing defectors (I) compared to non-punishing defectors (D) is small enough that it can be compensated by the additional punishment efforts contributed by moralists.
- To explain the well-known coexistence of D and C [11], it is useful to remember that defectors can be crowded out by cooperators, when the synergy factor exceeds a critical value (even when punishment is not considered). Slightly below this threshold, neither cooperators nor defectors have a sufficient advantage to get rid of the other strategy, which results in a coexistence of both strategies.

Generally, a coexistence of strategies occurs, when the payoffs at the interface between clusters of different strategies are balanced. In order to understand why the coexistence is possible in a certain parameter area rather than just for an infinitely small parameter set, it is important to consider that typical cluster sizes vary with the parameter values. This also changes the typical radius of the interface between the coexisting strategies and, thereby, the typical number of neighbors applying the same strategy or a different one. In other words, a change in the shape of a cluster can partly counter-balance payoff differences between two strategies by varying the number of "friends" and "enemies" involved in the battle at the interface between spatial areas with different strategies (see Fig. 8.4).

Finally, we would like to discuss the robustness of our observations. It is well-known that the level of cooperation in the public goods game is highest in *small* groups [10]. However, we have found that moralists can crowd out non-punishing cooperators also for group sizes of $k + 1 = 9, 13, 21$, or 25 interacting individuals, for example. In the limiting case of *large* groups, where everybody interacts with everybody else, we expect the outcome of the well-mixed case, which corresponds to defection by everybody (if other mechanisms like reputation effects [11] or abstaining are not considered [20]). That is, the same mechanisms that can create cooperation among friends may *fail* to establish shared moral standards, when

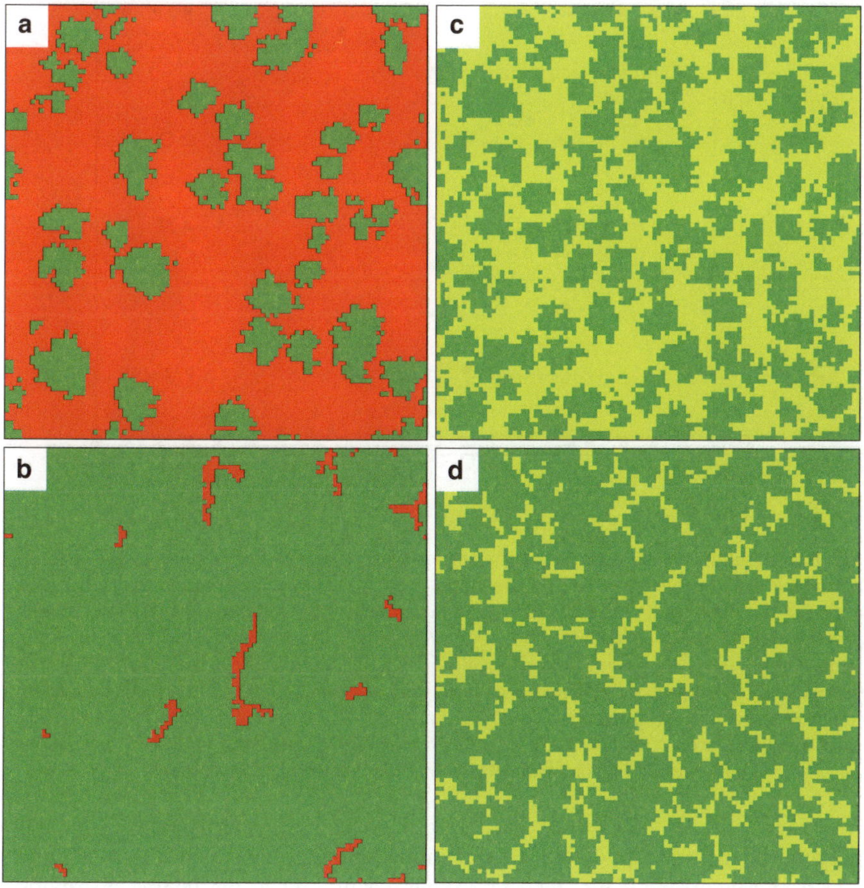

Fig. 8.4 Dependence of cluster shapes on the punishment fine β in the stationary state, supporting an adaptive balance between the payoffs of two different strategies at the interface between competing clusters. Snapshots in the top row were obtained for low punishment fines, while the bottom row depicts results obtained for higher values of β. (**a**) Coexistence of moralists and defectors for a synergy factor $r = 3.5$, punishment cost $\gamma = 0.20$, and punishment fine $\beta = 0.25$. (**b**) Same parameters, apart from $\beta = 0.4$. (**c**) Coexistence of moralists and immoralists for $r = 3.5$, $\gamma = 0.05$, and $\beta = 0.12$. (**d**) Same parameters, apart from $\beta = 0.25$. A similar change in the cluster shapes is found for the coexistence of cooperators and defectors, if the synergy factor r is varied

spatial interactions are negligible. It would therefore be interesting to study, whether the fact that interactions in the financial system are global, has contributed to the financial crisis. Typically, when social communities exceed a certain size, they need sanctioning institutions to stabilize cooperation (such as laws, an executive system, and police).

Note that our principal discoveries are not expected to change substantially for spatial interactions within *irregular* grids (i.e. neighborhoods different from Moore

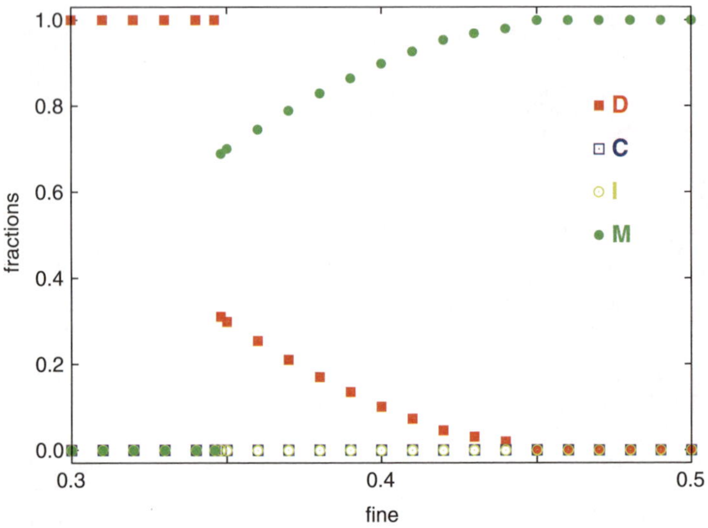

Fig. 8.5 Resulting fractions of the four strategies C, D, I, and M, for random regular graphs as a function of the punishment fine β. The graphs were constructed by rewiring links of a square lattice of size 400×400 with probability Q, thereby preserving the degree distribution (i.e. every player has four nearest neighbors) [49]. For small values of Q, small-world properties result, while for $Q \to 1$, we have a random regular graph. By keeping the degree distribution fixed, we can study the impact of randomness in the network structure independently of other effects. An inhomogeneous degree distribution can further promote cooperation [37]. The results displayed here are averages over ten simulation runs for the model parameters $r = 3.5$, $\gamma = 0.05$, and $Q = 0.99$. Similar results can be obtained also for other parameter combinations

neighborhoods) [48]. In case of *network* interactions, we have checked that small-world or random networks lead to similar results, when the degree distribution is the same (see Fig. 8.5). A *heterogeneous* degree distribution is even expected to *reduce* free-riding [37] (given the average degree is the same). Finally, adding other cooperation-promoting mechanisms to our model such as direct reciprocity (a shadow of the future through repeated interactions [50]), indirect reciprocity [17] (trust and reputation effects [11, 18]), abstaining from a joint enterprize [19–23], or success-driven migration [51], will strengthen the victory of moralists over conventional and second-order free-riders.

In order to test the robustness of our observations, we have also checked the effect of randomness ("noise") originating from the possibility of strategy mutations. It is known that mutations may promote cooperation [52]. According to the numerical analysis of the spatial public goods game with punishment, the introduction of rare mutations does not significantly change the final *outcome* of the competition between moralists and non-punishing cooperators. Second-order free-riders will always be a negligible minority in the end, if the fine-to-cost ratio and mutation rate allows moralists to spread. While a large mutation rate naturally causes a uniform distribution of strategies, a low level of strategy mutations can be even beneficial for moralists. Namely, by permanently generating a number of defectors, small mutation rates can considerably accelerate the spreading of moralists, i.e. the slow logarithmic coarsening is replaced by another kind of dynamics [53]. Defectors created by mutations play the same role as in the $D + M$ phase (see Figs. 8.1–8.2). They put moralists into an advantage over non-punishing cooperators, resulting in a faster spreading of the moralists (which facilitates the elimination of second-order free-riders over realistic time periods). In this way, the presence of a few "bad guys" (defectors) can accelerate the spreading of moral standards. Metaphorically speaking, we call this "lucifer's positive side effect".

The current study paves the road for several interesting extensions. It is possible, for example, to study *antisocial* punishment [54], considering also strategies which punish cooperators [55]. The conditions for the survival or spreading of antisocial punishers can be identified by the *same* methodology, but the larger number of strategies creates new phases in the parameter space. While the added complexity transcends what can be discussed here, the current study demonstrates clearly how differentiated the moral dynamics in a society facing public goods problems can be and how it depends on a variety of factors (such as the punishment cost, punishment fine, and synergy factor). Going one step further, evolutionary game theory may even prove useful to understand how moral feelings have evolved.

Furthermore, it would be interesting to investigate the *emergence* of punishment within the framework of a coevolutionary model [56–58], where both, individual strategies and punishment levels are simultaneously spread. Such a model could, for example, assume that individuals show some exploration behavior [52] and stick to successful punishment levels for a long time, while they quickly abandon unsuccessful ones. In the beginning of this coevolutionary process, costly punishment would not pay off. However, after a sufficiently long time, mutually fitting punishment strategies are expected to appear in the same neighborhood by coincidence [51]. Once an over-critical number of successful punishment strategies have appeared in some area of the simulated space, they are eventually expected to spread. The consideration of success-driven migration should strongly support this process [51]. Over many generations, genetic-cultural coevolution could finally inherit costly punishment as a behavioral trait, as is suggested by the mechanisms of strong reciprocity [59].

Appendix A: Methods

We study the public goods game with punishment. Cooperative individuals (C and M) make a contribution of 1 to the public good, while defecting individuals (D and I) contribute nothing. The sum of all contributions is multiplied by r and the resulting amount equally split among the $k + 1$ interacting individuals. A defecting individual (D or I) suffers a fine β/k by each punisher among the interaction partners, and each punishment requires a punisher (M or I) to spend a cost γ/k on each defecting individual among the interaction partners. In other words, only defectors and punishing defectors (immoralists) are punished, and the overall punishment is proportional to the sum of moralists and immoralists among the k neighbors. The scaling by k serves to make our results comparable with models studying different groups sizes.

Denoting the number of so defined cooperators, defectors, moralists, and immoralists among the k interaction partners by N_C, N_D, N_M and N_I, respectively, an individual obtains the following payoff: If it is a cooperator, it gets $P_C = r(N_C + N_M + 1)/(k + 1) - 1$, if a defector, the payoff is $P_D = r(N_C + N_M)/(k + 1) - \beta(N_M + N_I)/k$, a moralist receives $P_M = P_C - \gamma(N_D + N_I)/k$, and an immoralist obtains $P_I = P_D - \gamma(N_D + N_I)/k$. Our model of the spatial variant of this game studies interactions in a simple social network allowing for clustering. It assumes that individuals are distributed on a square lattice with periodic boundary conditions and play a public goods game with $k = 4$ neighbors. We work with a fully occupied lattice of size $L \times L$ with $L = 200 \ldots 1,200$ in Fig. 8.1 and $L = 100$ in Figs. 8.2–8.4 (the lattice size must be large enough to avoid an accidental extinction of a strategy). The initial strategies of the L^2 individuals are equally and uniformly distributed. Then, we perform a random sequential update. The individual at the randomly chosen location x belongs to five groups. (It is the focal individual of a Moore neighborhood and a member of the Moore neighborhoods of four nearest neighbors). It plays the public goods game with the k interaction partners of a group g, and obtains a payoff P_x^g in all five groups it belongs to. The overall payoff is $P_x = \sum_g P_x^g$. Next, one of the four nearest neighbors is randomly chosen. Its location shall be denoted by y and its overall payoff by P_y. This neighbor imitates the strategy of the individual at location x with probability $q = 1/\{1+\exp[(P_y - P_x)/K]\}$ [45]. That is, individuals tend to imitate better performing strategies in their neighborhood, but sometimes deviate (due to trial-and-error behavior or mistakes) [31]. Realistic noise levels lie between the two extremes $K \to 0$ (corresponding to unconditional imitation by the neighbor, whenever the overall payoff P_x is higher than P_y) and $K \to \infty$ (where the strategy is copied with probability 1/2, independently of the payoffs). For the noise level $K = 0.5$ chosen in our study, the evolutionary selection pressure is high enough to eventually eliminate poorly performing strategies in favor of strategies with a higher overall payoff. This implies that the resulting frequency distribution of strategies in a large enough lattice is independent of the specific initial condition after a sufficiently long transient time. Close to the separating line between M and D+M in Fig. 8.1, the equilibration may require up to 10^7 iterations (involving L^2 updates each).

Acknowledgements D.H. would like to thank for useful comments by Carlos P. Roca, Moez Draief, Stefano Balietti, Thomas Chadefaux, and Sergi Lozano.

Author Contributions Conceived and designed the experiments: DH AS MP GS. Performed the experiments: DH AS MP GS. Wrote the paper: DH AS MP GS.

References

1. G. Hardin, The tragedy of the commons. Science **162**, 1243–1248 (1968)
2. E. Fehr, S. Gächter, Altruistic punishment in humans. Nature **415**, 137–140 (2002)
3. E. Fehr, U. Fischbacher, The nature of human altruism. Nature **425**, 785–791 (2003)
4. P. Hammerstein (ed.), *Genetic and Cultural Evolution of Cooperation* (MIT Press, Cambridge, MA, 2003)
5. M. Nakamaru, Y. Iwasa, The evolution of altruism and punishment: Role of selfish punisher. J. Theor. Biol. **240**, 475–488 (2006)
6. C.F. Camerer, E. Fehr, When does "economic man" dominate social behavior. Science **311**, 47–52 (2006)
7. O. Gurerk, B. Irlenbusch, B. Rockenbach, The competitive advantage of sanctioning institutions. Science **312**, 108–111 (2006)
8. K. Sigmund, C. Hauert, M.A. Nowak, Reward and punishment. Proc. Natl. Acad. Sci. USA **98**, 10757–10762 (2001)
9. J. Henrich, R. Boyd, Why people punish defectors. J. Theor. Biol. **208**, 79–89 (2001)
10. R. Boyd, H. Gintis, S. Bowles, P.J. Richerson, The evolution of altruistic punishment. Proc. Natl. Acad. Sci. USA **100**, 3531–3535 (2003)
11. H. Brandt, C. Hauert, K. Sigmund, Punishing and reputation in spatial public goods games. Proc. R. Soc. Lond. Ser. B **270**, 1099–1104 (2003)
12. T. Yamagishi, The provision of a sanctioning system as a public good. J. Pers. Soc. Psychol. **51**, 110–116 (1986)
13. E. Fehr, Don't lose your reputation. Nature **432**, 449–450 (2004)
14. A.M. Colman, The puzzle of cooperation. Nature **440**, 744–745 (2006)
15. J.H. Fowler, Second-order free-riding problem solved? Nature **437**, E8-E8 (2005)
16. K. Panchanathan, R. Boyd, Reply. Nature **437**, E8-E9 (2005)
17. K. Panchanathan, R. Boyd, Indirect reciprocity can stabilize cooperation without the second-order free rider problem. Nature **432**, 499–502 (2004)
18. M. Milinski, D. Semmann, H.-J. Krambeck, Reputation helps to solve the "tragedy of the commons". Nature **415**, 424–426 (2002)
19. J.H. Fowler, Altruistic punishment and the origin of cooperation. Proc. Natl. Acad. Sci. USA **102**, 7047–7049 (2005)
20. H. Brandt, C. Hauert, K. Sigmund, Punishing and abstaining for public goods. Proc. Natl. Acad. Sci. USA **103**, 495–497 (2006)
21. C. Hauert, A. Traulsen, H. Brandt, M.A. Nowak, K. Sigmund, Via freedom to coercion: The emergence of costly punishment. Science **316**, 1905–1907 (2007)
22. C. Hauert, S. De Monte, J. Hofbauer, K. Sigmund, Volunteering as red queen mechanism for cooperation in public goods game. Science **296**, 1129–1132 (2002)
23. D. Semmann, H.-J. Krambeck, M. Milinski, Volunteering leads to rock-paper-scissors dynamics in a public goods game. Nature **425**, 390–393 (2003)
24. M. Milinski, R.D. Sommerfeld, H.J. Krambeck, F.A. Reed, J. Marotzke, The collective-risk social dilemma and the prevention of simulated dangerous climate change. Proc. Natl. Acad. Sci. USA **105**, 2291–2294 (2008)
25. K. Sigmund, *The Calculus of Selfishness* (Princeton University Press, Princeton, 2010)
26. D.D. Heckathorn, The dynamics and dilemmas of collective action. Am. Soc. Rev. **61**, 250–277 (1996)
27. O.T. Eldakar, D.S. Wilson, Selfishness as second-order altruism. Proc. Natl. Acad. Sci. USA **109**, 6982–6986 (2008)
28. M. Nakamaru, Y. Iwasa, The evolution of altruism by costly punishment in the lattice structured population: Score-dependent viability versus score-dependent fertility. Evol. Ecol. Res. **7**, 853–870 (2005)
29. T. Sekiguchi, M. Nakamaru, Effect of the presence of empty sites on the evolution of cooperation by costly punishment in spatial games. J. Theor. Biol. **256**(2), 297–304 (2009)

30. M.A. Nowak, Five rules for the evolution of cooperation. Science 314, 1560–1563 (2006)
31. A. Traulsen, et al., Human strategy updating in evolutionary games. Proc. Natl. Acad. Sci. USA **107**, 2962–2966 (2010)
32. M.A. Nowak, C.E. Tarnita, T. Antal, Evolutionary dynamics in structured populations. Phil. Trans. R. Soc. B **365**, 19–30 (2010)
33. M.D. Hauser, *Moral Minds: How Nature Designed Our Universal Sense of Right and Wrong* (Ecco, New York, 2006)
34. A. Falk, E. Fehr, U. Fischbacher, Driving forces behind informal sanctions. Econometrica **73**, 2017–2030 (2005)
35. M. Shinada, T. Yamagishi, Y. Ohmura, False friends are worse than bitter enemies: "Altruistic" punishment of in-group members. Evol. Hum. Behav. **25**, 379–393 (2004)
36. G. Szabó, C. Hauert, Phase transitions and volunteering in spatial public goods games. Phys. Rev. Lett. **89**, 118101 (2002)
37. F.C. Santos, M.D. Santos, J.M. Pacheco, Social diversity promotes the emergence of cooperation in public goods games. Nature **454**, 213–216 (2008)
38. J.Y. Wakano, M.A. Nowak, C. Hauert, Spatial dynamics of ecological public goods. Proc. Natl. Acad. Sci. USA **106**, 7910–7914 (2009)
39. M.A. Nowak, R.M. May, Evolutionary games and spatial chaos. Nature **359**, 826–829 (1992)
40. M.A. Nowak, S. Bonhoeffer, R.M. May, More spatial games. Int. J. Bifurcat. Chaos **4**, 33–56 (1994)
41. M.A. Nowak, S. Bonhoeffer, R.M. May, Spatial games and the maintenance of cooperation. Proc. Natl. Acad. Sci. USA **91**, 4877–4881 (1994)
42. M.A. Nowak, S. Bonhoeffer, R.M. May, Robustness of cooperation. Nature **379**, 125–126 (1996)
43. C.G. Nathanson, C.E. Tarnita, M.A. Nowak, Calculating evolutionary dynamics in structured populations. PLoS Comput. Biol. **5**, e1000615 (2009)
44. J.M. Pacheco, F.L. Pinheiro, F.C. Santos, Population structure induces a symmetry breaking favoring the emergence of cooperation. PLoS Comput. Biol. **5**, e1000596 (2009)
45. G. Szabó, C. Tőke, Evolutionary prisoner's dilemma game on a square lattice. Phys. Rev. E **58**, 69–73 (1998)
46. D. Helbing, A. Szolnoki, M. Perc, G. Szabó, Punish, but not too hard: how costly punishment spreads in the spatial public goods game. New J. Phys. **12**(8), 083005 (2010)
47. I. Dornic, H. Chaté, J. Chave, H. Hinrichsen, Critical coarsening without surface tension: The universality class of the voter model. Phys. Rev. Lett. **87**, 045701 (2001)
48. A. Flache, R. Hegselmann, Do irregular grids make a difference? Relaxing the spatial regularity assumption in cellular models of social dynamics. J. Artif. Soc. Soc. Simul. **4**, 4 (2001). See http://www.soc.surrey.ac.uk/JASSS/4/4/6.html
49. G. Szabó, A. Szolnoki, R. Izsák, Rock-scissors-paper game on regular small-world networks. J. Phys. A: Math. Gen. **37**, 2599–2609 (2004)
50. R. Axelrod, *The Evolution of Cooperation* (Basic Books, New York, 1984)
51. D. Helbing, W. Yu, The outbreak of cooperation among success-driven individuals under noisy conditions. Proc. Natl. Acad. Sci. USA **106**, 3680–3685 (2009)
52. A. Traulsen, C. Hauert, H.D. Silva, M.A. Nowak, K. Sigmund, Exploration dynamics in evolutionary games. Proc. Natl. Acad. Sci. USA **106**, 709–712 (2009)
53. D. Helbing, A. Szolnoki, M. Perc, G. Szabó, Defector-accelerated cooperativeness and punishment in public goods games with mutations. Phys. Rev. E **81**(5), 057104 (2010)
54. B. Herrmann, C. Thöni, S. Gächter, Antisocial punishment across societies. Science **319**, 1362–1367 (2008)
55. D.G. Rand, H. Ohtsuki, M.A. Nowak, Direct reciprocity with costly punishment: Generous tit-for-tag prevails. J. Theor. Biol. **256**, 45–57 (2009)
56. G. Szabó, A. Szolnoki, V. Jeromos, Selection of dynamical rules in spatial Prisoner's Dilemma games. EPL **87**, 18007 (2009)

57. F.C. Santos, J.M. Pacheco, T. Lenaerts, Cooperation prevails when individuals adjust their social ties. PLoS Comput. Biol. **2**, 1284–1291 (2006)
58. M. Perc, A. Szolnoki, Coevolutionary games - A mini review. BioSystems **99**, 109–125 (2010)
59. S. Bowles, H. Gintis, The evolution of strong reciprocity: Cooperation in heterogeneous populations. Theor. Popul. Biol. **65**, 17–28 (2004)

Chapter 9
Coordination and Competitive Innovation Spreading in Social Networks*

9.1 Introduction

The analysis of percolation in random media has become a very popular framework over the last decades to address a wide variety of phenomena in disordered systems, such as oil mining in porous reservoirs, fire spreading in forests, fracture patterns in rocks, electromagnetic properties of composite materials, etc [1]. More recently, it has also been applied to shed light on social phenomena, namely the diffusion of opinions [2] and innovations [3] in social networks. All of the aforementioned systems can be modeled as percolation problems. More precisely, they can be abstracted as a network of nodes representing the topology of the random medium, wherein nodes can be either "empty" or "occupied", depending on the state of their neighbors. Starting from an initial condition where some nodes are occupied, an occupied node becomes empty if the number of its occupied neighbors goes below a threshold k, the index of the percolation process ($k = 2$ for standard percolation [4] and $k \geq 3$ for bootstrap or k-core percolation [5, 6]). The underlying switching dynamics is therefore assumed to be *unidirectional*.

Here we introduce a percolation model that generalizes this powerful theoretical approach. Our extension assumes that nodes are of two types A or B, and that a node changes type when the number of neighbors of the same type is less than k. Consequently both changes A-to-B and B-to-A are possible, i.e. we are considering a *bi-directional* percolation dynamics instead. Figure 9.1 provides an example which illustrates the fundamental difference between both percolation processes. The problem we want to address is the competition between innovations [7]. Competition between products, tools or technical standards is ubiquitous. Well-known examples are railway gauges, keyboard layouts, computer operating systems, high-definition video standards, e-book readers, etc. The reasons that determine

*This chapter has been prepared by *C. Roca, Moez Draief,* and D. Helbing under the project title "Percolate or die: Multi-percolation decides the struggle between competing innovations".

D. Helbing (ed.), *Social Self-Organization*, Understanding Complex Systems, DOI 10.1007/978-3-642-24004-1_9, © Springer-Verlag Berlin Heidelberg 2012

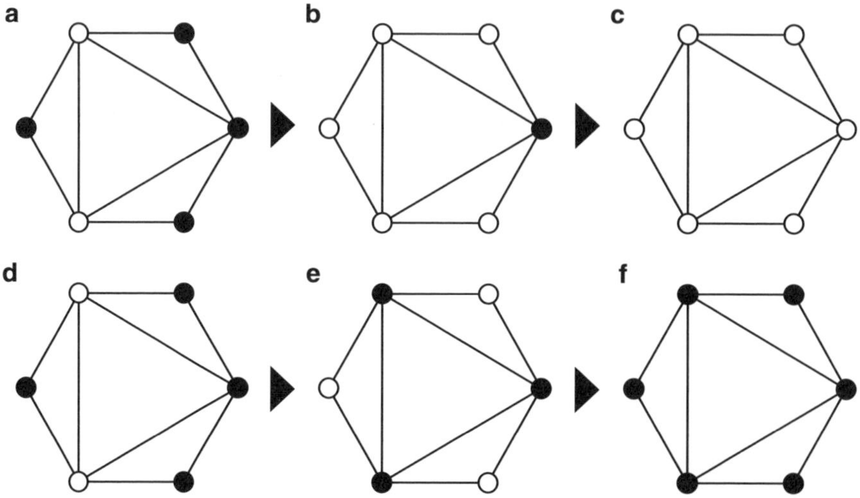

Fig. 9.1 Comparison of unidirectional (panels **A**–**C**) vs bi-directional percolation (panels **D**–**F**). In unidirectional percolation, occupied nodes (*in black*) become empty (*in white*) when they have less than k occupied neighbors (in this example $k = 2$). In the end there is no occupied node that survives the percolation pruning. With bi-directional percolation, both white and black nodes switch color when they have less than $k = 2$ neighbors of their same color. The end result in this case is an all-black graph, with no white nodes surviving the competitive bi-directional percolation process. All black nodes end up with two black neighbors and they are connected, hence they form a percolating cluster. Notice that although both cases have the same initial condition and percolation index k, the outcome is opposite

the outcome of these fierce competitions have puzzled researchers of different disciplines for a long time [8, 9]. Previous work has highlighted the combined influence of intrinsic benefits of each option together with costs incurred due to switching [10]. In addition, it has been suggested that social structure, i.e. the network of social relationships in a group or population, would play a crucial role [11]. So far, however, there has been little analytical work that elucidates the outcome of such competitions. In this work we show that the competition between innovations can be understood as a bi-directional percolation process, which ultimately determines the fate of the options in contest.

9.2 Model

To start with, let us consider a simple model with two competing options, A and B (for example Blu-ray Disc vs HD DVD), whose benefits to individuals depend on intrinsic factors as well as on the acceptance by others in a certain social neighborhood. This can be modeled as a coordination problem [12, 13], in which individuals choosing one of the two options A or B obtain a payoff $\pi_{\mathrm{A}} = q\tilde{x}_{\mathrm{A}}$ and

$\pi_B = (1-q)\tilde{x}_B$ respectively. The relative advantage of one option over the other is represented by the parameter q, where $0 \leq q \leq 1$. Quantities $\tilde{x}_A$ and $\tilde{x}_B$ give, respectively, the proportion of people adhering to option A or B among the social acquaintances who have an influence on the individual's decision, such us family members, friends, co-workers, etc ($\tilde{x}_A + \tilde{x}_B = 1$ for every individual). In addition, we consider that changing option entails some switching cost, which is called c_A for a follower of option A who changes to B, and c_B in the opposite case. Thus, A- and B-individuals have the following effective payoff matrices

$$A: \begin{pmatrix} q & 0 \\ -c_A & 1-q-c_A \end{pmatrix} \qquad B: \begin{pmatrix} q-c_B & -c_B \\ 0 & 1-q \end{pmatrix}, \tag{9.1}$$

where we follow the standard convention for symmetric games: rows represent own strategy and columns that of the interaction partner.

For the moment, we assume that individuals are able to assess to a good extent the benefits and drawbacks of options A and B, and also the degree of penetration of each option in their social neighborhood, i.e. we assume a low level of uncertainty in the decision-making process (more on this important point later). Therefore, individuals choose a best response to the current state of their social context according to the payoffs expected from (9.1). As a consequence, A-individuals change to option B if the proportion of A-neighbors is below a certain threshold, namely $\tilde{x}_A < 1-q-c_A$, while B-individuals switch if the proportion of B-neighbors is less than certain value, $\tilde{x}_B < q - c_B$.

This defines an evolutionary game [14, 15], which consists in a dynamical system with state variable x, the global density of the followers of one of the options. We set $x = x_B$ without loss of generality. Disregarding the effect of social structure for the moment, the evolution of x_B can easily be calculated assuming a well-mixed population [16], equivalent to the mean field hypothesis in physics or the representative agent in economics. It posits that every individual has, in her social neighborhood, a proportion of A- or B-individuals equal to the respective global densities, i.e. for every individual and at any time $\tilde{x}_B = x_B$. Under this assumption, the population rapidly reaches an equilibrium with stationary value $x_B^* = \lim_{t\to\infty} x_B(t)$

$$x_B^* = \begin{cases} 0, & \text{if } x_B^0 < q - c_B, \\ x_B^0, & \text{if } q - c_B \leq x_B^0 \leq q + c_A, \\ 1, & \text{if } x_B^0 > q + c_A, \end{cases} \tag{9.2}$$

where x_B^0 represents the initial density of individuals following option B. Equation (9.2) shows that under well-mixed conditions switching costs induce the appearance of a heterogeneous state, in which both competing options keep a share of the population. If costs are left out ($c_A = c_B = 0$), then we find the standard solution of a coordination game, with an unstable equilibrium at $x_B^* = x_B^0$, which separates the basins of attraction of the stable equilibria $x_B^* = 0$ and $x_B^* = 1$ [17].

Let us now consider this model embedded in a social network [18, 19], that is, in a network of social relationships which determines who interacts with whom [14,20,21]. Here we use regular random networks [22], which are networks where nodes are linked at random and where each node has the same number of neighbors, or degree, z. Such networks are known to have the most neutral effect on evolutionary games [23], just fixing neighborhood size and preserving the social context of individuals. They avoid particular effects that some topological features, such as clustering or degree heterogeneity, may have [24], which could obscure the processes we want to reveal here.

9.3 Results

Figure 9.2 displays simulation results for this model, showing the stationary density of B-individuals x_{B}^{*} as a function of their initial density x_{B}^{0} (see the Materials and Methods section for full details about the simulations). Notably, there are large differences to mean field theory, which predicts a heterogeneous population for a much wider range of initial conditions. In order to understand this deviation we have to consider the time evolution of the model. To that purpose, it is better to start with a simpler case, setting one of the switching costs with so large a value that it prevents the switching of individuals following that strategy. For example, let us set $c_{\mathrm{A}} \geq 1 - q$, so that only B-individuals can change to option A. This switching takes place when the proportion of B-individuals in the neighborhood of the focal individual satisfies $\tilde{x}_{\mathrm{B}} < q - c_{\mathrm{B}}$. Hence the subsequent dynamics exactly coincides with the pruning of nodes of a standard site percolation process with unidirectional dynamics (see Fig. 9.1), A- and B-individuals corresponding to empty and occupied nodes, respectively.

When B-nodes become A-nodes, they leave other B-nodes with fewer B-neighbors. The process repeats until it stabilizes to a subset of B-nodes, all of which have $(q - c_{\mathrm{B}})$ or more B-neighbors. When the size of this subset is a non-negligible fraction of the size of the full graph, or infinite in the case of infinite graphs, then percolation is said to occur [4]. The appearance of a percolating cluster constitutes a phase transition, and it takes place when the initial density of occupied nodes is larger than a critical density. In our case, the index of the percolation process of B-individuals switching to option A, called k_{B}, is given by

$$k_{\mathrm{B}} = \lceil z(q - c_{\mathrm{B}}) \rceil. \tag{9.3}$$

Herein, $\lceil x \rceil$ denotes the smallest integer equal or larger than x. Conversely, considering only the transitions of A-individuals to option B, we have *another* percolation process with index k_{A}, whose value is given by

$$k_{\mathrm{A}} = \lceil z(1 - q - c_{\mathrm{A}}) \rceil. \tag{9.4}$$

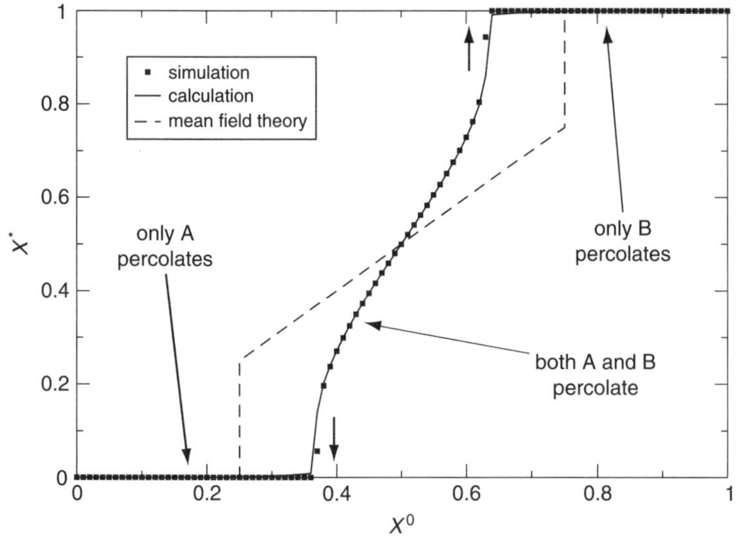

Fig. 9.2 The fate of two options in contest is determined by the underlying process of multi-percolation, taking place on the social network. The graph shows the stationary density of B-individuals x^* as a function of the initial density x^0 (simulation results, *black squares*). Parameter $q = 0.5$, so both options A and B are intrinsically equally good. Switching costs are also symmetric, with values $c_A = c_B = 0.25$. As a result, both percolation indices have the same value $k_A = k_B = 3$. Interactions take place in a regular random network of degree $z = 9$. The difference with the prediction of mean theory (*dashed line*) demonstrates the crucial role played by the social network. Labels indicate the possible regions of behavior, depending on the percolation of one option, the other or both. Notice that a heterogeneous population is sustainable only when both options percolate, but this case occurs for a significant range of initial conditions. *Small arrows* near abscissa axes mark the critical density to attain percolation of each strategy, as predicted by a calculation based on standard unidirectional percolation. The discrepancy with simulation results highlights the fact that the mutual interference between both percolation processes changes the percolation thresholds. This is confirmed by an analytical calculation that takes into account this interplay (*solid line*)

Note that $k = 0$ and $k = 1$ are degenerate cases, whereas $k = 2$ is the index of standard percolation and $k \geq 3$ corresponds to bootstrap or k-core percolation.

The actual dynamics of our model is given by the competition between these two percolation processes. The dynamics is therefore bi-directional, with both transitions A-to-B and B-to-A taking place simultaneously. A calculation based on standard unidirectional percolation, applied to each process separately, estimates the percolation thresholds only poorly, as the arrows in Fig. 9.2 show. It is also possible, however, to take into account the interference between both processes, with a recursive calculation on the switching times of individuals. Figure 9.3 shows the excellent agreement of this calculation with the computer simulations, and it clearly demonstrates that mutual interference between both percolation processes occurs. Interestingly, we find that this interplay supports the success of the dominated

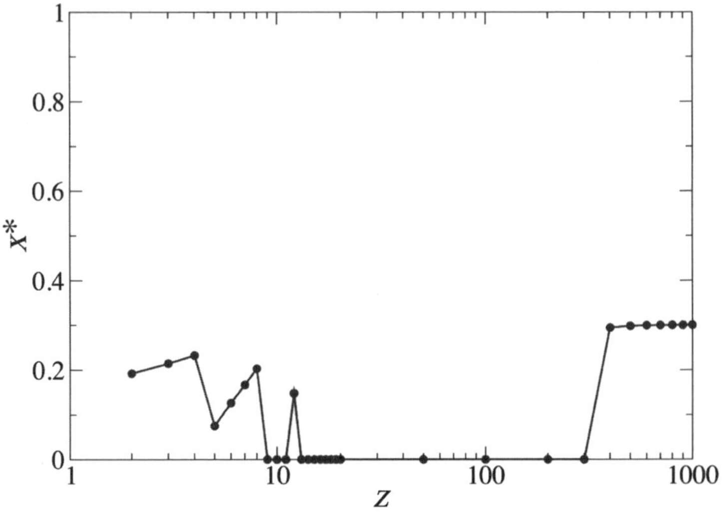

Fig. 9.3 Stationary density x^* of B-individuals for a coordination game on a regular random network of size $N = 10^4$ with switching costs $c_A = c_B = 0.25$, parameter $q = 0.5$, and initial condition $x^0 = 0.3$. The theoretically determined values according to formulas (9.12) and (9.20) match the values determined in computer simulations of the multi-percolation process perfectly well, despite the complicated, ragged nature of the curve

option, i.e. it allows some individuals following the minor option to percolate with initial conditions for which unidirectional percolation does not occur (range $0.36 \lesssim x_B^0 \lesssim 0.38$ for option B, and range $0.62 \lesssim x_B^0 \lesssim 0.64$ for option A). Individuals who have switched obviously promote percolation of the newly acquired option, as the switching increases the density of that option in the neighborhoods of adjacent nodes. The time scale of switching for the major option is much faster than for the minor one [25]. This implies that the pruning of nodes for the major option is virtually done by the time the pruning of the minor option proceeds, with the consequence that only changes of major to minor option have time to effectively foster percolation of the latter option. More importantly, this analytical theory confirms that the competition between options A and B gives rise to a bi-directional percolation process, which allows a simple rationale for the outcome: *In the context of competing innovations, percolation of an option means survival and, as long as it is the only one that percolates, it also implies dominance of the market.* We refer the reader to Sec. 9.4 for details of the analytical calculations and further discussion.

The joint influence of switching costs and social networks becomes most salient when one of the options is intrinsically superior to the other, i.e. when $q \neq 0.5$. Figure 9.4 shows an example, displaying again the asymptotic density of B-individuals x_B^* as a function of their initial density x_B^0 (see solid line and black squares). In this case, the asymmetry of the game results in different percolation indices for each option, namely $k_A = 4$ and $k_B = 1$ (see (9.3) and (9.4)), which

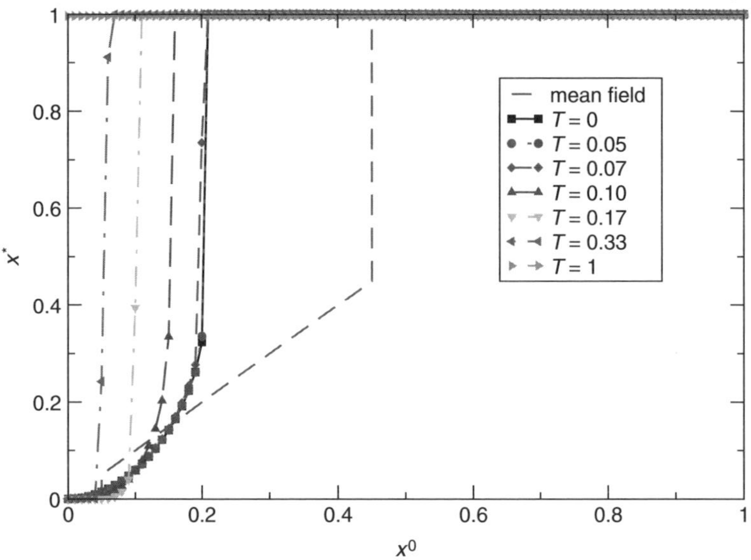

Fig. 9.4 Noise only has an effect on the multi-percolation process when the amount is large. The graph shows the stationary density of B-individuals x^* as a function of the initial density x^0, for different amounts of uncertainty or noise in the decision rule of individuals (simulation results, see line styles and symbols in legend). Compared to Fig. 9.2, in this case $q = 0.25$, so option B is superior to option A. Switching costs are equal for both options, with values $c_A = c_B = 0.2$. The social network is a regular random network of degree $z = 6$. For $T = 0$ (no noise), the asymmetry in intrinsic value between the options translates into different percolation indices, $k_A = 4$ and $k_B = 1$, which causes different kinds of transitions to homogeneous population (A- or B-dominated). This fact favors option B compared to the mean field prediction. Additional curves show results for non-zero amounts of noise. Moderate noise does not change the result qualitatively and, strikingly, larger amounts reinforce the superior option B rather than yielding an more balanced outcome

causes a continuous transition towards an A-dominated population ($x_B^0 \lesssim 0.1$), but a discontinuous one in the B-dominated case ($x_B^0 \gtrsim 0.2$). The difference between both transitions [26] originates from the characteristic transition of standard percolation, in the former case, versus that of bootstrap percolation, in the latter. Interestingly, the net effect of this imbalance between the two competing percolation processes is a fostering of the superior option B. Note that if the same game, with or without switching costs, took place on a well-mixed population, a symmetric outcome around $x_B^0 = q = 0.25$ would result instead (see (9.2) and dashed red line in Fig. 9.4). Let us finally address the issue of uncertainty or noise in the decision rule of individuals. To this end, we assume that individuals choose options stochastically, with a probability that follows a multi-nomial logit model [27, 28], which is also known in physics as Fermi rule [29, 30]. Specifically, if the expected variation in payoff resulting from a change of option is $\Delta\pi$, then the probability of switching strategy is assumed to be $1/(1 + \exp(-\Delta\pi/T))$. The parameter T determines the amount of noise in the decision process. In the limit $T \to 0$, noise disappears and we have the deterministic dynamics used so far. Additional curves in Fig. 9.4 display the influence of noise, showing that the qualitative behavior of the model remains the same for low to moderate amounts of noise. It is striking, however, that the evolution of the population is biased towards the superior option B, i.e. noise reinforces the imbalance between options rather than washing it out.

9.4 Discussion

In conclusion, we have shown that the competition between several options gives rise to bi-directional (or, more generally, multi-directional) percolation processes which can be analytically understood. Multi-percolation thus provides a powerful theoretical tool to understand the problem of competition between innovations in networks. It offers predictions about the survival and dominance of the options in contest, as well as insights into their dynamical evolution in time. Our general finding is that percolation of an option implies its survival and, if only one option percolates, it will be eventually supported by everyone. The latter may be favorable, for example when it promotes a shared technical standard. Nevertheless, it could also create monopolies or endanger pluralism. Our conclusions are expected to be also relevant to the results of marketing and political campaigns, and to the diffusion of opinions [11] and behavior [31]. Model variants or extensions may also describe the spread of health behavior, such as obesity, smoking, depression or happiness [32]. We recognize that the practical applicability of this theory requires a good knowledge of the social network and a sound modeling of the decision making process of individuals [33], but given the increasing availability of massive social data, such information may be available for example systems, soon.

Appendix A: Methods

All the simulation results reported in the main text have been obtained according to the procedures described in the following. Population size is 10^4 individuals. Regular random networks are generated by randomly assigning links between nodes, ensuring that each node ends up with exactly the same number of links. Results have been obtained with synchronous update. This means that, first, the next strategy is calculated for all individuals and, then, it is updated for them all at once. We have also checked the influence of asynchronous update, which assigns next strategies to individuals proceeding one by one in random order, but we have found no significant qualitative difference in the results. The times of convergence allowed for the population to reach a stationary state are 10^2 steps for the main model (best response rule) and 10^3 steps for the model with noise (multi-nomial logit or Fermi decision rule). We have verified that these are sufficiently large values.

In the first case, population reaches a frozen state and, in the second one, results do not change if a time of convergence 10^4 steps is used instead. Results plotted in the graphs correspond, for each data point, to an average of 100 realizations. Each realization is carried out with a newly generated random network, where strategies were randomly assigned to individuals in accordance with the initial densities of both strategies.

Appendix B: Analytical Theory

We study a coordination game with switching costs. A-individuals have the following payoff matrix

$$\begin{pmatrix} q & 0 \\ -c_{\mathrm{A}} & 1-q-c_{\mathrm{A}} \end{pmatrix}, \tag{9.5}$$

while B-individuals have

$$\begin{pmatrix} q-c_{\mathrm{B}} & -c_{\mathrm{B}} \\ 0 & 1-q \end{pmatrix}. \tag{9.6}$$

All three parameters satisfy $q, c_{\mathrm{A}}, c_{\mathrm{B}} \in [0, 1]$.

We call x^t the global density of B-individuals in the population at time step t, x^0 its initial value and x^* its asymptotic value ($x^* = \lim_{t\to\infty} x^t$). At time $t = 0$ individuals are randomly assigned options A or B, with probabilities $x^0_{\mathrm{A}} = (1 - x^0)$ or $x^0_{\mathrm{B}} = x^0$ respectively. Given an individual i, we call $\tilde{x}_i$ the proportion of B-individuals in her neighborhood. Then, an A-individual i changes to option B when her neighborhood satisfies

$$\tilde{x}_i > q + c_{\mathrm{A}}, \tag{9.7}$$

whereas a B-individual j switches strategy when

$$\tilde{x}_j < q - c_{\mathrm{B}}. \tag{9.8}$$

The above strategy switchings correspond to the removal of nodes (pruning) of two competing site percolation processes, whose indices are respectively

$$k_{\mathrm{A}} = \lceil z(1 - q - c_{\mathrm{A}})\rceil \tag{9.9}$$

and

$$k_{\mathrm{B}} = \lceil z(q - c_{\mathrm{B}})\rceil, \tag{9.10}$$

wherein $\lceil x\rceil$ denotes the smallest integer equal or larger than x.

In the pruning of nodes associated with a site percolation process, the nodes that remain in the stationary state are those which have k or more neighbors of the same strategy and hence they belong to a percolating cluster of that strategy. We call p_{A} and p_{B} the probabilities that a node belongs to a percolating cluster assuming *independent* percolation processes for transitions A $\to$ B and B $\to$ A, respectively.

Obviously, $p_{\mathrm{A}} \leq x_{\mathrm{A}} = 1 - x^0$ and $p_{\mathrm{B}} \leq x_{\mathrm{B}} = x^0$. Both probabilities can be calculated for infinite Bethe lattices [5,6]. They offer a rough approximation of the behavior of the model, assuming no interference between both processes, which yields the following prediction for the asymptotic density of B-individuals

$$x^* = 1 - x^0 + p_{\mathrm{B}} - p_{\mathrm{A}}. \tag{9.11}$$

This approximation obviously fails to account for the change in the percolation probabilities that the simulation results reflect. Therefore, we propose the following analytical theory, which takes into account the interplay between both competing percolation processes.

Let us first define a simplified version of the model, which we call the 1T-model. This model is the same as the original one, with the only difference that each node can only switch strategy once. That is, nodes change strategy according to (9.7) and (9.8), but once they have changed, they stick to the new strategy, no matter what (9.7) and (9.8) dictate. Note that this idea can be generalized to a nT-model, where each node is allowed to switch at most n times. Our original model could also be called ∞T-model, corresponding to an unbounded value of n.

In fact, we can assume in the original model that most nodes will switch only once or never. For a node to switch, it is required that the number of neighbors of the same strategy is below a given threshold k. Once it has switched, it will have a number of neighbors of the (new) strategy larger than $z - k$ (z is the degree of the network), which usually means a large number of them. So it is reasonable to expect that the node will be locked in the new strategy forever, as it would require a large number of changes in its neighborhood to switch back. The great similarity between the simulation results of the original ∞T-model and the 1T-model supports this intuition (see Figs. 9.5–9.8).

The 1T-model can be calculated for an infinite Bethe lattice, with a recursive procedure that we present in the following. From now on, let us denote by X one of the options, A or B, and by Y the other one, i.e. Y is the only element of the set $\{\mathrm{A}, \mathrm{B}\} - \{X\}$. The index of the percolation process with transitions $X \to Y$ is denoted k_X. The fundamental recursive property of the 1T-model is that the time of switching of any X-node is 1+ (the k_X-th greater time of switching among its X-neighbors). For example, with a percolation process for transitions $\mathrm{B} \to \mathrm{A}$ with index $k_{\mathrm{B}} = 3$, if a B-node has six neighbors that are all B-nodes and whose times of switching are $t = 2, 2, 3, 5, 7, 7$, respectively, then the node will switch at time $t = 6$. Notice that if an X-node belongs to a percolating cluster then its time of switching is unbounded, because it has k_X or more neighbors that also have unbounded switching times.

First, we calculate the switching probabilities of a node, conditioned to the event of being connected to a node of the same or the other type. Thus, we define r_X^t as the probability of a node being of type X and switching at time t, conditioned to the event of being connected to an X-node until time $t - 1$, with $t \geq 1$. Similarly, we define s_X^t as the probability of a node being of type X and switching at time t, conditioned to the event of being connected to a Y-node, i.e. of the opposite type, until time $t-1$, with $t \geq 1$.

Second, we calculate the probabilities of a node being of type X and switching at time t, which we represent by p_X^t, given the conditional probabilities of switching of its child nodes until time $t-1$, namely $r_X^1, \ldots, r_X^{t-1}$ and $s_Y^1, \ldots, s_Y^{t-1}$.

Third, in the stationary state X-nodes will be either nodes that have been X-nodes since the beginning or initial Y-nodes that have switched option. Hence the stationary density of B-nodes x^* can be expressed as

$$x^* = x^0 - \sum_{t=1}^{\infty} p_{\mathrm{B}}^t + \sum_{t=1}^{\infty} p_{\mathrm{A}}^t. \tag{9.12}$$

To calculate the probabilities $\{r_{\mathrm{A}}^t, s_{\mathrm{A}}^t, p_{\mathrm{A}}^t, r_{\mathrm{B}}^t, s_{\mathrm{B}}^t, p_{\mathrm{B}}^t\}$ we need to consider all the possible configurations of neighborhood of a node. To this end, we classify the neighbors at time t into one of these types:

[1] Nodes of the same type that switch at time $t-1$.
[2] Nodes of the same type that switch at time t or later.
[3] Nodes of the same type that switch at time $t-2$ or before.
[4] Nodes of the other type that switch at time t or later.
[5] Nodes of the other type that switch at time $t-1$ or before.

Note that neighbors of type [1] are the ones that trigger switching of the focal node at time t. For a given configuration, the number of neighbors of each type appears as an exponent in the corresponding combinatorial expression. In (9.13)–(9.21) below, we use the exponents i, j, k, m, n for types [1] to [5], respectively.

In addition, we define $R_X^0 = S_X^0 = 0$, and $R_X^t = \sum_{\tau=1}^{t} r_X^\tau$ and $S_X^t = \sum_{\tau=1}^{t} s_X^\tau$, for $t \geq 1$. The degree of the Bethe lattice is z. The calculation proceeds according to the following (9.13)–(9.21)

$$r_X^1 = \sum_{j=0}^{k_X-2} \binom{z-1}{j} (x_X^0)^{j+1} (x_Y^0)^{z-1-j}, \tag{9.13}$$

$$s_X^1 = \sum_{j=0}^{k_X-1} \binom{z-1}{j} (x_X^0)^{j+1} (x_Y^0)^{z-1-j}. \tag{9.14}$$

For $t > 1$

$$r_X^t = x_X^0 \sum_{(i,j,k,m,n)\in\mathscr{R}} \binom{z-1}{i,j,k,m,n} (r_X^{t-1})^i \, (x_X^0 - R_X^{t-1})^j \, (R_X^{t-2})^k \, (x_Y^0 - S_Y^{t-1})^m \, (S_Y^{t-1})^n, \tag{9.15}$$

$$s_X^t = x_X^0 \sum_{(i,j,k,m,n)\in\mathscr{S}} \binom{z-1}{i,j,k,m,n} (r_X^{t-1})^i \, (x_X^0 - R_X^{t-1})^j \, (R_X^{t-2})^k \, (x_Y^0 - S_Y^{t-1})^m \, (S_Y^{t-1})^n, \tag{9.16}$$

where the sets of exponents $\mathscr{R}$ and $\mathscr{S}$ are

$$\begin{aligned}\mathscr{R} = \{\, (i,j,k,m,n) \in \{0,1,2,\ldots,z-1\}^5 :\ & i+j+k+m+n = z-1,\\ & i+j+k \geq k_X - 1,\\ & i+j+n \geq k_X - 1,\\ & j+n \leq k_X - 2\},\end{aligned} \tag{9.17}$$

$$\begin{aligned}\mathscr{S} = \{\, (i,j,k,m,n) \in \{0,1,2,\ldots,z-1\}^5 :\ & i+j+k+m+n = z-1,\\ & i+j+k \geq k_X,\\ & i+j+n \geq k_X,\\ & j+n \leq k_X - 1\,\}.\end{aligned} \tag{9.18}$$

$$p_X^1 = \sum_{j=0}^{k_X-1} \binom{z}{j} (x_X^0)^{j+1} (x_Y^0)^{z-j}. \tag{9.19}$$

For $t > 1$

$$p_X^t q = x_X^0 \sum_{(i,j,k,m,n)\in\mathscr{P}} \binom{z}{i,j,k,m,n} (r_X^{t-1})^i (x_X^0 - R_X^{t-1})^j (R_X^{t-2})^k (x_Y^0 - S_Y^{t-1})^m (S_Y^{t-1})^n, \tag{9.20}$$

where the set of exponents $\mathscr{P}$ is

$$\begin{aligned}\mathscr{P} = \{\, (i,j,k,m,n) \in \{0,1,2,\ldots,z\}^5 :\ & i+j+k+m+n = z,\\ & i+j+k \geq k_X,\\ & i+j+n \geq k_X,\\ & j+n \leq k_X - 1\,\}.\end{aligned} \tag{9.21}$$

As the probabilities of switching decay exponentially with time, it is enough to calculate a finite number of terms in the sums of (9.12). For the results reported in this work 100 terms were used, which yields an excellent agreement with the computer simulations (see Figs. 9.5–9.8).

Notice that (9.15)–(9.18) assume that nodes of the other strategy that have switched (exponent n) have done so before or at the same time as those of the own strategy (exponent k). This simplification thus avoids considering the full set of possible histories of switchings in times previous to $t-1$. This assumption is based on the separation of time scales between the switchings of both percolation processes. For unidirectional percolation, the probability of a node switching at time t decreases exponentially with the difference between the initial density and the critical percolation density. That is, the nearer the systems starts to the percolation threshold, the slower is the pruning of nodes [25]. As a consequence,

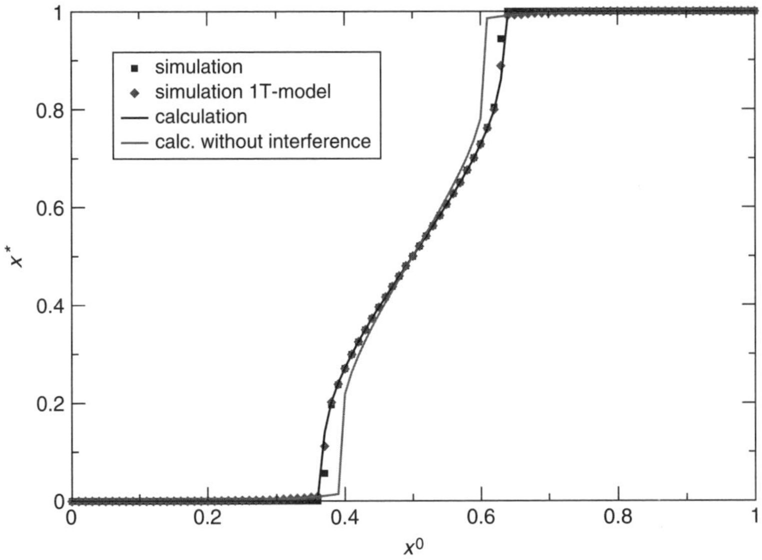

Fig. 9.5 Comparison of simulation with analytical results. Asymptotic density of B-nodes x^* as a function of the initial density x^0. Model parameters are the same as in Fig. 9.2: relative preference $q = 0.5$, switching costs $c_A = c_B = 0.25$, and network degree $z = 9$. The corresponding indices of the percolation processes are $k_A = k_B = 3$

with bidirectional percolation, when an option is near its percolation threshold the other is well beyond it, so that the switching of the latter will be exponentially faster, which supports the assumption above. It is interesting to point out that the proposed recursive scheme based on the switching times can be modified to carry out a calculation neglecting interference, then becoming equivalent to (9.11). The modification consists in setting $s_X^t = 0$, for any $t \geq 1$, instead of using (9.14) and (9.16).

Figures 9.5 and 9.6 show the results presented in Figs. 9.2 and 9.4 of the main text, but this time including also the simulation results for the 1T-model and the analytical results obtained with the calculation without interference. These figures show the close similarity between the original model and the 1T-model, which confirms that most nodes that switch in the original model do so only once. They also demonstrate the high accuracy of the proposed approximation, as compared to the calculation neglecting interference, which fails to reflect the actual percolation thresholds properly. Figures 9.7 and 9.8 display two additional examples.

Finally, we want to point out that, strictly speaking, the networks considered in the analytical calculations (namely, infinite Bethe lattices) differ from the ones used in the computer simulations (finite regular random networks). Apart from the different size, in the former case there are no closed paths or loops, whereas in the latter there exists a (low) number of them. Figures 9.5–9.8 show, however, that these

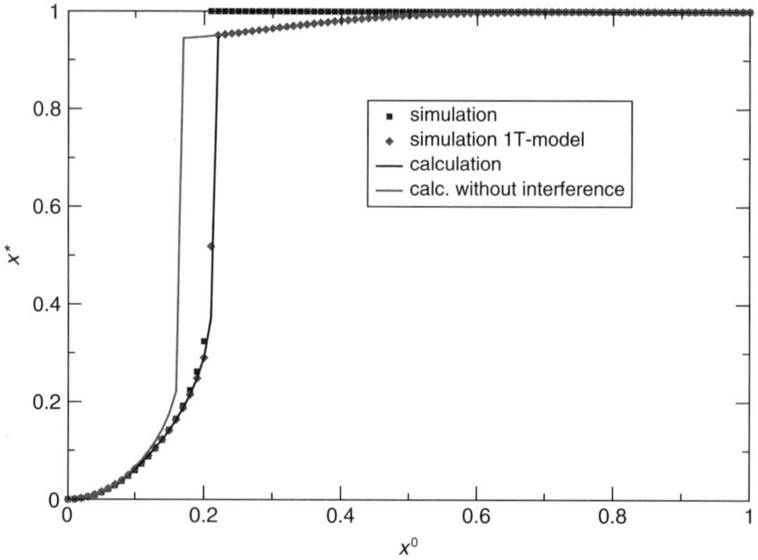

Fig. 9.6 Comparison of simulation with analytical results. Asymptotic density of B-nodes x^* as a function of the initial density x^0. Model parameters are the same as in Fig. 9.4: relative preference $q = 0.25$, switching costs $c_A = c_B = 0.2$, and network degree $z = 6$. The corresponding indices of the percolation processes are $k_A = 4$, $k_B = 1$

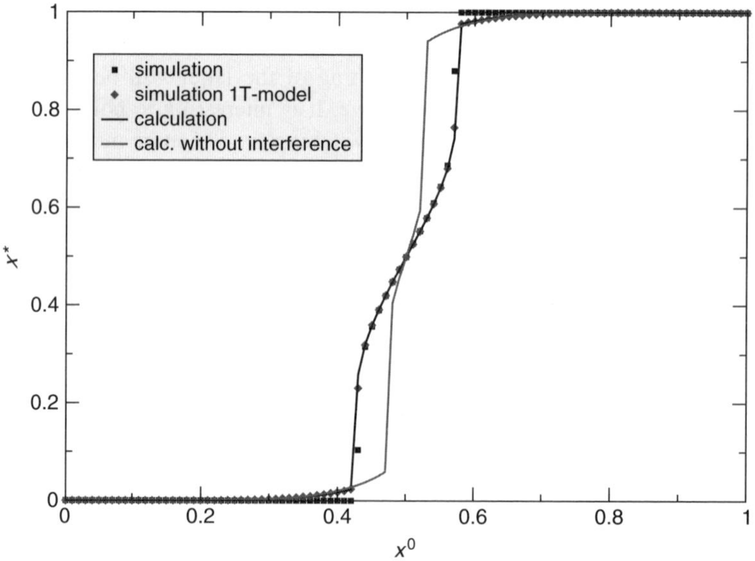

Fig. 9.7 Comparison of simulation with analytical results. Asymptotic density of B-nodes x^* as a function of the initial density x^0. Model parameters: relative preference $q = 0.5$, switching costs $c_A = c_B = 0.2$, and network degree $z = 11$. The indices of the percolation processes are $k_A = k_B = 4$

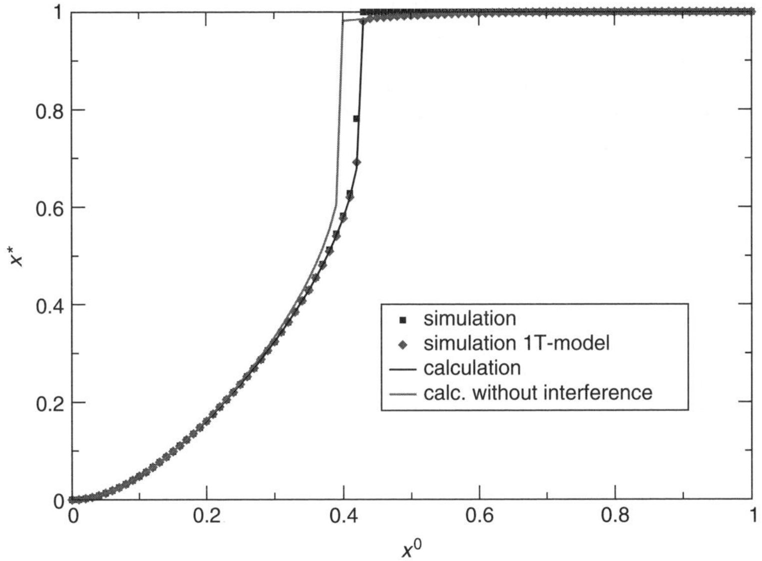

Fig. 9.8 Comparison of simulation with analytical results. Asymptotic density of B-nodes x^* as a function of the initial density x^0. Model parameters: relative preference $q = 0.3$, switching costs $c_A = c_B = 0.2$, and network degree $z = 6$. The indices of the percolation processes are $k_A = 3$, $k_B = 1$

differences in network topology do not produce a significative discrepancy between computational and analytical results.

Acknowledgements C. P. R. and D. H. were partially supported by the Future and Emerging Technologies programme FP7-COSI-ICT of the European Commission through project QLectives (grant no. 231200).

References

1. M. Sahimi, *Applications of Percolation Theory* (Taylor & Francis, PA, 1994)
2. J. Shao, S. Havlin, H.E. Stanley, Dynamic opinion model and invasion percolation. Phys. Rev. Lett. **103**, 018701 (2009)
3. J. Goldenberg, B. Libai, S. Solomon, N. Jan, D. Stauffer, Maketing percolation. Physica A **284**, 335–347 (2000)
4. D. Stauffer, A. Aharony, *Introduction to Percolation Theory*, second edn. (Taylor & Francis, Philadelphia, 1991)
5. J. Chalupa, P.L. Leath, G.R. Reich, Bootstrap percolation on a bethe lattice. J. Phys. C: Solid State Phys. **12**, L31–L35 (1979)
6. S.N. Dorogovtsev, A.V. Goltsev, J.F.F. Mendes, k-core organization of complex networks. Phys. Rev. Lett. **96**, 040601 (2006)
7. W.B. Arthur, Competing technologies, increasing returns, and lock-in by historical events. Econ. J. **99**, 116–131 (1989)

8. E.M. Rogers, *Diffusion of Innovations*, 5th edn. (Simon and Schuster, NY, 2003)
9. H. Amini, M. Draief, M. Lelarge, Marketing in a random network. Network Contr. Optim. **LNCS 5425**, 17–25 (2009)
10. P. Klemperer, Competition when consumers have switching costs: An overview with applications to industrial organization, macroeconomics, and international trade. Rev. Econ. Stud. **62**, 515–539 (1995)
11. M.S. Granovetter, Threshold models of collective behavior. Am. J. Sociol. **83**, 1420–1443 (1978)
12. R.B. Myerson, *Game Theory: Analysis of Conflict* (Harvard University Press, Cambridge, 1991)
13. B. Skyrms, *The Stag Hunt and the Evolution of Social Structure* (Cambridge University Press, Cambridge, 2003)
14. M.A. Nowak, R.M. May, Evolutionary games and spatial chaos. Nature **359**, 826–829 (1992)
15. H. Gintis, *Game Theory Evolving: A Problem-Centered Introduction to Modeling Strategic Interaction*, 2nd edn. (Princeton University Press, Princeton, 2009)
16. J. Hofbauer, K. Sigmund, *Evolutionary Games and Population Dynamics* (Cambridge University Press, Cambridge, 1998)
17. D. Helbing, A mathematical model for behavioral changes through pair interactions, in *Economic Evolution and Demographic Change* ed. by G. Hagg, U. Mueller, K.G. Troitzsch (Springer, Berlin, 1992), pp. 330–348
18. S. Wasserman, K. Faust, *Social Network Analysis: Methods and Applications* (Cambridge University Press, Cambridge, 1994)
19. F. Vega-Redondo, *Complex Social Networks* (Cambridge University Press, Cambridge, 2007)
20. M. Nakamaru, S.A. Levin, Spread of two linked social norms on complex interaction networks. J. Theor. Biol. **230**, 57–64 (2004)
21. A. Galeotti, S. Goyal, M.O. Jackson, F. Vega-Redondo, L. Yariv, Network games. Rev. Econ. Studies **77**, 218–244 (2010)
22. M.E.J. Newman, D.J. Watts, S.H. Strogatz, Random graph models of social networks. Proc. Natl. Acad. Sci. U.S.A. **99**, 2566–2572 (2002)
23. C.P. Roca, J.A. Cuesta, A. Sánchez, Evolutionary game theory: Temporal and spatial effects beyond replicator dynamics. Phys. Life Rev. **6**, 208–249 (2009)
24. G. Szabó, G. Fáth, Evolutionary games on graphs. Phys. Rep. **446**, 97–216 (2007)
25. G. Grimmett, *Percolation*, 2nd edn. (Springer, Berlin, 1999)
26. D. Achlioptas, R.M. D'Souza, J. Spencer, Explosive percolation in random networks. Science **323**, 1453–1455 (2009)
27. D. McFadden, Conditional logit analysis of qualitative choice behavior, in *Frontiers of Econometrics*, ed. by P. Zarembka (Academic Press, New York, 1974), pp. 105–142
28. J.K. Goeree, C.A. Holt, Stochastic game theory: For playing games, not just for doing theory. Proc. Natl. Acad. Sci. U.S.A. **96**, 10564–10567 (1999)
29. L.E. Blume, The statistical mechanics of strategic interaction. Games Econ. Behav. **5**, 387–424 (1993)
30. A. Traulsen, M.A. Nowak, J.M. Pacheco, Stochastic dynamics of invasion and fixation. Phys. Rev. E **74**, 011909 (2006)
31. D. Centola, The spread of behavior in an online social network experiment. Science **329**, 1194–1197 (2010)
32. K.P. Smith, N.A. Christakis, Social networks and health. Annu. Rev. Sociol. **34**, 405–429 (2008)
33. A. Traulsen, D. Semmann, R.D. Sommerfeld, H.-J. Krambeck, M. Milinski, Human strategy updating in evolutionary games. Proc. Natl. Acad. Sci. U.S.A. **107**, 2962–2966 (2010)

Chapter 10
Heterogeneous Populations: Coexistence, Integration, or Conflict*

10.1 Introduction

In order to gain a better understanding of factors preventing or promoting cooperation among individuals, biologists, economists, social scientists, mathematicians and physicists have intensively studied game theoretical problems such as the prisoner's dilemma and the snowdrift game (also known as chicken or hawk-dove game) [1–3]. These games have in common that a certain fraction of people or even everyone is expected to behave uncooperatively (see Fig. 10.1). Therefore, a large amount of research has focused on how cooperation can be supported by mechanisms such as [3] repeated interactions [1], reputation [4], clusters of cooperative individuals [5], costly punishment [6–12], or success-driven migration [13].

Unfortunately, comparatively little attention has been devoted to the problem of cooperation between groups with different preferences (e.g. people of different gender, status, age, or cultural background). Yet, what constitutes cooperative behavior for one group might be considered non-cooperative by another. For example, men and women appear to have different preferences many times, but they normally interact among and between each other on a daily basis. It is also more and more common that people with different religious beliefs live and work together, while their religions request some mutually incompatible behaviors (in terms of working days and free days, food one may eat or should avoid, headgear, or appropriate clothing, etc.). A similar situation applies, when people with different mother tongues meet, or businessmen from countries with different business practices make a deal. Is it possible to identify factors determining whether two such populations go their own way, find a common agreement, or end up in conflict? And what is

*The content of this chapter has some overlap with the following two papers, which should be cited instead: D. Helbing and *A. Johansson*, Evolutionary dynamics of populations with conflicting interactions: Classification and analytical treatment considering asymmetry and power. *Physical Review E* **81**, 016112 (2010); D. Helbing and A. Johansson, Cooperation, norms, and revolutions: A unified game-theoretical approach. *PLoS ONE* **5**(10): e12530.

D. Helbing (ed.), *Social Self-Organization*, Understanding Complex Systems, DOI 10.1007/978-3-642-24004-1_10,

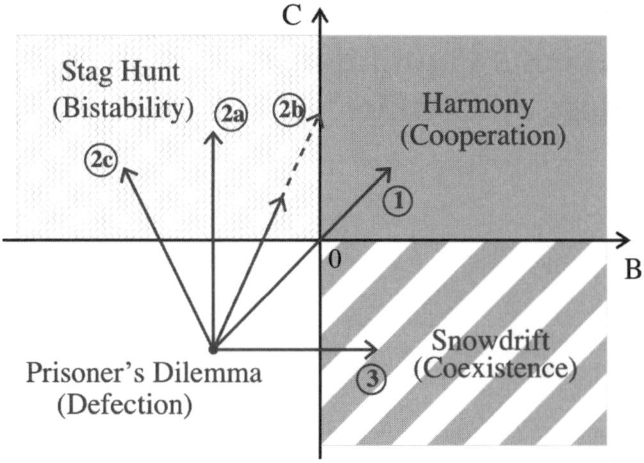

Fig. 10.1 Illustration of the parameter-dependent types and outcomes of symmetrical 2 × 2 games in a *single* population [14]. For prisoner's dilemmas (PD) and two-person public goods games, we have $B < 0$ and $C < 0$, and the expected outcome is defection by everybody. For snowdrift games (SD), which are also called chicken or hawk-dove games, we have $B > 0$ and $C < 0$. The stable stationary solution corresponds to a coexistence of a fraction $p_0 = |B|/(|B| + |C|)$ of cooperators with a fraction $1 - p_0$ of defectors (i.e. non-cooperative individuals). For harmony games (HG), which are sometimes called coordination games as well, we have $B > 0$ and $C > 0$, and everybody will eventually cooperate. Finally, for stag hunt games (SH), which are also called assurance games, we have $B < 0$ and $C > 0$, and there is a bistable situation: If the initial fraction $p(0)$ of cooperators is larger than $p_0 = |B|/(|B| + |C|)$, everybody is expected to cooperate in the end, otherwise everybody will eventually behave uncooperatively [15]. The *arrows* illustrate different routes to cooperation [3, 15]. Route 1 belongs to the way in which kin selection, network reciprocity, or group selection modify the payoff structure of the game. Route 2a corresponds to the effect of direct reciprocity (due to the "shadow of the future" through the likelihood of future interactions). Route 2b reflects the mechanism of indirect reciprocity (based on reputation effects), and route 2c corresponds to costly punishment. Route 3 results for certain kinds of network interactions [16]

the relevance of power in the rivalry of populations? Differences in power can, for example, result from different sizes of the interacting populations, their material resources (money, weapons, etc.), social capital (status, social influence, etc.), and other factors (charisma, moral persuasion, etc.).

10.2 Model

As a mathematical approach to this problem, we propose game-dynamical replicator equations for multiple populations [17, 18]. The crucial point is to adjust them in a way that reflects interactions between individuals with incompatible preferences (see Methods and [19]). These equations describe the time evolution of the proportions $p(t)$ and $q(t)$ of cooperative individuals in populations 1 and 2, respectively, as individuals imitate more successful behaviors in their own population. The success depends on the "payoffs" resulting from social interactions, i.e., on the own behavior and the behavior of the interaction partners.

In order to reflect incompatible interests of both populations, we assume that population 1 prefers behavior 1 (e.g. everybody should be undressed at the beach), while population 2 prefers behavior 2 (everybody should be properly dressed). If an interaction partner shows the behavior preferred by oneself, this behavior is called "cooperative", otherwise uncooperative. In other words, behavior 1 is cooperative from the viewpoint of population 1, but uncooperative from the viewpoint of population 2 (and vice versa). Furthermore, if an individual of population 1 interacts with an individual of population 2 and both display *same* behavior, we call this behavior "coordinated". Finally, if the *majority* of individuals in *both* populations shows a coordinated behavior, we speak of "normative behavior" or a "behavioral norm".

To establish a behavioral norm, the individuals belonging to *one* of the populations have to act against their own preferences, in particular as we assume that preferred behaviors and population sizes do not change. (Otherwise, identical behavior in both populations could simply result from the adaptation of preferences or group membership, which can, of course, further promote consensus in reality.)

It is very interesting to study under what conditions interactions within and between populations with incompatible preferences can lead to cooperation, conflict, or "normative behavior". To address this question, we adjust the theoretical framework of game-dynamical replicator equations to the case of two populations playing 2×2 games, which are represented by four payoffs T, R, P, and S. In the prisoner's dilemma, for example, the meaning of these parameters is "*T*emptation" to behave non-cooperatively, "*R*eward" for mutual cooperation, "*P*unishment" for mutual non-cooperative behavior and "*S*ucker's payoff" for a cooperative individual meeting an uncooperative one. The related game-dynamical replicator equations read

$$\frac{dp(t)}{dt} = p(t)[1 - p(t)]F\big(p(t), q(t)\big) \tag{10.1}$$

and

$$\frac{dq(t)}{dt} = q(t)[1 - q(t)]G\big(p(t), q(t)\big), \tag{10.2}$$

where the terms $p(1-p)$ and $q(1-q)$ can be interpreted as saturation factors. They make sure that the proportions $p(t)$ and $q(t)$ of individuals pursuing their preferred strategies stay within the range from 0 to 1. $F(p,q)$ and $G(p,q)$ are functions reflecting the interactions between individuals. They include terms describing "in-group" interactions ("self-interactions", reflecting encounters with individuals of the *same* population) and "out-group" interactions (among individuals belonging to *different* populations) (see Methods).

For simplicity, we will focus on the case where both populations play the same game. Then, the functions F and G only depend on the payoff-dependent parameters $B = S - P$, $C = R - T$, and the relative power f of population 1 (see Methods). Furthermore, we specify the power here by the relative population size (see [19] for details). The parameter C may be interpreted as gain of coordinating on one's own preferred behavior (if greater than zero, otherwise as loss). B may be viewed as gain when giving up coordinated, but non-preferred behavior. Despite the simplifications made, this model for two populations with incompatible preferences shows very interesting features, and it can be generalized in numerous ways (to consider more populations, to treat heterogeneous, but compatible interactions, to reflect that payoffs in out-group interactions may differ from in-group interactions, to consider migration between populations, spatial interactions, learning, punishment, etc.).

10.3 Results

We find that social interactions with incompatible interests do not necessarily produce conflict – they may even promote mutual coordination. Depending on the signs of B and C, which determine the character of the game, we have four archetypical situations: (1) The case $B < 0$ and $C < 0$ applies to the *multi-population prisoner's dilemma (MPD)*, (2) in the *multi-population harmony game (MHG)*, we have $B > 0$ and $C > 0$, (3) the *multi-population snowdrift game (MSD)* is characterized by $B > 0$ and $C < 0$, and (4) in the *multi-population stag hunt game (MSH)*, we have $B < 0$ and $C > 0$. In the multi-population harmony game, everybody shows cooperative behavior, while in the multi-population prisoner's dilemma, everybody is uncooperative in the end, as one may expect (see Fig. 10.2). However, as we will show in the following, the dynamics and outcome of the multi-population stag hunt and snowdrift games are in marked contrast to the one-population case. This can be demonstrated by mathematical analysis of the stationary solutions of (10.1) and (10.4) and their stability properties (see Methods and [19]). However, our results can be more intuitively illustrated by figures and movies showing the evolutionary equilibria of the games for various parameter values B and C, their basins of attraction, and representative flow lines. Details are discussed below and in the captions of Figs. 10.3 and 10.4.

10.3.1 Evolution of Normative Behavior in the Stag Hunt Game

The one-population stag hunt game is characterized by an equilibrium selection problem [20]: Everyone is finally expected to cooperate, if the initial fraction of cooperative individuals is above $p_0 = |B|/(|B| + |C|)$, otherwise everybody will behave uncooperative in the end [15]. The same applies to non-interacting

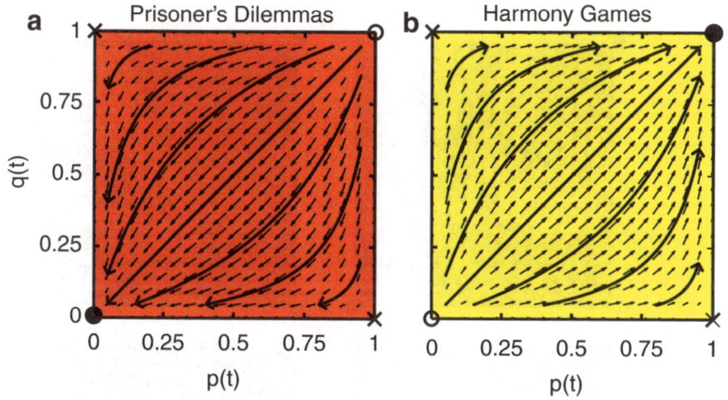

Fig. 10.2 Vector fields (*small arrows*), sample trajectories (*large arrows*) and phase diagrams (*colored areas*) for two interacting populations with incompatible preferences, when 80% of individuals belong to population 1 ($f = 0.8$). (**a**) If $B = C = -1$, individuals are facing prisoner's dilemma interactions, and everybody ends up with non-cooperative behavior in each population. (**b**) If $B = C = 1$, individuals are playing harmony games instead, and everybody will eventually behave cooperatively. The results look similar when the same two-population games are played with different values of f, $|B|$ or $|C|$

populations. However, in the multi-population stag hunt game with incompatible preferences and no self-interactions, it *never* happens that everybody or nobody cooperates in both populations (otherwise there should be yellow or red areas in the second part of Movie S2). Although both populations prefer different behaviors, all individuals end up coordinating themselves on a commonly shared behavior. This can be interpreted as self-organized evolution of a behavioral norm [21].

If self-interactions are taken into account as well, the case where everybody or nobody cooperates in both populations is possible, but rather exceptional (see Fig. 10.3). It requires that both populations have similar strengths ($f \approx 1/2$) and the initial levels of cooperation are comparable as well (see yellow area in Fig. 10.3b). Under such conditions, both populations may develop separate subcultures. Normally, however, both populations establish a commonly shared norm and either end up with behavior 1 (green area in Fig. 10.3) or with behavior 2 (blue area).

Due to the payoff structure of the multi-population stag hunt game, it can be profitable to coordinate oneself with the prevailing behavior in the other population. Yet, the establishment of a norm requires the individuals of one population to give up their *own* preferred behavior in favor of the one preferred by the *other* population. Therefore, it is striking that the preferred behavior of the *weaker* population can actually win through and finally establish the norm (see blue areas in Figs. 10.3a,c,d). *Who* adapts to the preferred strategy of the other population essentially depends on the *initial* fractions of behaviors. The majority behavior is likely to determine the resulting behavioral norm, but a powerful population is in a favorable position: The area of possible histories leading to an establishment of the

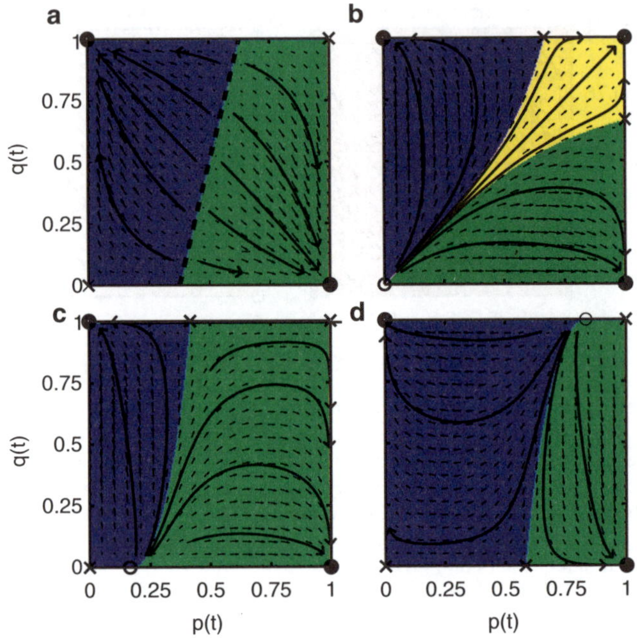

Fig. 10.3 Vector fields (*small arrows*), sample trajectories (*large arrows*) and phase diagrams (*colored areas*) for two interacting populations with incompatible preferences, playing stag hunt games with $B < 0$ and $C > 0$ (*right*). p is the fraction of individuals in population 1 showing their preferred, cooperative behavior 1, and q is the fraction of cooperative individuals in population 2 showing their preferred behavior 2. The vector fields show $(dp/dt, dq/dt)$, i.e. the direction and size of the expected temporal change of the behavioral distribution, if the fractions of cooperative individuals in populations 1 and 2 are $p(t)$ and $q(t)$. Sample trajectories illustrate some representative flow lines $(p(t), q(t))$ as time t passes. The flow lines move away from unstable stationary points (*empty circles* or *dashed lines*) and are attracted towards stable stationary points (*black circles* or *solid diagonal lines*). The colored areas represent the basins of attraction, i.e. all initial conditions $(p(0), q(0))$ leading to the same fix point [*yellow* = (1,1), *blue* = (0,1), *green* = (1,0)]. Saddle points (*crosses*) are attractive in one direction, but repulsive in another. The model parameters are as follows: (**a**) $|B| = |C| = 1$ and $f = 0.8$, i.e. 80% of all individuals belong to population 1, (**b**) $|C| = 2|B| = 2$ and $f = 1/2$, i.e. both populations are equally strong, (**c**) $|C| = 2|B| = 2$ and $f = 0.8$, (**d**) $2|C| = |B| = 2$ and $f = 0.8$. In the multi-population stag hunt game (MSH), due to the asymptotically stable fix points at (1,0) and (0,1), all individuals of both populations finally show the behavior preferred in population 1 (when starting in the *green area*) or the behavior preferred in population 2 (when starting in the blue area). This case can be considered to describe the evolution of a shared behavioral norm. Only for similarly strong populations ($f \approx 1/2$) and similar initial fractions $p(0)$ and $q(0)$ of cooperators in both populations (*yellow area*), both populations will end up with population-specific norms ("subcultures"), corresponding to the asymptotically stable point at (1,1). The route towards the establishment of a shared norm may be quite unexpected, as the flow line starting with the *white circle* shows: The fraction $q(t)$ of individuals in population 2 who are uncooperative from the viewpoint of population 1 grows in the beginning, but later on it goes down dramatically. Therefore, a momentary trend does not allow to easily predict the final outcome of the struggle between two interest groups

norm preferred by population 1 tends to increase with power f (compare the size of the green areas in Figs. 10.3b+c).

Note that the evolution of norms is one of the most fundamental challenges in the social and economic sciences. Norms are crucial for society, as they reduce uncertainty, bargaining efforts, and conflict in social interactions. They are like social forces guiding our interactions in numerous situations and subtle ways, creating an "invisible hand" kind of self-organization of society [21].

Researchers from various disciplines have worked on the evolution of norms, often utilizing game-theoretical concepts [22–26], but it has been hard to reveal the conditions under which behavioral consensus[27] is established. This is so, because cooperation norms, in contrast to coordination norms, are not self-enforcing. In other words, there are incentives for unilateral deviance. Considering the fact that norms require people to constrain self-interested behavior [28] and to perform socially prescribed roles, the ubiquity of norms is quite surprising. Yet, widespread cooperation-enhancing mechanisms such as group pressure can transform prisoner's dilemmas into stag hunt interactions [3,15,29] (see Fig. 10.1). This creates a natural tendency towards the formation of norms, whatever their content may be.

Our model sheds new light on the problem of whether, why and how a norm can establish. In particular, it reveals that the dynamics and finally resulting state of the system is not just determined by the payoff structure. It also crucially depends on the power of populations and even on the initial proportions of cooperative individuals (the "initial conditions").

10.3.2 Occurrence of Conflict in the Snowdrift Game

In the one-population *snowdrift game*, there is one stable stationary point, corresponding to a fraction $p_0 = |B|/(|B| + |C|)$ of cooperative individuals [15]. If this were transferable to the multi-population case, we should have $p = q = p_0$ in the limit of large times $t \to \infty$. Instead, we find a variety of different outcomes, depending on the values of the model parameters B, C, and f:

(a) *The interactions between both populations shift the fraction of cooperative individuals in each population to values different from p_0.* If $|B| = |C|$, we discover a line of *infinitely many* stationary points, and the actually resulting stationary solution uniquely depends on the initial condition (see Fig. 10.4a). This line satisfies the relation $q = p$ only if $f = 1/2$, while for most parameter combinations we have $q \neq p \neq p_0$. Nevertheless, the typical outcome in the case $|B| = |C|$ is characterized by a finite fraction of cooperative individuals in each population.

(b) *Conflicting interactions between two equally strong groups destabilize the stationary solution $q = p = p_0$ of the one-population case, and both populations lose control over the final outcome.* For $|B| \neq |C|$, all stationary points are *discrete* and located on the *boundaries*, and only *one* of these points is

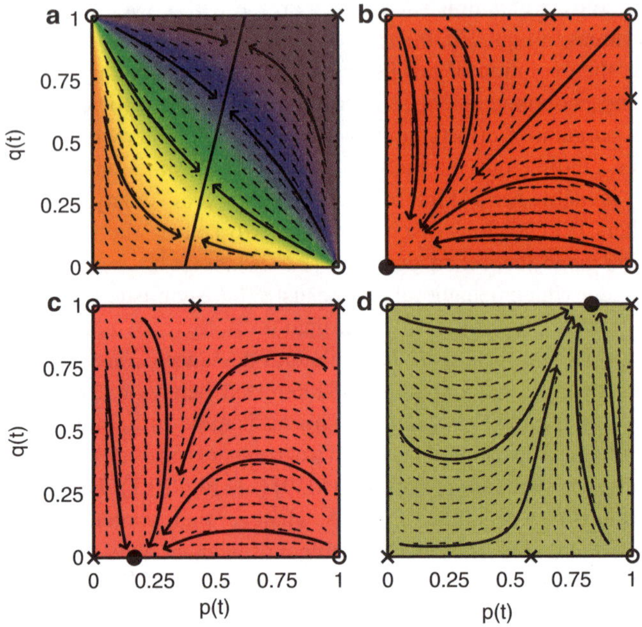

Fig. 10.4 Vector fields (*small arrows*), sample trajectories (*large arrows*) and phase diagrams (*colored areas*) for two interacting populations with incompatible preferences, playing snowdrift games with $B > 0$ and $C < 0$ (*right*). The representation is the same as in Fig. 10.3. In particular, the colored areas represent again the basins of attraction, i.e. all initial conditions $(p(0), q(0))$ leading to the same fix point [*red* = (0,0), *salmon* = $(u, 0)$, *mustard* = $(v, 1)$, *rainbow colors* = (u, v), with $0 < u, v < 1$]. The model parameters are as follows: (**a**) $|B| = |C| = 1$ and $f = 0.8$, i.e. 80% of all individuals belong to population 1, (**b**) $|C| = 2|B| = 2$ and $f = 1/2$, i.e. both populations are equally strong, (**c**) $|C| = 2|B| = 2$ and $f = 0.8$, (**d**) $2|C| = |B| = 2$ and $f = 0.8$. (**a**) In the multi-population snowdrift game (MSD), a mixture of cooperative and uncooperative behaviors results in both populations, if $|B| = |C|$. (**b**) For $|B| < |C|$ and equally strong populations, everybody ends up with non-cooperative behavior in each population. (**c**) For $|B| < |C|$ and $f - 1/2 \gg 0$, the weaker population 2 solidarizes with the minority of the stronger population 1 and opposes its majority. (**d**) Same as (**c**), but now, all individuals in the weaker population 2 show their own preferred behavior after the occurrence of a discontinuous ("revolutionary") transition of the evolutionary equilibrium from $(u, 0)$ to $(v, 1)$

an evolutionary equilibrium. If both populations have *equal* power ($f = 1/2$), we always end up with non-cooperative behavior by everybody (if $p_0 < 1/2$, see Fig. 10.4b), or everybody is cooperative (if $p_0 > 1/2$). Remarkably, there is no mixed stable solution between these two extremes.

(c) *The stronger population gains control over the weaker one, but a change of the model parameters may induce a "revolutionary" transition.* If $|B| \neq |C|$ and population 1 is much *stronger* than population 2 (i.e., $f - 1/2 \gg 0$), we find a finite fraction of cooperative individuals in the stronger population, while either 0% or 100% of the individuals are cooperative in the weaker population. A closer analysis reveals that the resulting overall fraction of cooperative individuals fits exactly the expectation p_0 of the stronger population (see Methods), while from the perspective of the weaker population, the overall fraction of cooperative individuals is largely different from $p_0 = |B|/(|B| + |C|)$. Note that the stronger population alone can not reach an overall level of cooperation of p_0. The desired outcome can only be produced by effectively controlling the behavior of the weaker population. This takes place in an unexpected way, namely by polarization: In the weaker population 2, *everyone* shows behavior 1 for $p_0 < 1/2$ (see Fig. 10.4c), otherwise everyone shows behavior 2 (see Fig. 10.4d). There is no solution in between these two extremes (apart from the special case $p_0 = 1/2$ for $|B| = |C|$).

It comes as a further surprise that the behavior in the weaker population is always coordinated with the *minority* behavior in the stronger population. Due to the payoff structure of the multi-population snowdrift game, it is profitable for the weaker population to oppose the majority of the stronger population, which creates a tacit alliance with its minority. Such antagonistic behavior is well-known from protest movements [30] and finds here a natural explanation.

Moreover, when $|C|$ changes from values greater than $|B|$ to values smaller than $|B|$, there is an unexpected, discontinuous transition in the weaker population 2 from a state in which everybody is cooperative from the point of view of population 1 to a state in which everybody shows the *own* preferred behavior 2 (see Movie S1 and Methods). History and science [31] have seen many abrupt regime shifts of this kind. Revolutions caused by class conflict provide ample empirical evidence for their existence. Combining the theory of phase transitions with "catastrophe theory" offers a quantitative scientific approach to interpret such revolutions as the outcome of social interactions [32]. Here, their recurrence becomes understandable in a unified and simple game-theoretical framework.

10.4 Discussion

Multi-population game-dynamical replicator equations provide an elegant and powerful approach to study the dynamics and outcomes expected for groups with incompatible interests. A detailed mathematical analysis reveals how interactions within and between groups can substantially change the dynamics of various game theoretical dilemmas. Generalizations to more than two behaviors or groups and to populations with heterogeneous preferences are easily possible.

When populations with incompatible preferences interact among and between each other, the signs of the payoff-dependent parameters B and C determine the character of the game. The snowdrift and stag hunt games show a particularly rich and interesting dynamics. For example, there is a discontinuous ("revolutionary") transition, when $1-|B|/|C|$ changes its sign. On top of this, the power f has a major influence on the outcome, and the initial distribution of behaviors can be crucial.

Note that such a rich system behavior is already found for the *simplest* setting of our model and that the concept of multi-population game-dynamical equations can be generalized in various ways to address a number of challenging questions in the future: How can we gain a better understanding of a clash of cultures, the outbreak of civil wars, or conflicts with ethnic or religious minorities? How can we analytically study migration and group competition? When do social systems become unstable and face a polarization of society? How can we understand the emergence of fairness norms in bargaining situations?

Another interesting aspect of our model is that it makes a variety of quantitative predictions. Therefore, it could be tested experimentally with iterated games in the laboratory, involving several groups of people with random matching and sufficiently many iterations. Suitable changes in the payoff matrices should be able to confirm the mathematical conditions under which different archetypical types of social phenomena or discontinuous transitions in the system behavior can occur (see Methods and [19]): (1) the breakdown of cooperation, (2) in-group cooperation (the formation of "sub-cultures"), (3) societal polarization and conflict with the possibility of discontinuous regime shifts ("revolutions"), and (4) the evolution of shared behavioral norms.

In the past, the problem of the evolution of norms has often been addressed by studying one-population prisoner's dilemma situations [21, 24], and by assuming cooperation-enhancing mechanisms such as repeated interactions (a "shadow of the future", as considered in the mechanism of direct reciprocity) [1], or the sanctioning of non-conforming behavior ("punishment") [7–12, 26, 33, 34] These mechanisms can, in fact, transform prisoner's dilemmas into stag hunt games [15, 23, 29] (see Fig. 10.1), which connects our approach with previous work addressing norms. However, our model goes beyond studying the circumstances under which people follow a *preset* norm, it considers situations where it is not clear from the beginning what behavior would eventually establish as a norm (or whether any of the behaviors would become a norm *at all*). When studying multi-population settings with incompatible interests, we do not only have the problem of how *cooperative* behavior can be promoted, as in the prisoner's dilemma. We also have a *normative* dilemma by the circumstance that *the establishment of a behavioral norm requires one population to adjust to the preferred behavior in the other population – against its own preference*.

Other cooperation-enhancing mechanisms such as kin selection (based on genetic relationship) and group selection tend to transform a prisoner's dilemma into a *harmony game* [15] (see Fig. 10.1). Therefore, our findings suggest that genetic relatedness and group selection are *not* ideal mechanisms to establish shared behavioral norms. They rather support the formation of subcultures. Moreover, the transformation of prisoner's dilemma interactions into a snowdrift game is expected to cause social conflict. Obviously, this has crucial implications for society, law and economics [20, 35], where conflicts need to be avoided or solved, and norms and standards are of central importance.

Take language as another example – probably the most distinguishing trait of humans [36, 37]. Successful communication requires the establishment of a norm, how words are used (the "evolution of meaning"). It will, therefore, be intriguing to study whether the explosive development of language and culture in humans is due to their ability to transform interactions into norm-promoting stag hunt interactions. From this point of view, repeated interactions due to human agglomeration in settlements, the development of reputation mechanisms, and the invention of sanctioning institutions should have largely accelerated cultural evolution [38].

Appendix A: Methods

Multi-population game-dynamical replicator equations describe the temporal evolution of the proportions $p_i^a(t)$ of individuals showing behavior i at time t in population a. They assume that more successful behaviors spread, as these are imitated by individuals of the same population at a rate proportional to the gain in the expected success. The expected success is determined from the frequency of interactions between two behaviors i and j, and by the associated payoffs A_{ij}^{ab} (see [19]). Focusing on the above-mentioned social dilemmas, in the case of two interacting populations $a, b \in \{1, 2\}$ and two behavioral strategies $i, j \in \{1, 2\}$, we assume the following for interactions within the *same* population a: If two interacting individuals show the same behavior i, both will either receive the payoff r_a or p_a. If we have $r_a \neq p_a$, we call the behavior with the larger payoff r_a "preferred" or "cooperative", the other behavior "non-cooperative" or "uncooperative". When one individual chooses the cooperative behavior and the interaction partner is uncooperative, the first one receives the payoff s_a and the second one the payoff t_a. To model conflicts of interests, we assume that population $a = 1$ prefers behavior $i = 1$ and population 2 prefers behavior 2. Therefore, if an individual of population 1 meets an individual belonging to population 2 and both show the same behavior $i = 1$, the first one will earn R_1 and the second one P_2, as behavior $i = 1$ is considered uncooperative in population 2. Analogously, for $i = 2$ they earn P_1 and R_2, respectively. If the interaction partners choose *different* behaviors i and j, they earn S_a, when the behavior corresponds to their cooperative behavior, otherwise they earn T_a.

Assuming constant preferences and fixed relative population strengths f_a, the resulting coupled game-dynamical replicator equations for the temporal evolution of the proportion $p(t) = p_1^1(t)$ of cooperative individuals in population 1 and the fraction $q(t) = p_2^2(t)$ of cooperative individuals in population 2 are given by (10.1) and (10.2) with

$$F(p, q) = b_1 f + (c_1 - b_1) f p(t) + C_1(1 - f) + (B_1 - C_1)(1 - f) q(t) \quad (10.3)$$

and

$$G(p, q) = b_2(1 - f) + (c_2 - b_2)(1 - f) q(t) + C_2 f + (B_2 - C_2) f p(t) \quad (10.4)$$

(see [19]). Here, we have used the abbreviation $f = f_1 = 1 - f_2$. Moreover, $b_a = s_a - p_a$, $B_a = S_a - P_a$, $c_a = r_a - t_a$, and $C_a = R_a - T_a$ are payoff-dependent model parameters, which can be positive, negative, or zero. Note that the above equations describe the *general* case for two populations with interactions and/or self-interactions playing *any* kind of 2×2 game. When setting $p_a = P_a = P$, $r_a = R_a = R$, $s_a = S_a = S$, and $t_a = T_a = T$ (i.e. $b_a = B_a = B$ and $c_a = C_a = C$), both populations play the same game. Moreover, the payoff depends on the own behavior i and the behavior j of the interaction partner only, but not on the population he/she belongs to. That is, in- and out-group interactions yield the same payoff, and the preference of the interaction partner does not matter for it.

Sanctioning of non-conforming behavior. Individuals often apply *group pressure* to support conformity and discourage uncoordinated behavior. This could be modeled by subtracting a value δ from the off-diagonal payoffs S and T or by adding δ to the diagonal elements R and P, resulting in the effective model parameters $b_a = B_a = B - \delta$ and $c_a = C_a = C + \delta$. Therefore, if the group pressure δ is large enough (namely, $\delta > |C|$), a prisoner's dilemma with $B < 0$ and $C < 0$ is transformed into a stag hunt game with $b_a = B_a < 0$ and $c_a = C_a > 0$.

Summary of the main analytical results. (see [19] for more general formulas). All of the multi-population games with interactions and self-interactions studied by us have the four stationary solutions $(p_1, q_1) = (0, 0)$, $(p_2, q_2) = (1, 1)$, $(p_3, q_3) = (1, 0)$ and $(p_4, q_4) = (0, 1)$, corresponding to the four corners of the p-q-space. Their stability properties depend on the eigenvalues $\lambda_l = (1 - 2p_l)F(p_l, q_l)$ and $\mu_l = (1 - 2q_l)G(p_l, q_l)$, where $l \in \{1, 2, 3, 4\}$. Stable (attractive) fix points require $\lambda_l < 0$ and $\mu_l < 0$, unstable (repulsive) fix points $\lambda_l < 0$ and $\mu_l < 0$. $\lambda_l \mu_l < 0$ implies a saddle point. In the *multi-population prisoner's dilemma (MPD)*, the only stable fix point is $(0, 0)$, while in the *multi-population harmony game (MHG)*, it is $(1, 1)$. In both games, $(1, 0)$ and $(0, 1)$ are always saddle points. For $B, C < 0$ (the MPD) and $B, C > 0$ (the MHG), no further stationary points exist (see Fig. 10.2). For the *multi-population stag hunt game (MSH)* with $B < 0$ and $C > 0$, we find:

- (0,1) and (1,0) are always stable fix points, see Fig. 10.3.
- (1,1) is a stable fix point for $C/|B| > \max[f/(1 - f), (1 - f)/f]$ (see Fig. 10.3b).
- (0,0) is a stable fix point for $C/|B| < \min[f/(1 - f), (1 - f)/f]$.

For the *multi-population snowdrift game (MSD)* with $B > 0$ and $C < 0$ we have:

- (1,0) and (0,1) are always unstable fix points (see Fig. 10.4).
- (0,0) is a stable fix point for $C/|B| > \max[f/(1 - f), (1 - f)/f]$ (see Fig. 10.4b).
- (1,1) is a stable fix point for $C/|B| < \min[f/(1 - f), (1 - f)/f]$.

Moreover, if B and C have different signs, further stationary points (p_l, q_l) with $l \in \{5, 6, 7, 8\}$ may occur on the boundaries, while inner points (p_9, q_9) with $0 < p_9 < 1$ and $0 < q_9 < 1$ can only occur for $B = -C$ (see Figs. 10.3a and 10.4a). As $|C|$ is increased from 0 to high values, we find the following additional stationary points for the MSH and MSD, where we use the abbreviations

$p_5 = [Bf + C(1-f)]/[(B-C)f]$, $p_6 = B/[(B-C)f]$, $q_7 = B/[(B-C)(1-f)]$, and $q_8 = [B(1-f) + Cf]/[(B-C)(1-f)]$:

- $(p_5, 0)$ and $(0, q_8)$, if $f \geq 1/2$ and $|C|/|B| \leq (1-f)/f$ or if $f \leq 1/2$ and $|C|/|B| \leq f/(1-f)$.
- $(p_5, 0)$ and $(p_6, 1)$, if $f \geq 1/2$ and $(1-f)/f < |C|/|B| < f/(1-f)$, or $(1, q_7)$ and $(0, q_8)$ if $f \leq 1/2$ and $f/(1-f) < |C|/|B| < (1-f)/f$ (see Figs. 10.3c+d and 10.4c+d).
- $(p_6, 1)$ and $(1, q_7)$, if $f \geq 1/2$ and $|C|/|B| \geq f/(1-f)$ or if $f \leq 1/2$ and $|C|/|B| \geq (1-f)/f$ (see Figs. 10.3b and 10.4b).

For $B < 0 < C$ (the MSH), these fix points are unstable or saddle points, while they are stable or saddle points for $C < 0 < B$ (the MSD). Obviously, there are transitions to a qualitatively different system behavior at the points $|C|/|B| = (1-f)/f$ and $|C|/|B| = f/(1-f)$. Moreover, there are discontinuous, "revolutionary" transitions, when $|C|$ crosses the value of $|B|$, as the stability properties of pairs of fix points are then *interchanged*. This can be followed from the fact that the dynamic behavior and final outcome for the case $|B| > |C|$ can be derived from the results for $|B| < |C|$, namely by applying the transformations $B \leftrightarrow -C$, $p \leftrightarrow (1-p)$, and $q \leftrightarrow (1-q)$, which do not change the game-dynamical equations (see (10.1)–(10.4)). Generally, *discontinuous transitions in the system behavior may occur, when the sign of* $1 - |B|/|C|$ *changes, or if the sign of* B *or* C *changes* (which modifies the character of the game, for example from a MPD to a MSH game).

Fraction of cooperators in the multi-population snowdrift game. When $(p_5, 0)$ is the stable stationary point, the *average* fraction of cooperative individuals in *both* populations from the perspective of the stronger population 1 can be determined as the fraction of cooperative individuals in population 1 times the relative size f of population 1, plus the fraction $1 - q_5 = 1$ of non-cooperative individuals in population 2 (who are cooperative from the point of view of population 1), weighted by its relative size $(1-f)$:

$$p_5 \cdot f + (1-q_5) \cdot (1-f) = \frac{Bf + C(1-f)}{(B-C)f} \cdot f + 1 \cdot (1-f) = \frac{B}{B-C}.$$

Considering $C < 0$, this corresponds to the expected fraction $p_0 = |B|/(|B| + C)$ of cooperative individuals in the one-population snowdrift game [15].

Acknowledgements The authors would like to thank for partial support by the EU Project QLectives and the ETH Competence Center "Coping with Crises in Complex Socio-Economic Systems" (CCSS) through ETH Research Grant CH1-01 08-2. They are grateful to Thomas Chadefaux, Ryan Murphy, Carlos P. Roca, Stefan Bechtold, Sergi Lozano, Heiko Rauhut, Wenjian Yu and further colleagues for valuable comments and to Sergi Lozano for drawing Fig. 10.1. D.H. thanks Thomas Voss for his insightful seminar on social norms.

References

1. R. Axelrod, *The Evolution of Cooperation* (Basic Books, New York, 1984)
2. H. Gintis, *Game Theory Evolving* (Princeton University, Princeton, NJ, 2000)
3. M.A. Nowak, Five rules for the evolution of cooperation. Science **314**, 1560–1563 (2006)
4. M. Milinski, D. Semmann, H.J. Krambeck, Reputation helps solve the "tragedy of the commons". Nature **415**, 424–426 (2002)
5. M.A. Nowak, R.M. May, Evolutionary games and spatial chaos. Nature **359**, 826–829 (1992)
6. E. Fehr, S. Gächter, Altruistic punishment in humans. Nature **415**, 137–140 (2002)
7. C. Hauert, A. Traulsen, H. Brandt, M.A. Nowak, K. Sigmund, Via freedom to coercion: The emergence of costly punishment. Science **316**, 1905–1907 (2007)
8. B. Rockenbach, M. Milinski, The efficient interaction of indirect reciprocity and costly punishment. Nature **444**, 718–723 (2006)
9. O. Gurerk, B. Irlenbusch, B. Rockenbach, The competitive advantage of sanctioning institutions. Science **312**, 108–111 (2006)
10. J. Henrich et al., Costly punishment across human societies. Science **312**, 1767–1770 (2006)
11. J.H. Fowler, Altruistic punishment and the origin of cooperation. Proc. Natl. Acad. Sci. USA **102**, 7047–7049 (2005)
12. R. Boyd, H. Gintis, S. Bowles, P.J. Richerson, The evolution of altruistic punishment. Proc. Natl. Acad. Sci. USA **100**, 3531–3535 (2003)
13. D. Helbing, W. Yu, The outbreak of cooperation among success-driven individuals under noisy conditions. Proc. Natl. Acad. Sci. USA **106**(8), 3680–3685 (2009)
14. J.W. Weibull, *Evolutionary Game Theory* (MIT Press, Cambridge, MA, 1996)
15. D. Helbing, S. Lozano, Phase transitions to cooperation in the prisoner's dilemma. Phys. Rev. E **81**(5), 057102 (2010)
16. H. Ohtsukia, M.A. Nowak, The replicator equation on graphs. J. Theor. Biol. **243**, 86–97 (2006)
17. P. Schuster, K. Sigmund, J. Hofbauer, R. Gottlieb, P. Merz, Selfregulation of behaviour in animal societies. III. Games between two populations with selfinteraction. Biol. Cyber. **40**, 17–25 (1981)
18. D. Helbing, A mathematical model for behavioral changes by pair interactions, in *Economic Evolution and Demographic Change* ed. by G. Haag, U. Mueller, K.G. Troitzsch (Springer, Berlin, 1992), pp. 330–348
19. D. Helbing, A. Johansson, Evolutionary dynamics of populations with conflicting interactions: Classification and analytical treatment considering asymmetry and power. Phys. Rev. E **81**, 016112 (2010)
20. L. Samuelson, Chap. 5: The Ultimatum Game, in *Evolutionary Games and Equilibrium Selection* (The MIT Press, Cambridge, 1998)
21. R. Axelrod, An evolutionary approach to norms. Am. Pol. Sci. Rev. **80**(4), 1095–1111 (1986)
22. M. Hechter, K.D. Opp (eds.), particularly Chap. 4 Game-theoretical perspectives on the emergence of social norms, in *Social Norms*, ed. by T. Voss (Russell Sage, New York, 2001), pp. 105–136
23. C. Bicchieri, R. Jeffrey, B. Skyrms (eds.), *The Dynamics of Norms* (Cambridge University, Cambridge, 2009)
24. J. Bendor, P. Swistak, The evolution of norms. Am. J. Sociol. **106**(6), 1493–1545 (2001)
25. F.A.C.C. Chalub, F.C. Santos, J.M. Pacheco, The evolution of norms. J. Theor. Biol. **241**, 233–240 (2006)
26. E. Ostrom, Collective action and the evolution of social norms. J. Econ. Perspect. **14**(3), 137–158 (2000)
27. P.R. Ehrlich, S.A. Levin, The evolution of norms. PLoS Biol. **3**(6), 0943–0948 (2005)
28. K. Keizer, S. Lindenberg, L. Steg, The spreading of disorder. Science **322**, 1681–1685 (2008)
29. B. Skyrms, *The Stag Hunt and the Evolution of Social Structure* (Cambridge University, Cambridge, 2003)

30. K.D. Opp, *Theories of Political Protest and Social Movements* (Routledge, London, 2009)
31. T.S. Kuhn, *The Structure of Scientific Revolutions* (University of Chicago, Chicago, 1962)
32. W. Weidlich, H. Huebner, Dynamics of political opinion formation including catastrophe theory. J. Econ. Behav. Organ. **67**, 1–26 (2008)
33. E. Fehr, U. Fischbacher, S. Gächter, Strong reciprocity, human cooperation, and the enforcement of social norms. Hum. Nat. **13**, 1–25 (2002)
34. A. Whiten, V. Horner, F.B.M. de Waal, Conformity to cultural norms of tool use in chimpanzees. Nature **437**, 737–740 (2005)
35. K. Binmore, *Natural Justice* (Oxford University, New York, 2005)
36. M.A. Nowak, N.L. Komarova, P. Niyogi, Computational and evolutionary aspects of language. Nature **417**, 611–617 (2002)
37. V. Loreto, L. Steels, Emergence of language. Nat. Phys. **3**, 758–760 (2007)
38. R. Boyd, P.J. Richerson, *The Origin and Evolution of Cultures* (Oxford University, Oxford, 2005)

Chapter 11
Social Experiments and Computing

11.1 Introduction

When Nowak and May published their computational study of spatial games in 1992, it soon became a scientific milestone [1]. They showed that altruistic ("cooperative") behavior would be able to survive through spatial clustering. This finding, also called "network reciprocity" [2], is enormously important, as cooperation is the essence that keeps societies together. It is the basis of solidarity and social order. When humans stop cooperating, this implies a war of everybody against everybody.

Understanding why and under what conditions humans cooperate is one of the grand challenges of science [3], particularly in social dilemma situations (where collective cooperation is beneficial, but individual free-riding is even more profitable). How should humans otherwise be able to create public goods (such as a shared culture or a public infrastructure), build up functioning social benefit systems, or fight global warming collectively in the future? From a theoretical point of view, Nowak and May's work demonstrates that the representative agent paradigm of economics (according to which interactions with others can be represented by the interaction with average individuals) can be quite misleading. This paradigm predicts that cooperation should completely disappear in social dilemma situations, leading to a "tragedy of the commons". If the world was really like this, social systems would not work.

---

*This chapter was written together with *Wenjian Yu*. It is an extended version of the following Commentary, which the reader is requested to cite instead: D. Helbing and *W. Yu*, The future of social experimenting. *Proceedings of the National Academy of Sciences USA* **107**(12), 5265–5266 (2010).

D. Helbing (ed.), *Social Self-Organization*, Understanding Complex Systems,
DOI 10.1007/978-3-642-24004-1_11,

However, when the same interactions take place in a spatial setting, they can cause correlations between the behaviors of neighboring individuals, which can dramatically change the outcome of the system (as long as the interactions are local rather than global). The effect is even more pronounced, when a success-driven kind of mobility is considered in the model [4]. Spatio-temporal pattern formation facilitates a co-evolution of the behaviors and the spatial organization of individuals, creating a "social milieu" that can encourage cooperative behavior. In fact, some long-standing puzzles in the social sciences find a natural solution, when spatial interactions (and mobility) are taken into account. This includes the higher-than-expected level of cooperation in social dilemma situations and the spreading of costly punishment (the eventual disappearance of defectors and "second-order free-riders", i.e. cooperators who abstain from the punishment of non-cooperative behaviors).

11.2 Experiment

Despite the importance of these topics, it took quite long until the effects of game-theoretic interactions in two-dimensional space were tested in laboratory experiments. A recent study of Traulsen et al. [5] now reports experiments of a spatial prisoner's dilemma game for the original setting of Nowak and May, while the size of the spatial grid, the number of interaction partners and the payoff parameters were modified for experimental reasons. According to their results, spatial interactions have no significant effect on the level of cooperation. The reason for this is that their experimental subjects did not show an unconditional imitation of neighbors with a higher payoff, as it is assumed in many game-theoretic models.

In fact, it is known that certain game-theoretic results are sensitive to details of the model such as the number of interaction partners, the inclusion of self-interactions or not, or significant levels of randomness (see Figs. 11.1–11.4). Moreover, researchers have proposed a considerable number of different strategy update rules, which matter as well. Besides unconditional imitation, these include the best response rule [6], multi-stage strategies such as tit for tat [7], win-stay-lose-shift rules [8] and aspiration-dependent rules [9], furthermore probabilistic rules such as the proportional imitation rule [10, 11], the Fermi rule [12], and the unconditional imitation rule with a superimposed randomness ("noise") [4]. In addition, there are voter [13] and opinion dynamics models [14] of various kinds, which assume social influence. According to these, individuals would imitate behavioral strategies, which are more frequent in their neighborhood. So, how do individuals really update their behavioral strategies?

Traulsen et al. find that the probability to cooperate increases with the number of cooperative neighbors as expected from the Asch experiment [15]. Moreover, the probability of strategy changes increases with the payoff difference in a way that can be approximated by the Fermi rule [12]. In the case of two behavioral strategies

Fig. 11.1 Snapshot of a computer simulation of the spatial prisoner's dilemma without self-interactions, illustrating the representative dynamics of strategy updating on a 49×49 lattice. Here, we assume an unconditional imitation of the best performing direct neighbor (given his/her payoff was higher). *Blue sites* correspond to cooperative individuals, *red sites* to defecting ones. The payoffs in the underlying prisoner's dilemma were assumed as in the paper by Traulsen et al. [5]. A video illustrating the dynamics of the game is available at http://www.soms.ethz.ch/research/socialexperimenting. It reveals that the level of cooperation decays quickly, and defectors prevail after a short time. Since the simulation assumes no randomness in the strategy updates, the spatial configuration "freezes" quickly, i.e. it does not change anymore after a few iterations

only, it corresponds to the well-known multi-nomial logit model of decision theory [16]. However, there is a discontinuity in the data as the payoff difference turns from positive to negative values, which may be an effect of risk aversion [17]. To describe the time-dependent level of cooperation, it is sufficient to assume unconditional imitation with a certain probability, otherwise strategy mutations. The mutation rate is surprisingly big in the beginning and exponentially decaying over time.

The most surprising fact is maybe not the high level of randomness, which is quite typical for social systems. While one may expect that a large noise level quickly reduces a high level of cooperation, it actually leads to more cooperation than the unconditional imitation rule predicts (see Fig. 2 of [5]). This goes along with a significantly higher average payoff than for the unconditional imitation rule. In other words, the random component of the strategy update is profitable for the experimental subjects. This suggests that noise in social systems may play a functional role.

Fig. 11.2 Snapshot of a computer simulation of the spatial prisoner's dilemma without self-interactions, illustrating the representative dynamics of strategy updating according to (3) of Traulsen et al. [5]. The lattice size, payoff parameters, and color coding are the same as before, but individuals are performing random strategy updates with an exponentially decaying probability, while unconditional imitation occurs only otherwise. Due to the presence of strategy mutations, the spatial configuration keeps changing. Compared to Fig. 11.1, the level of cooperation drops further, since metastable configurations are broken up by strategy mutations ("noise"). A related video is available at http://www.soms.ethz.ch/research/socialexperimenting

11.3 Discussion

Given that Traulsen et al. do not find effects of spatial interactions, do we have to say good bye to network reciprocity in social systems and to all the nice explanations that it offers? Comparing Fig. 11.6 with Fig. 11.5 suggests that this is not the case, since the experimental setting did not promote cooperative clusters for the occurring noise levels (see Fig. 11.2), while a scenario *with* self-interactions would have done so (see Fig. 11.4). Also the empirically confirmed spreading of obesity, smoking, happiness, and cooperation in social networks [18–22] suggests that effects of imitating neighbors (also friends or colleagues) are relevant, but probably over longer time periods than 25 interactions. In fact, in contrast to the scenario without self-interactions (see Fig. 11.5), according to formula (3) of Traulsen et al. one would expect a sudden spreading of cooperation when the mutation rate has decreased to low values (after about 40 iterations), given that self-interactions are taken into account (see Fig. 11.6). To make the effect observable experimentally, it

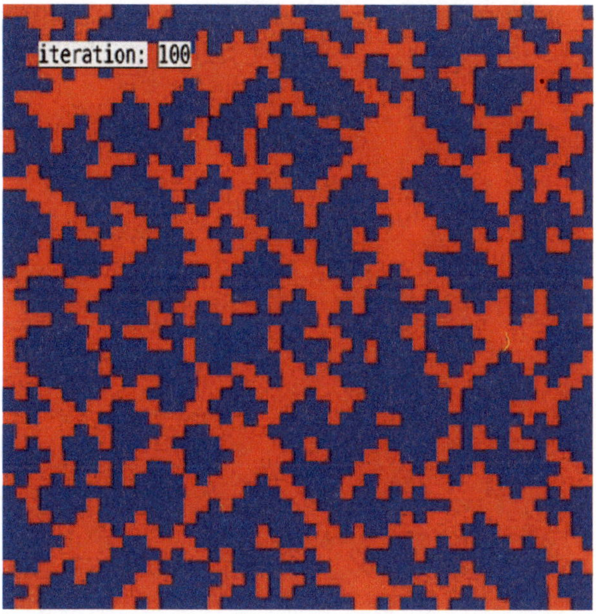

Fig. 11.3 Snapshot of a computer simulation of the spatial prisoner's dilemma assuming unconditional imitation. Compared to Fig. 11.1, we take self-interactions into account, which supports the spreading of cooperators. Since individuals are assumed to imitate unconditionally, there are no strategy mutations. As a consequence, the spatial configuration freezes after a few iterations. A related video is available at http://www.soms.ethz.ch/research/socialexperimenting

would be favorable to reduce the necessary number of iterations for its occurrence and to control the noise level.

The particular value of the work by Traulsen et al. [5] is that it facilitates more realistic computer simulations. Thereby it becomes possible to determine payoff values and other model parameters, which are expected to produce interesting effects (such as spatial correlations) after an experimentally accessible number of iterations. In fact, experimental games can have qualitatively different outcomes, which are hard to predict without extensive computer simulations scanning the parameter space (see [23] for an example and a variety of related "phase diagrams"). Such parameter dependencies could explain some of the apparent inconsistencies between empirical observations in different areas of the world [24] (at least when framing effects such as the expected level of reciprocity and their impact on the effective payoffs [2] are taken into account). The progress in the social sciences by understanding such parameter dependencies would be enormous. However, as the effort to experimentally determine phase diagrams is prohibitive, one can only check computationally predicted, parameter-dependent outcomes by targeted samples. The future of social experimenting lies in the combination of computational and experimental approaches, where computer simulations optimize the experimental

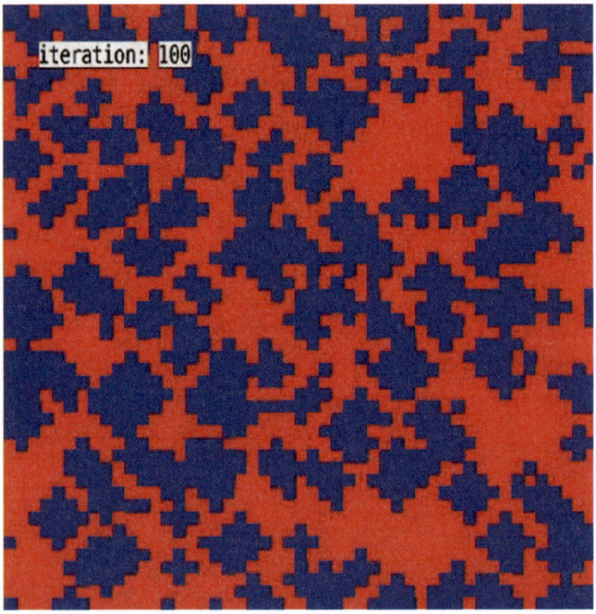

Fig. 11.4 Snapshot of a computer simulation of the spatial prisoner's dilemma with self-interactions, assuming strategy updates according to (3) of Traulsen et al. [5]. Initially, there is a high probability of strategy mutations, but it decreases exponentially. As a consequence, an interesting effect occurs: While the level of cooperation decays in the beginning, it manages to recover later and becomes almost as high as in the noiseless case represented by Fig. 11.3. A related video is available at http://www.soms.ethz.ch/research/socialexperimenting

setting and experiments are used to verify, falsify or improve the underlying model assumptions.

Besides selecting parameter values which maximize the signal-to-noise ratio and minimize the number of iterations after which the expected effect becomes visible, one could try to reduce the level of randomness by experimental noise control. For this, it would be useful to understand the origin and relevance of the observed randomness. Do the experimental subjects make mistakes and why? Do they try to optimize their behavioral strategies or do they apply simple heuristics (and which ones)? Do they use heterogeneous updating rules? Or do they just show exploratory behavior? [25,26] Is it useful to work with subjects who have some experience with behavioral experiments (without having a theoretical background in them)? How relevant is the homogeneity of the subject pool? What are potentials and dangers of framing effects? How can effects of the individual histories of experimental subjects be eliminated? Does it make sense to perform the experiment with a mixture of experimental subjects and computer agents (where the noise level can be reduced by implementing deterministic strategy updates of these agents)?

In view of the great theoretical importance of experiments with many iterations and spatial interactions, more large-scale experiments over long time horizons

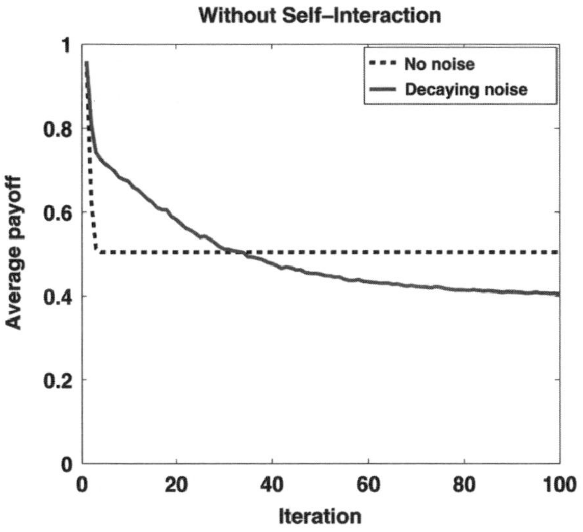

Fig. 11.5 Average payoff of all individuals in the spatial prisoner's dilemma without self-interactions, displayed over the number of iterations. It is clearly visible that the initial payoff drops quickly. In the noiseless case, it becomes constant after a few iterations, as the spatial configuration freezes (see *broken line*). In contrast, in the case of a decaying rate of strategy mutations according to (3) of Traulsen et al. [5], the average payoff keeps changing (see *solid line*). It is interesting that the average payoff is higher in the noisy case than in the noiseless one for approximately 30 iterations, particularly over the time period of the laboratory experiment by Traulsen et al. (covering 25 iterations). The better performance in the presence of strategy mutations could be a possible reason for the high level of strategy mutations observed by them

would be desirable. This calls for larger budgets (as they are common in the natural and engineering sciences), but also for new concepts. Besides connecting labs in different countries via internet, one may consider to perform experiments in "living labs" on the web itself [27]. It also seems worth exploring, how much we can learn from interactive games such as Second Life or Lords of Warcraft [28–30], which could be adapted for experimental purposes in order to create well-controlled environments. According to a recent replication of the Milgram experiment [31] with Avatars [32, 33], experiments with virtual humans may actually be surprisingly well transferable to real humans. One can furthermore hope that lab or web experiments will eventually become standardized measurement instruments to determine indices like the local "level of cooperation" as a function of time, almost like the gross domestic product is measured today. Knowing the "index of cooperativeness" would be good, as it reflects social capital. The same applies to the measurement of social norms, which are equally important for social order as cooperation, since they determine important factors such as coordination, adaptation, assimilation, integration, or conflict.

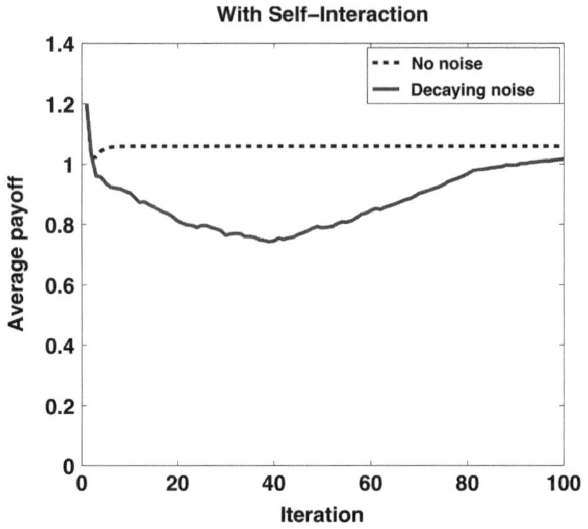

Fig. 11.6 Average payoff of all individuals in the spatial prisoner's dilemma with self-interactions, as a function of the number of iterations. Like in Fig. 11.5, the payoff drops considerably in the beginning. In the noiseless case, it stabilizes quickly (*broken line*). In comparison, when strategy mutations decay according to formula (3) of Traulsen et al. [5], the average payoff keeps decreasing for some time (see *solid line*) and falls significantly below the payoff of the noiseless case. However, after about 40 iterations, the average payoff recovers, which correlates with an increase in the level of cooperation. Due to the pronounced contrast to the case without self-interactions (see Fig. 11.5), it would be interesting to perform experiments *with* self-interactions. These should extend over significantly more than 25 iterations, or the payoff parameters would have to be changed in such a way that the average payoff recovers earlier. It is conceivable, however, that experimental subjects would show a lower level of strategy mutations under conditions where noise does not pay off (in contrast to the studied experimental setting *without* self-interactions represented in Fig. 11.5)

References

1. M.A. Nowak, R.M. May, Evolutionary games and spatial chaos. Nature **359**, 826–829 (1992)
2. M.A. Nowak, Five rules for the evolution of cooperation. Science **314**, 1560–1563, and supplementary information (2006)
3. E. Pennisi, How did cooperative behavior evolve? Science **309**, 93 (2005)
4. D. Helbing, W. Yu, The outbreak of cooperation among success-driven individuals under noisy conditions. Proc. Natl. Acad. Sci. USA **106**(8), 3680–3685 (2009)
5. A. Traulsen et al., Human strategy updating in evolutionary games. Proc. Natl. Acad. Sci. USA **107**, 2962–2966 (2010)
6. A. Matsui, Best response dynamics and socially stable strategies. J. Econ. Theor. **57**, 343–362 (1992)
7. R. Axelrod, *The Evolution of Cooperation* (Basic Books, New York, 1984)
8. M. Nowak, K. Sigmund, A strategy of win-stay, lose-shift that outperforms tit-for-tat in the Prisoner's Dilemma game. Nature **364**, 56–58 (1993)
9. M.W. Macy, A. Flache, Learning dynamics in social dilemmas. Proc. Natl. Acad. Sci. USA **99**(Suppl. 3), 7229–7236 (2002)

10. D. Helbing, in *Economic Evolution and Demographic Change*, ed. by G. Haag, U. Mueller, K.G. Troitzsch (Springer, Berlin, 1992), pp. 330–348
11. K.H. Schlag, Why imitate, and if so, how? A boundedly rational approach to multi- armed bandits. J. Econ. Theor. **78**(1), 130–156 (1998)
12. G. Szabo, C. Toke, Evolutionary prisoner's dilemma game on a square lattice. Phys. Rev. E **58**, 69–73 (1998)
13. I. Dornic, H. Chate, J. Chave, H. Hinrichsen, Critical coarsening without surface tension: The universality class of the voter model. Phys. Rev. Lett. **87**, 045701 (2001)
14. K. Sznajd-Weron, J. Sznajd, Opinion evolution in closed community. Int. J. Mod. Phys. C **11**(6), 1157–1165 (2000)
15. S.E. Asch, Studies of independence and conformity: A minority of one against a unanimous majority. Psychol. Monogr. **70**(9), 1–70 (1956)
16. D. McFadden, in *Frontiers of Econometrics*, ed. by P. Zarembka (Academic Press, New York, 1974), pp. 105–142
17. D. Kahnemann, A. Tversky, Prospect theory: An analysis of decision under risk. Econometrica **47**(2), 263–291 (1979)
18. K.P. Smith, N.A. Christakis, Social networks and health. Ann. Rev. Sociol. **34**, 405–429 (2008)
19. N.A. Christakis, J.H. Fowler, The spread of obesity in a large social network over 32 years. New Engl. J. Med. **357**(4), 370–379 (2007)
20. N.A. Christakis, J.H. Fowler, The collective dynamics of smoking in a large social network. New Engl. J. Med. **358**(21), 2249–2258 (2008)
21. J.H. Fowler, N.A. Christakis, Dynamic spread of happiness in a large social network. Br. Med. J. **337**, a2338 (2008)
22. J.H. Fowler, N.A. Christakis, Cooperative behavior cascades in human social networks. Proc. Natl. Acad. Sci. USA **107**(12), 5334–5338 (2010)
23. D. Helbing, A. Szolnoki, M. Perc, G. Szabo, Evolutionary establishment of moral and double moral standards through spatial interactions. Evolutionary establishment of moral and double moral standards through spatial interactions. PLoS Comput. Biol. **6**(4), e1000758 (2010). Supplementary videos are available at http://www.soms.ethz.ch/research/secondorder-freeriders and http://www.matjazperc.com/games/moral.html
24. B. Herrmann, C. Thoni, S. Gachter, Antisocial punishment across societies. Science **319**, 1362–1367 (2008)
25. D. Helbing, M. Schonhof, H.U. Stark, J.A. Holyst, How individuals learn to take turns. Adv. Complex Syst. **8**, 87–116 (2005)
26. A. Traulsen, C. Hauert, H.D. Silva, M.A. Nowak, K. Sigmund, Exploration dynamics in evolutionary games. Proc. Natl. Acad. Sci. USA **106**, 709–712 (2009)
27. M.J. Salganik, P.S. Dodds, D.J. Watts, Experimental study of inequality and unpredictability in an artificial cultural market. Science **311**, 854–856 (2006)
28. W.S. Bainbridge, The scientific research potential of virtual worlds. Science **317**, 472–476 (2007)
29. N.F. Johnson, C. Xu, Z. Zhao, N. Ducheneaut, N. Yee, G. Tita, P.M. Hui, Human group formation in online guilds and offline gangs driven by a common team dynamic. Phys. Rev. E **79**, 066117 (2009)
30. M. Szell, S. Thurner, Measuring social dynamics in a massive multiplayer online game. Social Networks **32**(4), 313–329 (2010)
31. S. Milgram, Behavioral study of obedience. J. Abnorm. Soc. Psychol. **67**(4), 371–378 (1963)
32. M. Slater, et al., A virtual reprise of the Stanley Milgram obedience experiments. PLoS One **1**, e39 (2006)
33. M. Cheetham, A.F. Pedroni, A. Antley, M. Slater, L. Jancke, Virtual Milgram: Empathic concern or personal distress? Front. Hum. Neurosci. **3**, 29, 1–13 (2009)

Chapter 12
Learning of Coordinated Behavior*

12.1 Introduction

Congestion is a burden of today's traffic systems, affecting the economic prosperity of modern societies. Yet, the optimal distribution of vehicles over alternative routes is still a challenging problem and uses scarce resources (street capacity) in an inefficient way. Route choice is based on interactive, but decentralized individual decisions, which cannot be well described by classical utility-based decision models [27]. Similar to the minority game [16, 39, 43], it is reasonable for different people to react to the same situation or information in *different* ways. As a consequence, individuals tend to develop characteristic response patterns or roles [26]. Thanks to this differentiation process, individuals learn to coordinate better in the course of time. However, according to current knowledge, selfish routing does not establish the system optimum of minimum overall travel times. It rather tends to establish the Wardrop equilibrium, a special user or Nash equilibrium characterized by equal travel times on all alternative routes chosen from a certain origin to a given destination (while routes with longer travel times are not taken) [71].

Since Pigou [53], it has been suggested to resolve the problem of inefficient road usage by congestion charges, but are they needed? Is the missing establishment of a system optimum just a problem of varying traffic conditions and changing origin-destination pairs, which make route-choice decisions comparable to one-shot games? Or would individuals in an *iterated* setting of a day-to-day route choice game with identical conditions spontaneously establish cooperation in order to increase their returns, as the folk theorem suggests [6]?

How would such a cooperation look like? Taking turns could be a suitable solution [62]. While simple symmetrical cooperation is typically found for the

---

*This chapter reprints a previous publication with kind permission of the copyright owner, World Scientific. It is requested to cite this work as follows: D. Helbing, *M. Schönhof, H.-U. Stark,* and *J. A. Holyst*, How individuals learn to take turns: Emergence of alternating cooperation in a congestion game and the prisoner's dilemma. *Advances in Complex Systems* **8**, 87–116 (2005).

D. Helbing (ed.), *Social Self-Organization*, Understanding Complex Systems,
DOI 10.1007/978-3-642-24004-1_12, © Springer-Verlag Berlin Heidelberg 2012

repeated Prisoner's Dilemma [2, 3, 44–46, 49, 52, 55, 59, 64, 67, 69], emergent alternating reciprocity has been recently discovered for the games Leader and Battle of the Sexes [11].[1] Note that such coherent oscillations are a time-dependent, but deterministic form of individual decision behavior, which can establish a persistent phase-coordination, while mixed strategies, i.e. statistically varying decisions, can establish cooperation only by chance or in the statistical average. This difference is particularly important when the number of interacting persons is small, as in the particular route choice game discussed below.

Note that oscillatory behavior has been found in iterated games before:

- In the rock-paper-scissors game [67], cycles are predicted by the game-dynamical equations due to unstable stationary solutions [28].
- Oscillations can also result by coordination problems [1,29,31,33], at the cost of reduced system performance.
- Moreover, blinker strategies may survive in repeated games played by a mixture of finite automata [5] or result through evolutionary strategies [11, 15, 16, 38, 39, 42, 43, 74].

However, these oscillation-generating mechanisms are clearly to be distinguished from the establishment of phase-coordinated alternating reciprocity we are interested in (coherent oscillatory cooperation to reach the system optimum).

Our paper is organized as follows: In Sect. 12.2, we will formally introduce the route choice game for N players, including issues like the Wardrop equilibrium [71] and the Braess paradox [10]. Section 12.3 will focus on the special case of the 2-person route choice game, compare it with the minority game [1, 15, 16, 38, 39, 42, 43, 74], and discuss its place in the classification scheme of symmetrical 2×2 games. This section will also reveal some apparent shortcomings of the previous game-theoretical literature:

- While it is commonly stated that among the 12 ordinally distinct, symmetrical 2×2 games [11, 57] only four archetypical 2×2 games describe a strategical conflict (the Prisoner's Dilemma, the Battle of the Sexes, Chicken, and Leader) [11,18,56], we will show that, for specific payoffs, the route choice game (besides Deadlock) also represents an interesting strategical conflict, at least for iterated games.
- The conclusion that conservative driver behavior is best, i.e. it does not pay off to change routes [7,65,66], is restricted to the special case of route-choice games with a system-optimal user equilibrium.
- It is only half the truth that cooperation in the iterated Prisoner's Dilemma is characterized by symmetrical behavior [11]. Phase-coordinated asymmetric reciprocity is possible as well, as in some other symmetrical 2×2 games [11].

New perspectives arise by less restricted specifications of the payoff values.

[1] See Fig. 12.2 for a specification of these games.

In Sect. 12.4, we will discuss empirical results of laboratory experiments with humans [12, 18, 32]. According to these, reaching a phase-coordinated alternating state is only one problem. Exploratory behavior and suitable punishment strategies are important to establish asymmetric oscillatory reciprocity as well [11, 20]. Moreover, we will discuss several coefficients characterizing individual behavior and chances for the establishment of cooperation. In Sect. 12.5, we will present multi-agent computer simulations of our observations, based on a novel win-stay, lose-shift [50, 54] strategy, which is a special kind of reinforcement learning strategy [40]. This approach is based on individual historical experience [13] and, thereby, clearly differs from the selection of the best-performing strategy in a set of hypothetical strategies as assumed in studies based on evolutionary or genetical algorithms [5, 11, 15, 16, 39, 42, 43]. The final section will summarize our results and discuss their relevance for game theory and possible applications such as data routing algorithms [35, 72], advanced driver information systems [8, 14, 30, 37, 41, 63, 70, 73], or road pricing [53].

12.2 The Route Choice Game

In the following, we will investigate a scenario with two alternative routes between a certain origin and a given destination, say, between two places or towns A and B (see Fig. 12.1). We are interested in the case where both routes have different capacities, say a freeway and a subordinate or side road. While the freeway is faster when it is empty, it may be reasonable to use the side road when the freeway is congested.

The "success" of taking route i could be measured in terms of its inverse travel time $1/T_i(N_i) = V_i(N_i)/L_i$, where L_i is the length of route i and $V_i(N_i)$ the average velocity when N_i of the N drivers have selected route i. One may roughly approximate the average vehicle speed V_i on route i by the linear relationship [24]

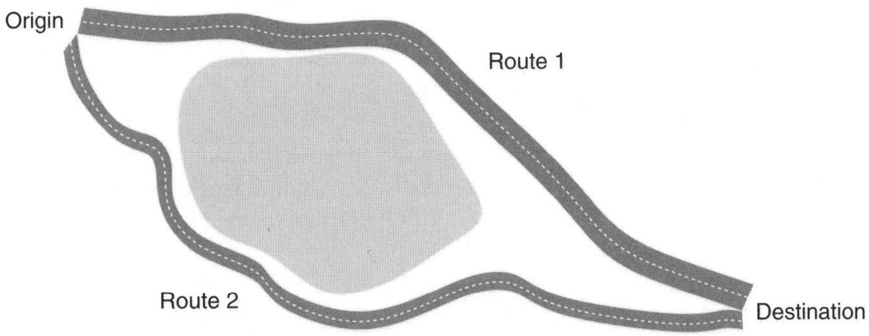

Fig. 12.1 Illustration of the investigated day-to-day route choice scenario. We study the dynamic decision behavior in a repeated route choice game, where a given destination can be reached from a given origin via two different routes, a freeway (route 1) and a side road (route 2)

$$V_i(N_i) = V_i^0 \left(1 - \frac{N_i(t)}{N_i^{\max}}\right), \tag{12.1}$$

where V_i^0 denotes the maximum velocity (speed limit) and $N_i^{\max}$ the capacity, i.e. the maximum possible number of vehicles on route i. With $A_i = V_i^0/L_i$ and $B_i = V_i^0/(N_i^{\max} L_i)$, the inverse travel time then obeys the relationship

$$1/T(N_i) = A_i - B_i N_i, \tag{12.2}$$

which is linearly decreasing with the road occupancy N_i. Other monotonously falling relationships $V_i(N_i)$ would make the expression for the inverse travel times non-linear, but they would probably not lead to qualitatively different conclusions.

The user equilibrium of equal travel times is found for a fraction

$$\frac{N_1^{\mathrm{e}}}{N} = \frac{B_2}{B_1 + B_2} + \frac{1}{N}\frac{A_1 - A_2}{B_1 + B_2} \tag{12.3}$$

of persons choosing route 1. In contrast, the system optimum corresponds to the maximum of the overall inverse travel times $N_1/T_1(N_1) + N_2/T_2(N_2)$ and is found for the fraction

$$\frac{N_1^{\mathrm{o}}}{N} = \frac{B_2}{B_1 + B_2} + \frac{1}{2N}\frac{A_1 - A_2}{B_1 + B_2} \tag{12.4}$$

of 1-decisions. The difference between both fractions vanishes in the limit $N \to \infty$. Therefore, only experiments with a few players allow to find out, whether the test persons adapt to the user equilibrium or to the system optimum. We will see that both cases have completely different dynamical implications: While the most successful strategy to establish the user equilibrium is to stick to the same decision in subsequent iterations [27, 65, 66], the system optimum can only be reached by a time-dependent strategy (at least, if no participant is ready to pay for the profits of others).

Note that alternative routes can reach comparable travel times only when the total number N of vehicles is large enough to fulfil the relationships $P_1(N) < P_2(0) = A_2$ and $P_2(N) < P_1(0) = A_1$. Our route choice game will address this traffic regime and additionally assume $N \leq N_i^{\max}$. The case $N_i = N_i^{\max}$ corresponds to a complete gridlock on route i.

Finally, it may be interesting to connect the previous quantities with the vehicle densities ρ_i and the traffic flows Q_i: If route i consists of I_i lanes, the relation with the average vehicle density is $\rho_i(N_i) = N_i/(I_i L_i)$, and the relation with the traffic flow is $Q_i(N_i) = \rho_i V_i(N_i) = N_i/[I_i T_i(N_i)]$.

In the following, we will linearly transform the inverse travel time $1/T_i(N_i)$ in order to define the so-called payoff

$$P_i(N_i) = C_i - D_i N_i \tag{12.5}$$

for choosing route i. The payoff parameters C_i and D_i depend on the parameters A_i, B_i, and N, but will be taken constant. We have scaled the parameters so that we have the payoff $P_i(N_i^e) = 0$ (zero payoff points) in the user equilibrium and the payoff $N_1 P_1(N_1^o) + N_2 P_2(N - N_1^o) = 100N$ (an average of 100 payoff points) in the system optimum. This serves to reach generalizable results and to provide a better orientation to the test persons.

Note that the investigation of social (multi-person) games with linearly falling payoffs is not new [33]. For example, Schelling [62] has discussed situations with "conditional externality", where the outcome of a decision depends on the independent decisions of potentially many others [62]. Pigou has addressed this problem, which has been recently focused on by Schreckenberg and Selten's project SURVIVE [7,65,66] and others [8,41,58].

The route choice game is a special congestion game [22,47,60]. More precisely speaking, it is a multi-stage symmetrical N-person single commodity congestion game [68]. Congestion games belong to the class of "potential games" [48], for which many theorems are available. For example, it is known that there always exists a Wardrop equilibrium [71] with essentially unique Nash flows [4]. This is characterized by the property that no individual driver can decrease his or her travel time by a different route choice. If there are several alternative routes from a given origin to a given destination, the travel times on all used alternative routes in the Wardrop equilibrium is the same, while roads with longer travel times are not used. However, the Wardrop equilibrium as expected outcome of selfish routing does not generally reach the system optimum, i.e. minimize the total travel times. Nash flows are often inefficient, and selfish behavior implies the possibility of decreased network performance.[2] This is particularly pronounced for the Braess paradox [10,61], according to which additional streets may sometimes increase the overall travel time and reduce the throughput of a road network. The reason for this is the possible existence of badly performing Nash equilibria, in which no single person can improve his or her payoff by changing the decision behavior.

In fact, recent laboratory experiments indicate that, in a "day-to-day route choice scenario" based on selfish routing, the distribution of individuals over the alternative routes is fluctuating around the Wardrop equilibrium [27,63]. Additional conclusions from the laboratory experiments by Schreckenberg, Selten et al. are as follows [65,66]:

- Most people, who change their decision frequently, respond to their experience on the previous day (i.e. in the last iteration).
- There are only a few different behavioral patterns: direct responders (44%), contrarian responders (14%), and conservative persons, who do not respond to the previous outcome.

[2]For more details see the work by T. Roughgarden.

- It does not pay off to react to travel time information in a sensitive way, as conservative test persons reach the smallest travel times (the largest payoffs) on average.
- People's reactions to short term travel forecasts can invalidate these. Nevertheless, travel time information helps to match the Wardrop equilibrium, so that excess travel times due to coordination problems are reduced.

A closer experimental analysis based on longer time series (i.e. more iterations) for smaller groups of test persons reveals a more detailed picture [26]:

- Individuals do not only show an adaptive behavior to the travel times on the previous day, but also change their response pattern in time [26, 34].
- In the course of time, one finds a differentiation process which leads to the development of characteristic, individual response patterns, which tend to be almost deterministic (in contrast to mixed strategies).
- While some test persons respond to small differences in travel times, others only react to medium-sized deviations, further people respond to large deviations, etc. In this way, overreactions of the group to deviations from the Wardrop equilibrium are considerably reduced.

Note that the differentiation of individual behaviors is a way to resolve the coordination problem to match the Wardrop equilibrium exactly, i.e. which participant should change his or her decision in the next iteration in order to compensate for a deviation from it. This implies that the fractions of specific behavioral response patterns should depend on the parameters of the payoff function. A certain fraction of "stayers", who do not respond to travel time information, can improve the coordination in the group, i.e. the overall performance. However, stayers can also prevent the establishment of a system optimum, if alternating reciprocity is needed, see (12.14).

12.3 Classification of Symmetrical 2 × 2 Games

In contrast to previous laboratory experiments, we have studied the route choice game not only with a very high number of repetitions, but also with a small number $N \in \{2, 4\}$ of test persons, in order to see whether the system optimum or the Wardrop equilibrium is established. Therefore, let us shortly discuss how the 2-person game relates to previous game-theoretical studies.

Iterated symmetrical two-person games have been intensively studied [12, 18], including Stag Hunt, the Battle of the Sexes, or the Chicken Game (see Fig. 12.2). They can all be represented by a payoff matrix of the form $\mathbf{P} = (P_{ij})$, where P_{ij} is the success ("payoff") of person 1 in a one-shot game when choosing strategy $i \in \{1, 2\}$ and meeting strategy $j \in \{1, 2\}$. The respective payoffs of the second person are given by the symmetrical values P_{ji}. Figure 12.2 shows a systematics of the previously mentioned and other kinds of symmetrical two-person games [21].

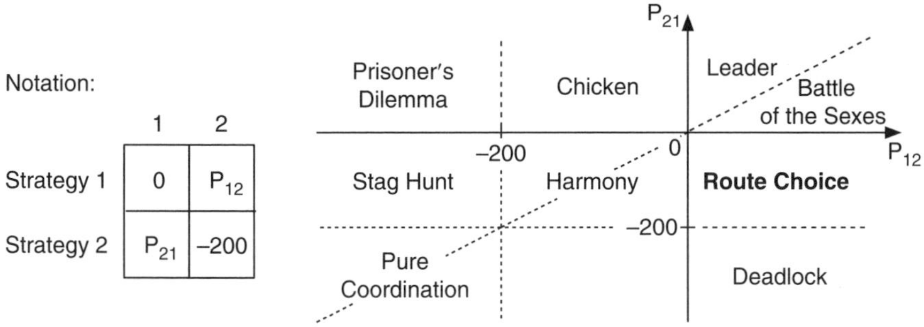

Fig. 12.2 Classification of symmetrical 2 × 2 games according to their payoffs P_{ij}. Two payoff values have been kept constant as payoffs may be linearly transformed and the two strategies of the one-shot game renumbered. Our choice of $P_{11} = 0$ and $P_{22} = -200$ was made to define a payoff of 0 points in the user equilibrium and an average payoff of 100 in the system optimum of our investigated route choice game with $P_{12} = 300$ and $P_{21} = -100$

a

	1	2
Strategy 1	0	P_{12}
Strategy 2	P_{21}	–200

b

	Coop.	Def.
Cooperation	0	–300
Defection	100	–200

c

	Coop.	Def.
Cooperation	0	–300
Defection	500	–200

d

	1	2
Route 1	0	300
Route 2	–100	–200

Fig. 12.3 Payoff specifications of the symmetrical 2 × 2 games investigated in this paper. (**a**) General payoff matrix underlying the classification scheme of Fig. 12.2. (**b**), (**c**) Two variants of the Prisoner's Dilemma. (**d**) Route choice game with a strategical conflict between the user equilibrium and the system optimum

The relations

$$P_{21} > P_{11} > P_{22} > P_{12}, \tag{12.6}$$

for example, define a Prisoner's Dilemma. In this paper, however, we will mainly focus on the 2-person route choice game defined by the conditions

$$P_{12} > P_{11} > P_{21} > P_{22} \tag{12.7}$$

(see Fig. 12.3). Despite some common properties, this game differs from the minority game [16, 39, 43] or El Farol bar problem [1] with $P_{12}, P_{21} > P_{11}, P_{22}$, as a minority decision for alternative 2 is less profitable than a majority decision for alternative 1. Although oscillatory behavior has been found in the minority game as well [9, 15, 16, 36, 43], an interesting feature of the route choice experiments discussed in the following is the regularity and phase-coordination (coherence) of the oscillations.

The 2-person route choice game fits well into the classification scheme of symmetrical 2 × 2 games. In Rapoport and Guyer's taxonomy of 2 × 2 games [57], the 2-person route choice game appears on page 211 as game number 7 together with four other games with strongly stable equilibria. Since then, the game has almost been forgotten and did not have a commonly known interpretation or name.

Therefore, we suggest to name it the 2-person "route choice game". Its place in the extended Eriksson-Lindgren scheme of symmetrical 2×2 games is graphically illustrated in Fig. 12.2.

According to the game-theoretical literature, there are 12 ordinally distinct, symmetric 2×2 games [57], but after excluding strategically trivial games in the sense of having equilibrium points that are uniquely Pareto-efficient, there remain four archetypical 2×2 games: the Prisoner's Dilemma, the Battle of the Sexes, Chicken (Hawk-Dove), and Leader [56]. However, this conclusion is only correct, if the four payoff values P_{ij} are specified by the four values $\{1, 2, 3, 4\}$. Taking different values would lead to a different conclusion: If we name subscripts so that $P_{11} > P_{22}$, a strategical conflict between a user equilibrium and the system optimum results when

$$P_{12} + P_{21} > 2P_{11}. \tag{12.8}$$

Our conjecture is that players tend to develop alternating forms of reciprocity if this condition is fulfilled, while symmetric reciprocity is found otherwise. This has the following implications (see Fig. 12.2):

- If the 2×2 games Stag Hunt, Harmony, or Pure Coordination are repeated frequently enough, we expect always a symmetrical form of cooperation.
- For Leader and the Battle of the Sexes, we expect the establishment of asymmetric reciprocity, as has been found by Browning and Colman with a computer simulation based on a genetic algorithm incorporating mutation and crossing-over [11].
- For the games Route Choice, Deadlock, Chicken, and Prisoner's Dilemma both, symmetric (simultaneous) and asymmetric (alternating) forms of cooperation are possible, depending on whether condition (12.8) is fulfilled or not. Note that this condition cannot be met for some games, if one restricts to ordinal payoff values $P_{ij} \in \{1, 2, 3, 4\}$ only. Therefore, this interesting problem has been largely neglected in the past (with a few exceptions, e.g. [51]). In particular, convincing experimental evidence of alternating reciprocity is missing. The following sections of this paper will, therefore, not only propose a simulation model, but also focus on an experimental study of this problem, which promises interesting new results.

12.4 Experimental Results

Altogether we have carried out more than 80 route choice experiments with different experimental setups, all with different participants. In the 24 two-person [12 four-person] experiments evaluated here (see Figs. 12.4–12.15), test persons were instructed to choose between two possible routes between the same origin and destination. They knew that route 1 corresponds to a "freeway" (which may be fast or congested), while route 2 represents an alternative route (a "side road"). Test persons were also informed that, if two [three] participants would choose

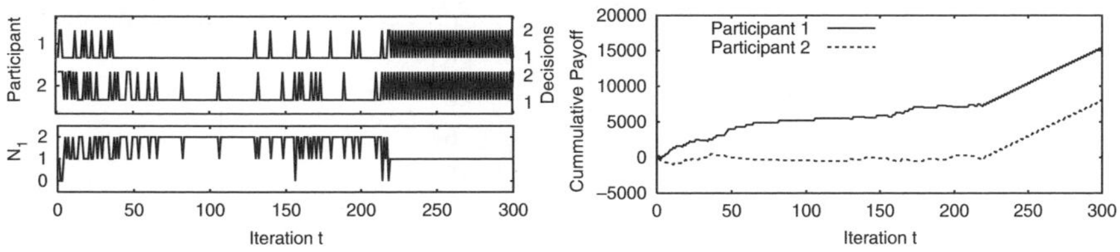

Fig. 12.4 Representative example for the emergence of coherent oscillations in a 2-person route choice experiment with the parameters specified in Fig. 12.3d. *Top left:* Decisions of both participants over 300 iterations. *Bottom left:* Number $N_1(t)$ of 1-decisions over time t. Note that $N_1 = 1$ corresponds to the system optimum, while $N_1 = 2$ corresponds to the user equilibrium of the one-shot game. *Right:* Cumulative payoff of both players in the course of time t (i.e. as a function of the number of iterations). Once the coherent oscillatory cooperation is established ($t > 220$), both individuals have high payoff gains on average

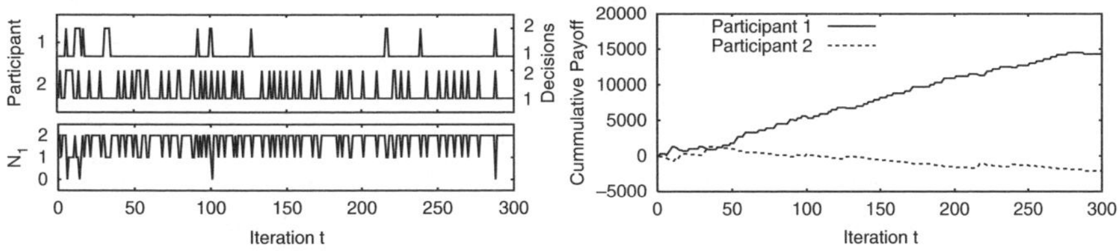

Fig. 12.5 Representative example for a 2-person route choice experiment, in which no alternating cooperation was established. Due to the small changing frequency of participant 1, there were not enough cooperative episodes that could have initiated coherent oscillations. *Top left:* Decisions of both participants over 300 iterations. *Bottom left:* Number $N_1(t)$ of 1-decisions over time t. *Right:* The cumulative payoff of both players in the course of time t shows that the individual with the smaller changing frequency has higher profits

route 1, everyone would receive 0 points, while if half of the participants would choose route 1, they would receive the maximum average amount of 100 points, but 1-choosers would profit at the cost of 2-choosers. Finally, participants were told that everyone could reach an average of 100 points per round with variable, situation-dependent decisions, and that the (additional) individual payment after the experiment would depend on their cumulative payoff points reached in at least 300 rounds (100 points = 0.01 EUR).

Let us first focus on the two-person route-choice game with the payoffs $P_{11} = P_1(2) = 0$, $P_{12} = P_1(1) = 300$, $P_{21} = P_2(1) = -100$, and $P_{22} = P_2(2) = -200$ (see Fig. 12.3d), corresponding to $C_1 = 600$, $D_1 = 300$, $C_2 = 0$, and $D_2 = 100$. For this choice of parameters, the best individual payoff in each iteration is obtained by choosing route 1 (the "freeway") and have the co-player(s) choose route 2. Choosing route 1 is the dominant strategy of the one-shot game, and players are tempted to use it. This produces an initial tendency towards the "strongly stable" user equilibrium [57] with 0 points for everyone. However, this decision behavior is not Pareto efficient in the repeated game. Therefore, after many iterations, the players often learn to establish the Pareto optimum of the multi-stage supergame by

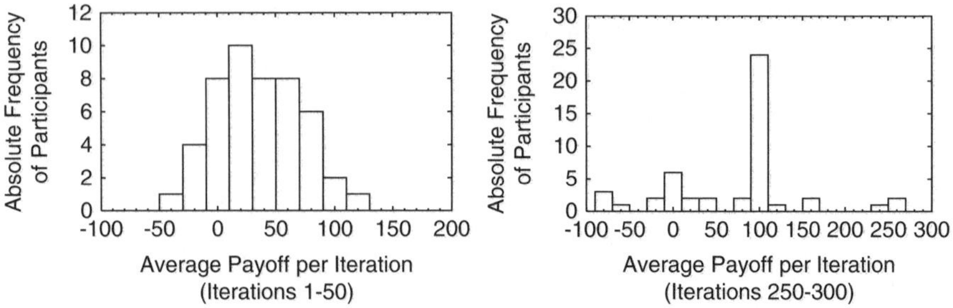

Fig. 12.6 Frequency distributions of the average payoffs of the 48 players participating in our 24 two-person route choice experiments. *Left:* Distribution during the first 50 iterations. *Right:* Distribution between iterations 250 and 300. The initial distribution with a maximum close to 0 points (*left*) indicates a tendency towards the user equilibrium corresponding to the dominant strategy of the one-shot game. However, after many iterations, many individuals learn to establish the system optimum with a payoff of 100 points (*right*)

selecting route 1 in turns (see Fig. 12.4). As a consequence, the experimental payoff distribution shows a maximum close to 0 points in the beginning and a peak at 100 points after many iterations (see Fig. 12.6), which clearly confirms that the choice behavior of test persons tends to change over time. Nevertheless, in 7 out of 24 two-person experiments, persistent cooperation did not emerge during the experiment. Later on, we will identify reasons for this.

12.4.1 Emergence of Cooperation[3] *and Punishment*

In order to reach the system optimum of $(-100+300)/2 = 100$ points per iteration, one individual has to leave the freeway for one iteration, which yields a reduced payoff of –100 in favour of a high payoff of +300 for the other individual. To be profitable also for the first individual, the other one should reciprocate this "offer" by switching to route 2, while the first individual returns to route 1. Establishing this oscillatory cooperative behavior yields 100 extra points on average. If the other individual is not cooperative, both will be back to the user equilibrium of 0 points only, and the uncooperative individual has temporarily profited from the offer by the other individual. This makes "offers" for cooperation and, therefore, the establishment of the system optimum unlikely.

Hence, the innovation of oscillatory behavior requires intentional or random changes ("trial-and-error behavior"). Moreover, the consideration of multi-period decisions is helpful. Instead of just 2 one-stage (i.e. one-period) alternative decisions 1 and 2, there are 2^n different n-stage (n-period) decisions. Such multi-

[3]The term cooperation is used here, because a coordination in time is only part of the problem. Individuals are also facing a dilemma situation, in which selfish behavior tends to prevent cooperation and temporal coordination.

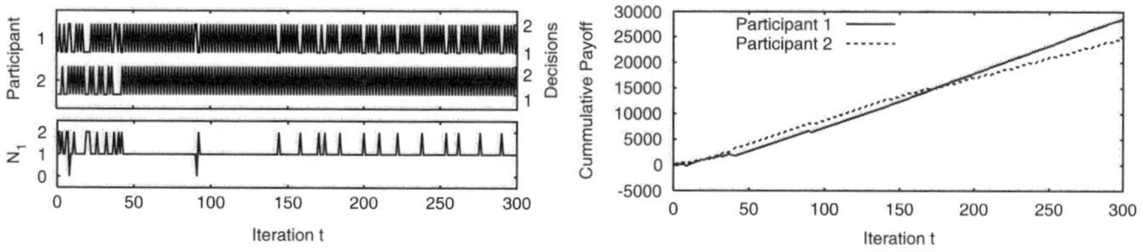

Fig. 12.7 Representative example for a 2-person route choice experiment, in which participant 1 leaves the pattern of oscillatory cooperation temporarily in order to make additional profits. Note that participant 2 does not "punish" this selfish behavior, but continues to take routes in an alternating way. *Top left:* Decisions of both participants over 300 iterations. *Bottom left:* Number $N_1(t)$ of 1-decisions over time t. *Right:* Cumulative payoff of both players as a function of the number of iterations. The different slopes indicate an unfair outcome despite of high average payoffs of both players

	1	2
Route 1	0	300
Route 2	–100	–200

	11	12	21	22
11	0	300	300	600
12	–100	–200	200	100
21	–100	200	–200	100
22	–200	–300	–300	–400

Fig. 12.8 Illustration of the concept of higher-order games defined by n-stage strategies. *Left:* Payoff matrix $\mathbf{P} = (P_{ij})$ of the one-shot 2×2 route choice game. *Right:* Payoff matrix $(P^{(2)}_{(i_1 i_2),(j_1 j_2)}) = (P_{i_1 j_1} + P_{i_2 j_2})$ of the 2nd-order route choice game defined by 2-stage decisions (*right*). The analysis of the one-shot game (*left*) predicts that the user equilibrium (with both persons choosing route 1) will establish and that no single player could increase the payoff by another decision. For two-period decisions (*right*), the system optimum (strategy 12 meeting strategy 21) corresponds to a fair solution, but one person can increase the payoff at the cost of the other (see *arrow* 1), if the game is repeated. A change of the other person's decision can reduce losses and punish this egoistic behavior (*arrow* 2), which is likely to establish the user equilibrium with payoff 0. In order to leave this state again in favour of the system optimum, one person will have to make an "offer" at the cost of a reduced payoff (*arrow* 3). This offer may be due to a random or intentional change of decision. If the other person reciprocates the offer (*arrow* 4), the system optimum is established again. The time-averaged payoff of this cycle lies below the system optimum

stage strategies can be used to define higher-order games and particular kinds of supergame strategies. In the two-person 2nd-order route choice game, for example, an encounter of the two-stage decision 12 with 21 establishes the system optimum and yields equal payoffs for everyone (see Fig. 12.8). Such an optimal and fair solution is not possible for one-stage decisions. Yet, the encounter of 12 with 21 ("cooperative episode") is not a Nash equilibrium of the two-stage

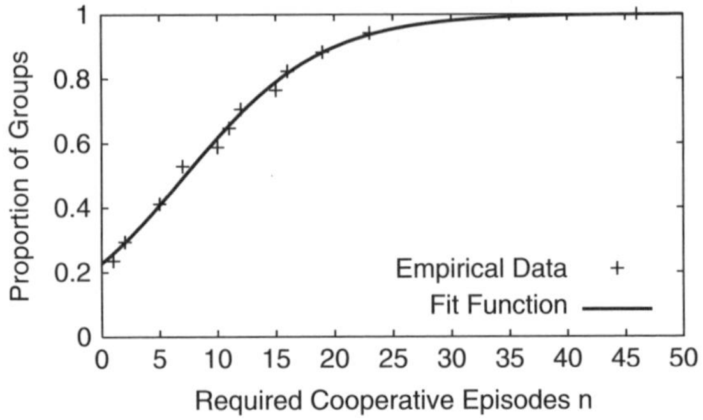

Fig. 12.9 Cumulative distribution of required cooperative episodes until persistent cooperation was established, given that cooperation occurred during the duration of the game as in 17 out of 24 two-person experiments. The experimental data are well approximated by the logistic curve (12.9) with the fit parameters $c_2 = 3.4$ and $d_2 = 0.17$

game, as an individual can increase his or her own payoff by selecting 11 (see Fig. 12.8). Probably for this reason, the first cooperative episodes in a repeated route choice game (i.e. encounters of 12-decisions with 21-decisions in two subsequent iterations) do often not persist (see Fig. 12.9). Another possible reason is that cooperative episodes may be overlooked. This problem, however, can be reduced by a feedback signal that indicates when the system optimum has been reached. For example, we have experimented with a green background color. In this setup, a cooperative episode could be recognized by a green background that appeared in two successive iterations together with two different payoff values.

The strategy of taking route 1 does not only dominate on the first day (in the first iteration). Even if a cooperative oscillatory behavior has been established, there is a temptation to leave this state, i.e. to choose route 1 several times, as this yields more than 100 points on average for the uncooperative individual at the cost of the participant continuing an alternating choice behavior (see Figs. 12.7 and 12.8). That is, the conditional changing probability $p_l(2|1, N_1 = 1; t)$ of individuals l from route 1 to route 2, when the system optimum in the previous iteration was established (i.e. $N_1 = 1$) tends to be small initially. However, oscillatory cooperation of period 2 needs $p_l(2|1, N_1 = 1; t) = 1$. The required transition in the decision behavior can actually be observed in our experimental data (see Fig. 12.10, left). With this transition, the average frequency of 1-decisions goes down to 1/2 (see Fig. 12.10, right). Note, however, that alternating reciprocity does not necessarily require oscillations of period 2. Longer periods are possible as well (see Fig. 12.11), but have occurred only in a few cases (namely, 3 out of 24 cases).

How does the transition to oscillatory cooperation come about? The establishment of alternating reciprocity can be supported by a suitable punishment strategy: If the other player should have selected route 2, but has chosen route 1 instead, he or she can be punished by changing to route 1 as well, since this causes an average payoff of less than 100 points for the other person (see Fig. 12.8). Repeated punishment of uncooperative behavior can, therefore, reinforce

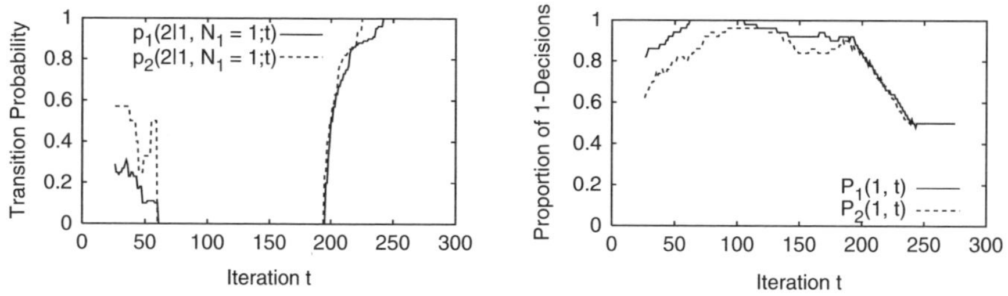

Fig. 12.10 *Left:* Conditional changing probability $p_l(2|1, N_1 = 1; t)$ of person l from route 1 (the "freeway") to route 2, when the other person has chosen route 2, averaged over a time window of 50 iterations. The transition from initially small values to 1 (for $t > 240$) is characteristic and illustrates the learning of cooperative behavior. In this particular group (cf. Fig. 12.4) the values started even at zero, after a transient time period of $t < 60$. *Right:* Proportion $P_l(1, t)$ of 1-decisions of both participants l in the two-person route choice experiment displayed in Fig. 12.4. While the initial proportion is often close to 1 (the user equilibrium), it reaches the value 1/2 when persistent oscillatory cooperation (the system optimum) is established

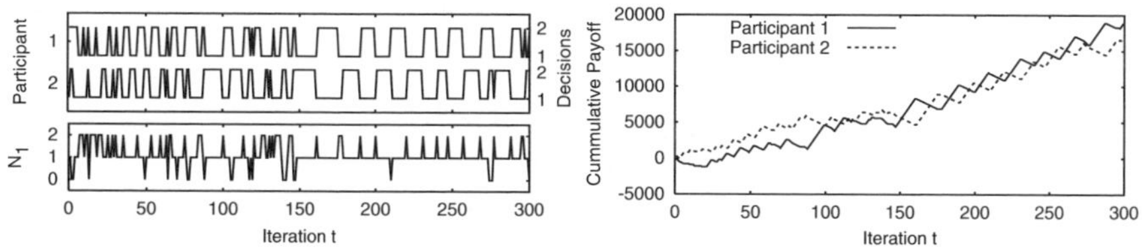

Fig. 12.11 Representative example for a 2-person route choice experiment with phase-coordinated oscillations of long (and varying) time periods larger than 2. *Top left:* Decisions of both participants over 300 iterations. *Bottom left:* Number $N_1(t)$ of 1-decisions over time t. *Right:* Cumulative payoff of both players as a function of the number of iterations. The sawtooth-like increase in the cumulative payoff indicates gains by phase-coordinated alternations with long oscillation periods

cooperative oscillatory behavior. However, the establishment of oscillations also requires costly "offers" by switching to route 2, which only pay back in case of alternating reciprocity. It does not matter whether these "offers" are intentional or due to exploratory trial-and-error behavior.

Due to punishment strategies and similar reasons, persistent cooperation is often established after a number n of cooperative episodes. In the 17 of our 24 two-person experiments, in which persistent cooperation was established, the cumulative distribution of required cooperative episodes could be mathematically described by the logistic curve

$$F_N(n) = 1/[1 + c_N \exp(-d_N n)] \qquad (12.9)$$

(see Fig. 12.9). Note that, while we expect that this relationship is generally valid, the fit parameters c_N and d_N may depend on factors like the distribution of participant intelligence, as oscillatory behavior is apparently difficult to establish (see below).

12.4.2 Preconditions for Cooperation

Let us focus on the time period before persistent oscillatory cooperation is established and denote the occurrence probability that individual l chooses alternative $i \in \{1, 2\}$ by $P_l(i)$. The quantity $p_l(j|i)$ shall represent the conditional probability of choosing j in the next iteration, if i was chosen by person l in the present one. Assuming stationarity for reasons of simplicity, we expect the relationship

$$p_l(2|1)P_l(1) = p_l(1|2)P_l(2), \quad (12.10)$$

i.e. the (unconditional) occurrence probability $P_l(1,2) = p_l(2|1)P_l(1)$ of having alternative 1 in one iteration and 2 in the next agrees with the joint occurrence probability $P_l(2,1) = p_l(1|2)P_l(2)$ of finding the opposite sequence 21 of decisions:

$$P_l(1,2) = P_l(2,1). \quad (12.11)$$

Moreover, if r_l denotes the average changing frequency of person l until persistent cooperation is established, we have the relation

$$r_l = P_l(1,2) + P_l(2,1). \quad (12.12)$$

Therefore, the probability that all N players simultaneously change their decision from one iteration to the next is $\prod_{l=1}^{N} r_l$. Note that there are 2^N such realizations of N decision changes 12 or 21, which have all the same occurrence probability because of (12.11). Among these, only the ones where $N/2$ players change from 1 to 2 and the other $N/2$ participants change from 2 to 1 establish cooperative episodes, given that the system optimum corresponds to an equal distribution over both alternatives. Considering that the number of different possibilities of selecting $N/2$ out of N persons is given by the binomial coefficient, the occurrence probability of cooperative events is

$$P_c = \frac{1}{2^N} \binom{N}{N/2} \prod_{l=1}^{N} r_l \quad (12.13)$$

(at least in the ensemble average). Since the expected time period T until the cooperative state incidentally occurs equals the inverse of P_c, we finally find the formula

$$T = \frac{1}{P_c} = 2^N \frac{(N/2)!^2}{N!} \prod_{l=1}^{N} \frac{1}{r_l}. \quad (12.14)$$

This formula is well confirmed by our 2-person experiments (see Fig. 12.12). It gives the lower bound for the expected value of the minimum number of required iterations until persistent cooperation can spontaneously emerge (if already the first cooperative episode is continued forever).

Obviously, the occurrence of oscillatory cooperation is expected to take much longer for a large number N of participants. This tendency is confirmed by our 4-person experiments compared to our 2-person experiments. It is also in agreement

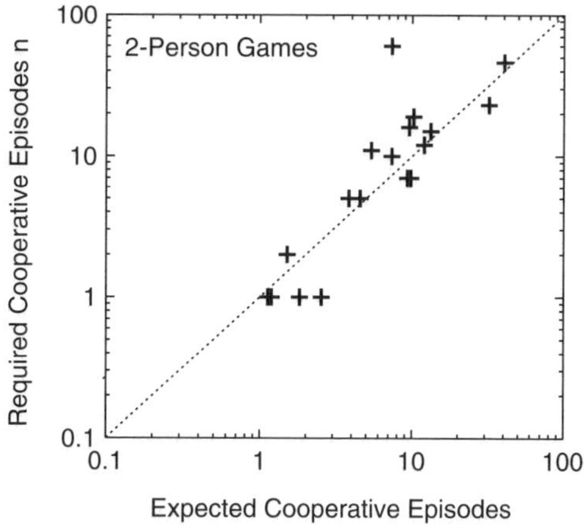

Fig. 12.12 Comparison of the required number of cooperative episodes y with the expected number x of cooperative episodes (approximated as occurrence time of persistent cooperation, divided by the expected time interval T until a cooperative episode occurs by chance). Note that the data points support the relationship $y = x$ and, thereby, formula (12.14)

with intuition, as coordination of more people is more difficult. (Note that mean first passage or transition times in statistical physics tend to grow exponentially in the number N of particles as well.)

Besides the number N of participants, another critical factor for the cooperation probability are the changing frequencies r_l: They are needed for the exploration of innovative strategies, coordination and cooperation. Although the instruction of test persons would have allowed them to conclude that taking turns would be a good strategy, the changing frequencies r_l of some individuals was so small that cooperation within the duration of the respective experiment did not occur, in accordance with formula (12.14). The unwillingness of some individuals to vary their decisions is sometimes called "conservative" [7, 65, 66] or "inertial behavior" [9]. Note that, if a player never reciprocates "offers" by other players, this may discourage further "offers" and reduce the changing frequency of the other player(s) as well (see the decisions 50 through 150 of player 2 in Fig. 12.4).

Our experimental time series show that most individuals initially did not know a periodic decision behavior would allow them to establish the system optimum. This indicates that the required depth of strategic reasoning [19] and the related complexity of the game for an average person are already quite high, so that intelligence may matter. Compared to control experiments, the hint that the maximum average payoff of 100 points per round could be reached "by variable, situation-dependent decisions", increased the average changing frequency (by 75 percent) and with this the occurrence frequency of cooperative events. Thereby, it also increased the chance that persistent cooperation established during the duration of the experiment.

Note that successful cooperation requires not only coordination [9], but also innovation: In their first route choice game, most test persons discover the oscillatory cooperation strategy only by chance in accordance with formula (12.14). The changing frequency is, therefore, critical for the establishment of innovative strategies: It determines the exploratory trial-and-error behavior. In contrast, cooperation is easy when test persons *know* that the oscillatory strategy is successful: When two

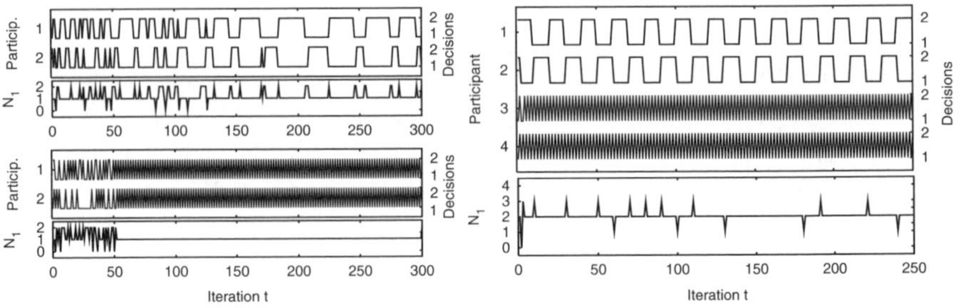

Fig. 12.13 Experimentally observed decision behavior when two groups involved in two-person route choice experiments afterwards played a four-person game with $C_1 = 900$, $D_1 = 300$, $C_2 = 100$, $D_2 = 100$. *Left:* While oscillations of period 2 emerged in the second group (*bottom*), another alternating pattern corresponding to n-period decisions with $n > 2$ emerged in the first group (*top*). *Right:* After all persons had learnt oscillatory cooperative behavior, the four-person game just required coordination, but not the invention of a cooperative strategy. Therefore, persistent cooperation was quickly established (in contrast to four-person experiments with new participants). It is clearly visible that the test persons continued to apply similar decision strategies (*right*) as in the previous two-person experiments (*left*)

teams, who had successfully cooperated in 2-person games, had afterwards to play a 4-person game, cooperation was *always* and quickly established (see Fig. 12.13). In contrast, unexperienced co-players suppressed the establishment of oscillatory cooperation in 4-person route choice games.

12.4.3 Strategy Coefficients

In order to characterize the strategic behavior of individuals and predict their chances of cooperation, we have introduced some strategy coefficients. For this, let us introduce the following quantities, which are determined from the iterations before persistent cooperation is established:

- c_l^k = relative frequency of a *changed* subsequent decision of individual l if the payoff was negative ($k = -$), zero ($k = 0$), or positive ($k = +$).
- s_l^k = relative frequency of individual l to *stay* with the previous decision if the payoff was negative ($k = -$), zero ($k = 0$), or positive ($k = +$).

The Yule-coefficient

$$Q_l = \frac{c_l^- s_l^+ - c_l^+ s_l^-}{c_l^- s_l^+ + c_l^+ s_l^-} \tag{12.15}$$

with $-1 \le Q_l \le 1$ was used by Schreckenberg, Selten et al. [65] to identify direct responders with $0.5 < Q_l \approx 1$ (who change their decision after a negative payoff and stay after a positive payoff), and contrarian responders with $-0.5 > Q_l \approx -1$ (who change their decision after a positive payoff and stay after a negative one).

A random decision behavior would correspond to a value $Q_l \approx 0$. However, a problem arises if one of the variables c_l^-, s_l^+, c_l^+, or s_l^- assumes the value 0. Then, we have $Q_l \in \{-1, 1\}$, independently of the other three values. If two of the variables become zero, Q_l is sometimes even undefined. Moreover, if the values are small, the resulting conclusion is not reliable. Therefore, we prefer to use the percentage difference

$$S_l = \frac{c_l^-}{c_l^- + s_l^l} - \frac{c_l^+}{c_l^+ + s_l^+} \tag{12.16}$$

for the assessment of strategies. Again, we have $-1 \leq S_l \leq 1$. Direct responders correspond to $S_l > 0.25$ and contrarian responders to $S_l < -0.25$. For $-0.25 \leq S_l \leq 0.25$, the response to the previous payoff is rather random.

In addition, we have introduced the Z-coefficient

$$Z_l = \frac{c_l^0}{c_l^0 + s_l^0}, \tag{12.17}$$

for which we have $0 \leq Z_l \leq 1$. This coefficient describes the likely response of individual l to the user equilibrium. $Z_l = 0$ means that individual l does not change routes, if the user equilibrium was reached. $Z_l = 1$ implies that person l always changes, while $Z_l \approx 0.5$ indicates a random response.

Figure 12.14 shows the result of the 2-person route choice experiments (cooperation or not) as a function of S_1 and S_2, and as a function of Z_1 and Z_2. Moreover, Fig. 12.15 displays the result as a function of the average strategy coefficients

$$Z = \frac{1}{N}\sum_{l=1}^{N} Z_l \tag{12.18}$$

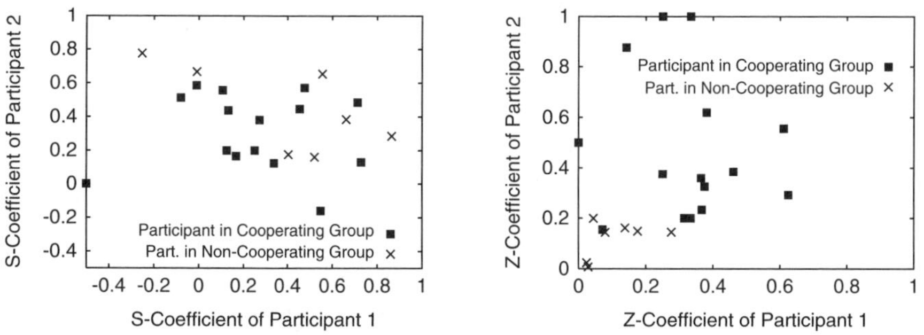

Fig. 12.14 Coefficients S_l and Z_l of both participants l in all 24 two-person route choice games. The values of the S-coefficients (i.e. the individual tendencies towards direct or contrarian responses) are not very significant for the establishment of persistent cooperation, while large enough values of the Z-coefficient stand for the emergence of oscillatory cooperation

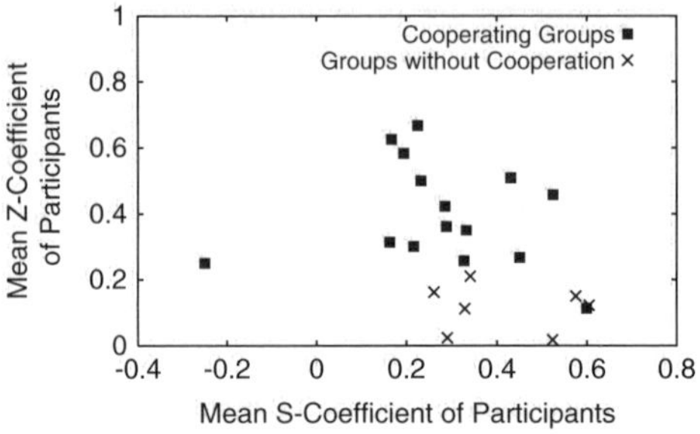

Fig. 12.15 S- and Z-coefficients averaged over both participants in all 24 two-person route choice games. The mainly small, but positive values of S indicate a slight tendency towards direct responses. However, the S-coefficient is barely significant for the emergence of persistent oscillations. A good indicator for their establishment is a sufficiently large Z-value

and

$$S = \frac{1}{N} \sum_{l=1}^{N} S_l . \tag{12.19}$$

Our experimental data indicate that the Z-coefficient is a good indicator for the establishment of cooperation, while the S-coefficient seems to be rather insignificant (which also applies to the Yule coefficient).

12.5 Multi-agent Simulation Model

In a first attempt, we have tried to reproduce the observed behavior in our 2-person route choice experiments by game-dynamical equations [28]. We have applied these to the 2×2 route choice game and its corresponding two-, three- and four-stage higher-order games (see Sect. 12.4.1). Instead of describing patterns of alternating cooperation, however, the game dynamical equations predicted a preference for the dominant strategy of the one-shot game, i.e. a tendency towards choosing route 1.

The reason for this becomes understandable through Fig. 12.8. Selecting routes 2 and 1 in an alternating way is not a stable strategy, as the other player can get a higher payoff by choosing two times route 1 rather than responding with 1 and 2. Selecting route 1 all the time even guarantees that the own payoff is never below the one by the other player. However, when both players select route 1 and establish the related user equilibrium, no player can improve his or her payoff in the next iteration by changing the decision. Nevertheless, it is possible to improve the long-term outcome, if *both* players change their decisions, and if they do it in a coordinated way. Note, however, that a strict alternating behavior of period 2 is an optimal strategy only in infinitely repeated games, while it is unstable to perturbations in finite games.

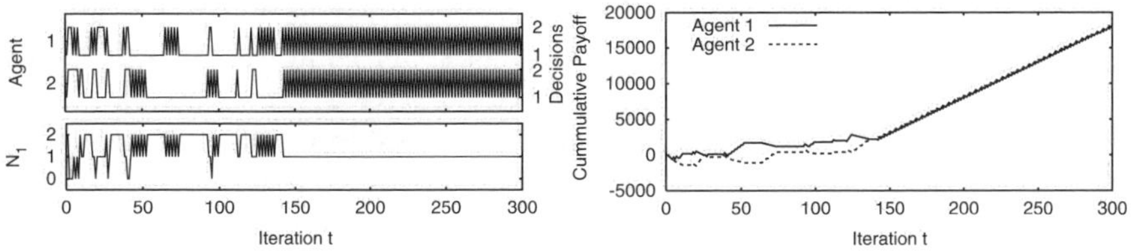

Fig. 12.16 Representative example for a 2-person route choice simulation based on our proposed multi-agent reinforcement learning model with $P_{\text{av}}^{\max} = 100$ and $P_{\text{av}}^{\min} = -200$. The parameter ν_l^1 has been set to 0.25. The other model parameters are specified in the text. *Top left:* Decisions of both agents over 300 iterations. *Bottom left:* Number $N_1(t)$ of 1-decisions over time t. *Right*: Cumulative payoff of both agents as a function of the number of iterations. The emergence of oscillatory cooperation is comparable with the experimental data displayed in Fig. 12.4

It is known that cooperative behavior may be explained by a "shadow of the future" [2, 3], but it can also be established by a "shadow of the past" [40], i.e. experience-based learning. This will be the approach of the multi-agent simulation model proposed in this section. As indicated before, the emergence of phase-coordinated strategic alternation (rather than a statistically independent application of mixed strategies) requires an almost deterministic behavior (see Fig. 12.16). Nevertheless, some weak stochasticity is needed for the establishment of asymmetric cooperation, both for the exploration of innovative strategies and for phase coordination. Therefore, we propose the following reinforcement learning model, which could be called a generalized win-stay, lose-shift strategy [50, 54].

Let us presuppose that an individual approximately memorizes or has a good feeling of how well he or she has performed on average in the last n_l iterations and since he or she has last responded with decision j to the situation (i, N_1). In our success- and history-dependent model of individual decision behavior, $p_l(j\,|i, N_1; t)$ denotes agent l's conditional probability of taking decision j at time $t + 1$, when i was selected at time t and $N_1(t)$ agents had chosen alternative 1. Assuming that p_l is either 0 or 1, $p_l(j\,|i, N_1; t)$ has the meaning of a deterministic response strategy: $p_l(j\,|i, N_1; t) = 1$ implies that individual l will respond at time $t + 1$ with the decision j to the situation (i, N_1) at time t.

Our reinforcement learning strategy can be formulated as follows: The response strategy $p_l(j\,|i, N_1, t)$ is switched with probability $q_l > 0$, if the average individual payoff since the last comparable situation with $i(t') = i(t)$ and $N_1(t') = N_1(t)$ at time $t' < t$ is less than the average individual payoff $\overline{P}_l(t)$ during the last n_l iterations. In other words, if the time-dependent aspiration level $\overline{P}_l(t)$ [40, 54] is not reached by the agent's average payoff since his or her last comparable decision, the individual is assumed to substitute the response strategy $p_l(j\,|i, N_1; t)$ by

$$p_l(j\,|i, N_1; t + 1) = 1 - p_l(j\,|i, N_1; t) \qquad (12.20)$$

with probability q_l. The replacement of dissatisfactory strategies orients at historical long-term profits (namely, during the time period $[t', t]$). Thereby, it avoids short-sighted changes after temporary losses. Moreover, it does not assume a comparison of the performance of the actually applied strategy with hypothetical ones as in most evolutionary models. A readiness for altruistic decisions is also not required, while exploratory behavior ("trial and error") is necessary. In order to reflect this, the decision behavior is randomly switched from $p_l(j|i, N_1; t+1)$ to $1 - p_l(j|i, N_1; t+1)$ with probability

$$\nu_l(t) = \max\left(\nu_l^0, \nu_l^1 \frac{P_{\text{av}}^{\max} - \overline{P}_l(t)}{P_{\text{av}}^{\max} - P_{\text{av}}^{\min}}\right) \ll 1. \tag{12.21}$$

Herein, $P_{\text{av}}^{\min}$ and $P_{\text{av}}^{\max}$ denote the minimum and maximum average payoff of all N agents (simulated players). The parameter ν_l^1 reflects the mutation frequency for $\overline{P}_l(t) = P_{\text{av}}^{\min}$, while the mutation frequency is assumed to be $\nu_l^0 \leq \nu_l^1$ when the time-averaged payoff $\overline{P}_l$ reaches the system optimum $\overline{P}_{\text{av}}^{\max}$.

In our simulations, no emergent cooperation is found for $\nu_l^0 = \nu_l^1 = 0$. $\nu_l^0 > 0$ or odd values of n_l may produce intermittent breakdowns of cooperation. A small, but finite value of ν_l^1 is important to find a transition to persistent cooperation. Therefore, we have used the parameter value $\nu_l^1 = 0.25$, while the simplest possible specification has been chosen for the other parameters, namely $\nu_l^0 = 0$, $q_l = 1$, and $n_l = 2$.

The initial conditions for the simulation of the route choice game were specified in accordance with the dominant strategy of the one-shot game, i.e. $P_l(1,0) = 1$ (everyone tends to choose the freeway initially), $p_l(2|1, N_1; 0) = 0$ (it is not attractive to change from the freeway to the side road) and $p_l(1|2, N_1; 0) = 1$ (it is tempting to change from the side road to the freeway). Interestingly enough, agents learnt to acquire the response strategy $p_l(2|1, N_1 = 1; t) = 1$ in the course of time, which established oscillatory cooperation with higher profits (see Figs. 12.16 and 12.17).

Fig. 12.17 *Left:* Conditional changing probability $p_l(2|1, N_1 = 1; t)$ of agent l from route 1 (the "freeway") to route 2, when the other agent has chosen route 2, averaged over a time window of 50 iterations. The transition from small values to 1 for the computer simulation displayed in Fig. 12.16 is characteristic and illustrates the learning of cooperative behavior. *Right:* Proportion $P_l(1,t)$ of 1-decisions of both participants l in the two-person route choice experiment displayed in Fig. 12.16. While the initial proportion is often close to 1 (the user equilibrium), it reaches the value 1/2 when persistent oscillatory cooperation (the system optimum) is established. The simulation results are compatible with the essential features of the experimental data (see, for example, Fig. 12.10)

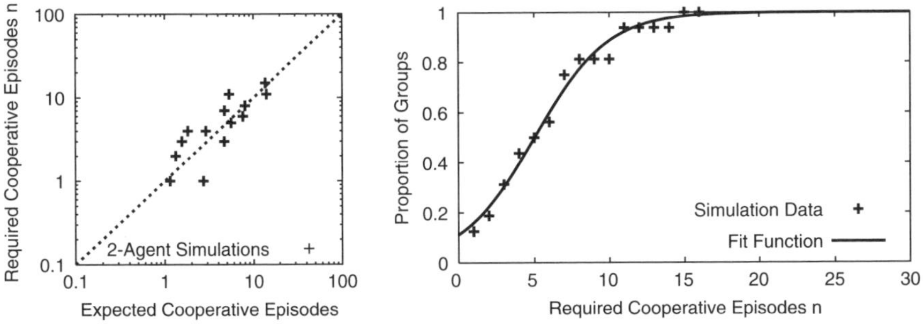

Fig. 12.18 *Left:* Comparison of the required number of cooperative episodes with the expected number of cooperative episodes in our multi-agent simulation of decisions in the route choice game. Note that the data points support formula (12.14). *Right:* Cumulative distribution of required cooperative episodes until persistent cooperation is established in our 2-person route choice simulations, using the simplest specification of model parameters (not calibrated). The simulation data are well approximated by the logistic curve (12.9) with the fit parameters $c_2 = 7.9$ and $d_2 = 0.41$

Note that the above described reinforcement learning model [40] responds only to the own previous experience [13]. Despite its simplicity (e.g. the neglection of more powerful, but probably less realistic k-move memories [11]), our "multi-agent" simulations reproduce the emergence of asymmetric reciprocity of two or more players, if an oscillatory strategy of period 2 can establish the system optimum. This raises the question why previous experiments of the N-person route choice game [27, 63] have observed a clear tendency towards the Wardrop equilibrium [71] with $P_1(N_1) = P_2(N_2)$ rather than phase-coordinated oscillations? It turns out that the payoff values must be suitably chosen (see (12.8)) and that several hundred repetitions are needed. In fact, the expected time interval T until a cooperative episode among $N = N_1 + N_2$ participants occurs in our simulations by chance is well described by formula (12.14), see Fig. 12.18. The empirically observed transition in the decision behavior displayed in Fig. 12.10 is qualitatively reproduced by our computer simulations as well (see Fig. 12.17). The same applies to the frequency distribution of the average payoff values (compare Fig. 12.19 with Fig. 12.6) or to the number of expected and required cooperative episodes (compare Fig. 12.18 with Figs. 12.9 and 12.12).

12.5.1 *Simultaneous and Alternating Cooperation in the Prisoner's Dilemma*

Let us finally simulate the dynamic behavior in the two different variants of the Prisoner's Dilemma indicated in Fig. 12.3b,c with the above experience-based reinforcement learning model. Again, we will assume $P_{11} = 0$ and $P_{22} = -200$.

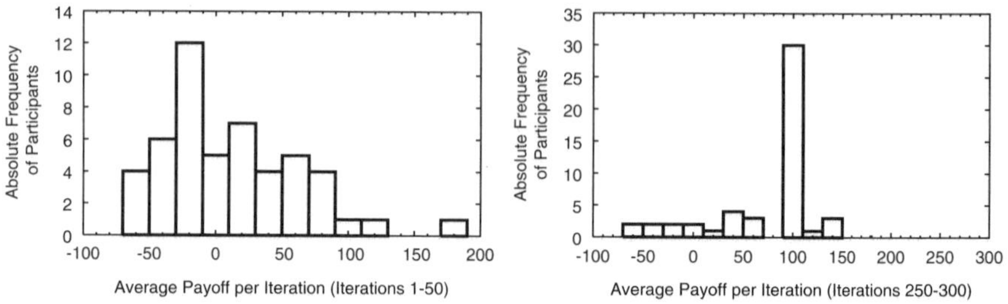

Fig. 12.19 Frequency distributions of the average payoffs in our computer simulations of the 2-person route choice game. *Left:* Distribution during the first 50 iterations. *Right:* Distribution between iterations 250 and 300. Our simulation results are compatible with the experimental data displayed in Fig. 12.6

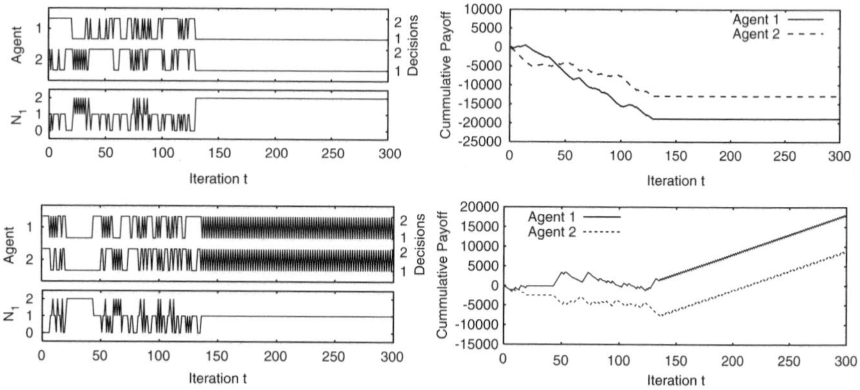

Fig. 12.20 Representative examples for computer simulations of the two different forms of the Prisoner's Dilemma specified in Fig. 12.3b,c. The parameter ν_l^1 has been set to 0.25, while the other model parameters are specified in the text. *Top:* Emergence of simultaneous, symmetrical cooperation, where decision 2 corresponds to defection and decision 1 to cooperation. The system optimum corresponds to $P_{\text{av}}^{\max} = 0$ payoff points, and the minimum payoff to $P_{\text{av}}^{\min} = -200$. *Bottom:* Emergence of alternating, asymmetric cooperation with $P_{\text{av}}^{\max} = 100$ and $P_{\text{av}}^{\min} = -200$. *Left:* Time series of the agents' decisions and the number $N_1(t)$ of 1-decisions. *Right:* Cumulative payoffs as a function of time t

According to (12.8), a simultaneous, symmetrical form of cooperation is expected for $P_{12} = -300$ and $P_{21} = 100$, while an alternating, asymmetric cooperation is expected for $P_{12} = -300$ and $P_{21} = 500$. Figure 12.20 shows simulation results for the two different cases of the Prisoner's Dilemma and confirms the two predicted forms of cooperation. Again, we varied only the parameter ν_l^1, while we chose the simplest possible specification of the other parameters $\nu_l^0 = 0$, $q_l = 1$, and $n_l = 2$. The initial conditions were specified in accordance with the expected non-cooperative outcome of the one-shot game, i.e. $P_l(1,0) = 0$ (everyone defects in the beginning), $p_l(2|2, N_1; 0) = 0$ (it is tempting to continue defecting), $p_l(1|1, N_1 = 1; 0) = 0$ (it is unfavourable to be the only cooperative player), and $p_l(1|1, N_1 = 2; 0) = 1$ (it is good to continue cooperating, if the other player cooperates). In the course of time, agents learn to acquire the response strategy $p_l(2|2, N_1 = 0; t) = 0$ when simultaneous cooperation evolves, but $p_l(2|2, N_1 = 1; t) = 0$ when alternating cooperation is established.

12.6 Summary, Discussion, and Outlook

In this paper, we have investigated the N-person day-to-day route-choice game. This special congestion game has not been thoroughly studied before in the case of small groups, where the system optimum can considerably differ from the user equilibrium. The 2-person route choice game gives a meaning to a previously uncommon repeated symmetrical 2×2 game and shows a transition from the dominating strategy of the one-shot game to coherent oscillations, if $P_{12} + P_{21} > 2P_{11}$. However, a detailed analysis of laboratory experiments with humans reveals that the establishment of this phase-coordinated alternating reciprocity, which is expected to occur in other 2×2 games as well, is quite complex. It needs either strategic experience or the invention of a suitable strategy. Such an innovation is driven by the potential gains in the average payoffs of all participants and seems to be based on exploratory trial-and-error behavior. If the changing frequency of one or several players is too low, no cooperation is established in a long time. Moreover, the emergence of cooperation requires certain kinds of strategies, which can be characterized by the Z-coefficient (12.18). These strategies can be acquired by means of reinforcement learning, i.e. by keeping response patterns which have turned out to be better than average, while worse response patterns are being replaced. The punishment of uncooperative behavior can help to enforce cooperation. Note, however, that punishment in groups of $N > 2$ persons is difficult, as it is hard to target the uncooperative person, and punishment hits everyone. Nevertheless, computer simulations and additional experiments indicate that oscillatory cooperation can still emerge in route choice games with more than two players after a long time period (rarely within 300 iterations) (see Fig. 12.21).

Altogether, spontaneous cooperation takes a long time. It is, therefore, sensitive to changing conditions reflected by time-dependent payoff parameters. As a consequence, emergent cooperation is unlikely to appear in real traffic systems. This is the reason why the Wardrop equilibrium tends to occur. However, cooperation could be rapidly established by means of advanced traveler information systems (ATIS) [8, 14, 30, 37, 41, 63, 70, 73], which would avoid the slow learning process described by (12.14). Moreover, while we do not recommend conventional congestion charges, a charge for *unfair* usage patterns would support the compliance with individual route choice recommendations. It would supplement the inefficient individual punishment mechanism.

Different road pricing schemes have been proposed, each of which has its own advantages and disadvantages or side effects. Congestion charges, for example, could discourage to take congested routes, which is actually required to reach

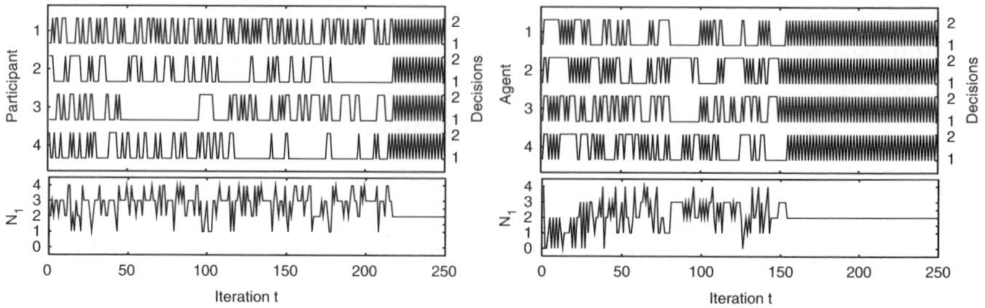

Fig. 12.21 Emergence of phase-coordinated oscillatory behavior in the 4-person route choice game with the parameters specified in Fig. 12.13. *Left*: Experimental data of the decisions of four unexperienced participants over 300 iterations. *Right*: Computer simulation with the reinforcement learning model

minimum *average* travel times. Conventional tolls and road pricing may reduce the trip frequency due to budget constraints, which potentially interferes with economic growth and fair chances for everyone's mobility.

In order to activate capacity reserves, we therefore propose an automated route guidance system based on the following principles: After specification of their destination, drivers should get individual (and, on average, fair) route choice recommendations in agreement with the traffic situation and the route choice proportions required to reach the system optimum. If an individual selects a faster route instead of the recommended route it should use, it will have to pay an amount proportional to the decrease in the overall inverse travel time compared to the system optimum. Moreover, drivers not in a hurry should be encouraged to take the slower route i by receiving the amount of money corresponding to the related increase in the overall inverse travel time. Altogether, such an ATIS could support the system optimum while allowing for some flexibility in route choice. Moreover, the fair usage pattern would be cost-neutral for everyone, i.e. traffic flows of potential economic relevance would not be suppressed by extra costs.

In systems with many similar routing decisions, a Pareto optimum characterized by asymmetric alternating cooperation may emerge even spontaneously. This could help to enhance the routing in data networks [72] and generally to resolve Braess-like paradoxes in networks [17].

Finally, it cannot be emphasized enough that taking turns is a promising strategy to distribute scarce resources in a fair and optimal way. It could be applied to a huge number of real-life situations due to the relevance for many strategical conflicts, including Leader, the Battle of the Sexes, and variants of Route Choice, Deadlock, Chicken, and the Prisoner's Dilemma. The same applies to their N-person generalizations, in particular social dilemmas [23, 25, 40]. It will also be interesting to find out whether and where metabolic pathways, biological supply networks, or information flows in neuronal and immune systems use alternating strategies to avoid the wasting of costly resources.

Acknowledgements D.H. is grateful for the warm hospitality of the Santa Fe Institute, where the Social Scaling Working Group Meeting in August 2003 inspired many ideas of this paper. The results shall be presented during the workshop on "Collectives Formation and Specialization in Biological and Social Systems" in Santa Fe (April 20–22, 2005).

References

1. W.B. Arthur, Inductive reasoning and bounded rationality. Am. Econ. Rev. **84**, 406–411 (1994)
2. R. Axelrod, D. Dion, The further evolution of cooperation. Science **242**, 1385–1390 (1988)
3. R. Axelrod, W.D. Hamilton, The evolution of cooperation. Science **211**, 1390–1396 (1981)
4. M. Beckmann, C.B., McGuire, C.B. Winsten, *Studies in the Economics of Transportation* (Yale University Press, New Haven, 1956)
5. K.G. Binmore, Evolutionary stability in repeated games played by finite automata. J. Econ. Theor. **57**, 278–305 (1992)
6. K. Binmore, *Fun and Games: A Text on Game Theory* (Heath, Lexington, MA, 1992), pp. 373–377
7. U. Bohnsack, Uni DuE: Studie SURVIVE gibt Einblicke in das Wesen des Autofahrers, Press release by *Informationsdienst Wissenschaft* (January 21, 2005)
8. P.W. Bonsall, T. Perry, Using an interactive route-choice simulator to investigate driver's compliance with route guidance information. Transpn. Res. Rec. **1306**, 59–68 (1991)
9. G. Bottazzi, G. Devetag, Coordination and self-organization in minority games: Experimental evidence, Working Paper 2002/09, Sant'Anna School of Advances Studies, May 2002
10. D. Braess, Über ein Paradoxon der Verkehrsplanung [A paradox of traffic assignment problems]. *Unternehmensforschung* **12**, 258–268 (1968). For about 100 related references see http://homepage.ruhr-uni-bochum.de/Dietrich.Braess/#paradox
11. L. Browning, A.M. Colman, Evolution of coordinated alternating reciprocity in repeated dyadic games. J. Theor. Biol. **229**, 549–557 (2004)
12. C.F. Camerer, *Behavioral Game Theory: Experiments on Strategic Interaction* (Princeton University Press, Princeton, 2003)
13. C.F. Camerer, T.-H. Ho, J.-K. Chong, Sophisticated experience-weighted attraction learning and strategic teaching in repeated games. J. Econ. Theor. **104**, 137–188 (2002)
14. N. Cetin, K. Nagel, B. Raney, A. Voellmy, Large scale multi-agent transportation simulations. Comput. Phys. Comm. **147**(1–2), 559–564 (2002)
15. D. Challet, M. Marsili, Relevance of memory in minority games. Phys. Rev. E **62**, 1862–1868 (2000)
16. D. Challet, Y.-C. Zhang, Emergence of cooperation and organization in an evolutionary game. Physica A **246**, 407–418 (1997)
17. J.E. Cohen, P. Horowitz Paradoxial behaviour of mechanical and electrical networks. Nature **352**, 699–701 (1991)
18. A.M. Colman, *Game Theory and its Applications in the Social and Biological Sciences*, 2nd edn. (Butterworth-Heinemann, Oxford, 1995)
19. A.M. Colman, Depth of strategic reasoning in games. Trends Cognit. Sci. **7**(1), 2–4 (2003)
20. P.H. Crowley, Dangerous games and the emergence of social structure: evolving memory-based strategies for the generalized hawk-dove game. Behav. Ecol. **12**, 753–760 (2001)
21. A. Eriksson, K. Lindgren Cooperation in an unpredictable environment, in *Proceedings of Artificial Life VIII*, ed. by R.K. Standish, M.A. Bedau, H.A. Abbass (MIT Press, Sidney, 2002), pp. 394–399 and poster available at http://frt.fy.chalmers.se/cs/people/eriksson.html
22. C.B. Garcia, W.I. Zangwill, *Pathways to Solutions, Fixed Points, and Equilibria* (Prentice Hall, New York, 1981)
23. N.S. Glance, B.A. Huberman, The outbreak of cooperation. J. Math. Sociol. **17**(4), 281–302 (1993)

24. B.D. Greenshield, A study of traffic capacity, in *Proceedings of the Highway Research Board*, Vol. 14 (Highway Research Board, Washington, D.C., 1935), pp. 448–477
25. G. Hardin, The tragedy of the commons. Science **162**, 1243–1248 (1968)
26. D. Helbing, Dynamic decision behavior and optimal guidance through information services: Models and experiments, in *Human Behaviour and Traffic Networks*, ed. by M. Schreckenberg, R. Selten (Springer, Berlin, 2004), pp. 47–95
27. D. Helbing, M. Schönhof, D. Kern, Volatile decision dynamics: Experiments, stochastic description, intermittency control, and traffic optimization. New J. Phys. **4**, 33.1–33.16 (2002)
28. J. Hofbauer, K. Sigmund, *The Theory of Evolution and Dynamical Systems* (Cambridge University Press, Cambridge, 1988)
29. T. Hogg, B.A. Huberman, Controlling chaos in distributed systems. IEEE Trans. Syst. Man Cybern. **21**(6), 1325–1333 (1991)
30. T.-Y. Hu, H.S. Mahmassani, Day-to-day evolution of network flows under real-time information and reactive signal control. Transport. Res. C **5**(1), 51–69 (1997)
31. S. Iwanaga, A. Namatame, The complexity of collective decision. Nonlinear Dynam. Psychol. Life Sci. **6**(2), 137–158 (2002)
32. J.H. Kagel, A.E. Roth (eds.), *The Handbook of Experimental Economics* (Princeton University, Princeton, NJ, 1995)
33. J.O. Kephart, T. Hogg, B.A. Huberman, Dynamics of computational ecosystems. Phys. Rev. A **40**(1), 404–421 (1989)
34. F. Klügl, A.L.C. Bazzan, Route decision behaviour in a commuting scenario: Simple heuristics adaptation and effect of traffic forecast. JASSS **7**(1) (2004)
35. Y.A. Korilis, A.A. Lazar, A. Orda Avoiding the Braess-paradox in non-cooperative networks. J. Appl. Probab. **36**, 211–222 (1999)
36. P. Laureti, P. Ruch, J. Wakeling, Y.-C. Zhang, The interactive minority game: a Web-based investigation of human market interactions. Physica A **331**, 651–659 (2004)
37. K. Lee, P.M. Hui, B.H. Wang, N.F. Johnson, Effects of announcing global information in a two-route traffic flow model. J. Phys. Soc. Japan **70**, 3507–3510 (2001)
38. T.S. Lo, H.Y. Chan, P.M. Hui, N.F. Johnson, Theory of networked minority games based on strategy pattern dynamics. Phys. Rev. E **70**, 056102 (2004)
39. T.S. Lo, P.M. Hui, N.F. Johnson, Theory of the evolutionary minority game. Phys. Rev. E **62**, 4393–4396 (2000)
40. M.W. Macy, A. Flache, Learning dynamics in social dilemmas. Proc. Natl. Acad. Sci. USA **99**(Suppl. 3), 7229–7236 (2002)
41. H.S. Mahmassani, R.C. Jou, Transferring insights into commuter behavior dynamics from laboratory experiments to field surveys. Transport. Res. A **34**, 243–260 (2000)
42. R. Mansilla, Algorithmic complexity in the minority game. Phys. Rev. E **62**, 4553–4557 (2000)
43. M. Marsili, R. Mulet, F. Ricci-Tersenghi, R. Zecchina, Learning to coordinate in a complex and nonstationary world. Phys. Rev. Lett. **87**, 208701 (2001)
44. J.M. McNamara, Z. Barta, A.I. Houston, Variation in behaviour promotes cooperation in the prisoner's dilemma game. Nature **428**, 745–748 (2004)
45. F. Michor, M.A. Nowak, The good, the bad and the lonely. Nature **419**, 677–679 (2002)
46. M. Milinski, D. Semmann, H.-J. Krambeck, Reputation helps solve the 'tragedy of the commons'. Nature **415**, 424–426 (2002)
47. D. Monderer, L.S. Shapley, Fictitious play property for games with identical interests. J. Econ. Theor. **1**, 258–265 (1996)
48. D. Monderer, L.S. Shapley, Potential games. Game. Econ. Behav. **14**, 124–143 (1996)
49. M.A. Nowak, A. Sasaki, C. Taylor, D. Fudenberg, Emergence of cooperation and evolutionary stability in finite populations. Nature **428**, 646–650 (2004)
50. M. Novak, K. Sigmund, A strategy of win-stay, lose-shift that outperforms tit-for-tat in the Prisoner's Dilemma game. Nature **364**, 56–58 (1993)
51. M.A. Nowak, K. Sigmund The alternating prisoner's dilemma. J. theor. Biol. **168**, 219–226 (1994)

52. M.A. Nowak, K. Sigmund, Evolution of indirect reciprocity by image scoring. Nature **393**, 573–577 (1998)
53. A.C. Pigou, *The Economics of Welfare* (Macmillan, London, 1920)
54. M. Posch, Win-stay, lose-shift strategies for repeated games—Memory length, aspiration levels and noise. J. Theor. Biol. **198**, 183–195 (1999)
55. D.C. Queller, Kinship is relative. Nature **430**, 975–976 (2004)
56. A. Rapoport, Exploiter, leader, hero, and martyr: the four archtypes of the 2 × 2 game. Behav. Sci. **12**, 81–84 (1967)
57. A. Rapoport, M. Guyer, A taxonomy of 2 × 2 games. Gen. Syst. **11**, 203–214 (1966)
58. P.D.V.G. Reddy, et al., Design of an artificial simulator for analyzing route choice behavior in the presence of information system. J. Math. Comp. Mod. **22**, 119–147 (1995)
59. R.L. Riolo, M.D. Cohen, R. Axelrod, Evolution of cooperation without reciprocity. Nature **414**, 441–443 (2001)
60. R.W. Rosenthal, A class of games possessing pure-strategy Nash equilibria. Int. J. Game Theor. **2**, 65–67 (1973)
61. T. Roughgarden, E. Tardos, How bad is selfish routing? J. ACM **49**(2), 236–259 (2002)
62. T.C. Schelling, *Micromotives and Macrobehavior* (WW Norton and Co, New York, 1978), pp. 224–231+237
63. M. Schreckenberg, R. Selten (eds.), *Human Behaviour and Traffic Networks* (Springer, Berlin, 2004)
64. F. Schweitzer, L. Behera, H. Mühlenbein, Evolution of cooperation in a spatial prisoner's dilemma. Adv. Complex Syst. **5**(2/3), 269–299 (2002)
65. R. Selten, et al., Experimental investigation of day-to-day route-choice behaviour and network simulations of autobahn traffic in North Rhine-Westphalia, in *Human Behaviour and Traffic Networks*, ed. by M. Schreckenberg, R. Selten (Springer, Berlin, 2004), pp. 1–21
66. R. Selten, M. Schreckenberg, T. Pitz, T. Chmura, S. Kube, Experiments and simulations on day-to-day route choice-behaviour. See http://papers.ssrn.com/sol3/papers.cfm?abstract_id=393841
67. D. Semmann, H.-J. Krambeck, M. Milinski, Volunteering leads to rock-paper-scissors dynamics in a public goods game. Nature **425**, 390–393 (2003)
68. P. Spirakis, Algorithmic aspects of congestion games, Invited talk at the 11th Colloquium on Structural Information and Communication Complexity (Smolenice Castle, Slovakia, June 21-23, 2004)
69. G. Szabó, C. Hauert, Phase transitions and volunteering in spatial public goods games. Phys. Rev. Lett. **89**, 118101 (2002)
70. J. Wahle, A.L.C. Bazzan, F. Klügl, M. Schreckenberg, Decision dynamics in a traffic scenario. Physica A **287**, 669–681 (2000)
71. J.G. Wardrop, Some theoretical aspects of road traffic research, in *Proceedings of the Institution of Civil Engineers II*, Vol. 1 (1952), pp. 325–378
72. D.H. Wolpert, K. Tumer, Collective intelligence, data routing and Braess' paradox. J. Artif. Intell. Res. **16**, 359–387 (2002)
73. T. Yamashita, K. Izumi, K. Kurumatani, Effect of using route information sharing to reduce traffic congestion. Lect. Notes Comput. Sci. **3012**, 86–104 (2004)
74. B. Yuan, K. Chen, Evolutionary dynamics and the phase structure of the minority game. Phys. Rev. E **69**, 067106 (2004)

Chapter 13
Response to Information*

Optimal route guidance strategies in overloaded traffic networks, for example, require reliable traffic forecasts (see Fig. 13.1). These are extremely difficult for two reasons: First of all, traffic dynamics is very complex, but after more than 50 years of research, it is relatively well understood [23]. The second and more serious problem is the invalidation of forecasts by the driver reactions to route choice recommendations. Nevertheless, some keen scientists hope to solve this long-standing problem by means of an iteration scheme [1, 2, 4–6, 8, 36, 42, 45]: If the driver reaction was known from experiments [10, 11, 15, 20, 28–30, 32, 35, 37, 43], the resulting traffic situation could be calculated, yielding improved route choice recommendations, etc. Given this iteration scheme converges, it would facilitate optimal recommendations and reliable traffic forecasts anticipating the driver reactions. Based on empirically determined transition and compliance probabilities, the new procedure developed in the following would even allow us to reach the optimal traffic distribution in one single step and in harmony with the forecast. Let us now quantify the success or payoff P_i of road users in terms of their inverse travel times. If one approximates the average vehicle speed V_i on route i by the linear relationship

$$V_i(n_i) = V_i^0 \left(1 - \frac{n_i(t)}{n_i^{\max}}\right), \tag{13.1}$$

the inverse travel times obey the payoff relations $P_i(n_i) = P_i^0 - P_i^1 n_i$ with

$$P_i^0 = \frac{V_i^0}{L_i} \quad \text{and} \quad P_i^1 = \frac{V_i^0}{n_i^{\max} L_i}. \tag{13.2}$$

---

*This chapter reprints parts of a previous publication with kind permission of the copyright owner, Springer Publishers. It is requested to cite this work as follows: D. Helbing, Dynamic decision behavior and optimal guidance through information services: Models and experiments. Pages 47–95 in: M. Schreckenberg and R. Selten (eds.) *Human Behaviour and Traffic Networks* (Springer, Berlin, 2004).

D. Helbing (ed.), *Social Self-Organization*, Understanding Complex Systems,
DOI 10.1007/978-3-642-24004-1_13, © Springer-Verlag Berlin Heidelberg 2012

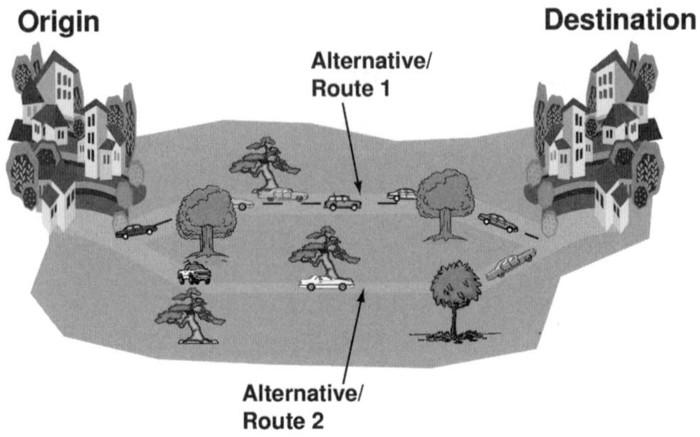

Fig. 13.1 Schematic illustration of the day-to-day route choice scenario (from [21]). Each day, the drivers have to decide between two alternative routes, 1 and 2. Note that, due to the different number of lanes, route 1 has a higher capacity than route 2. The latter is, therefore, used by less cars

Herein, V_i^0 denotes the maximum velocity (speed limit), n_i the number of drivers on route i, L_i its length, and $n_i^{\max}$ its capacity, i.e. the maximum possible number of vehicles on route i. For an improved approach to determine the travel times in road networks see [22]. Note that alternative routes can reach comparable payoffs (inverse travel times) only when the total number $N(t)$ of vehicles is large enough to fulfil the relations $P_1(N(t)) < P_2(0) = P_2^0$ and $P_2(N(t)) < P_1(0) = P_1^0$. Our route choice experiment will address this traffic regime. Furthermore, we have the capacity restriction $N(t) < n_1^{\max} + n_2^{\max}$. $N(t) = n_1^{\max} + n_2^{\max}$ would correspond to a complete gridlock.

13.1 Experimental Setup and Previous Results

To determine the route choice behavior, Schreckenberg, Selten et al. [43] have recently carried out a decision experiment (see Fig. 13.2). N test persons had to repeatedly decide between two alternatives 1 and 2 (the routes) and should try to maximize their resulting payoffs (describing something like the speeds or inverse travel times). To reflect the competition for a limited resource (the road capacity), the received payoffs

$$P_1(n_1) = P_1^0 - P_1^1 n_1 \quad \text{and} \quad P_2(n_2) = P_2^0 - P_2^1 n_2 \tag{13.3}$$

went down with the numbers of test persons n_1 and $n_2 = N - n_1$ deciding for alternatives 1 and 2, respectively. The *user equilibrium* corresponding to equal payoffs for both alternative decisions is found for a fraction

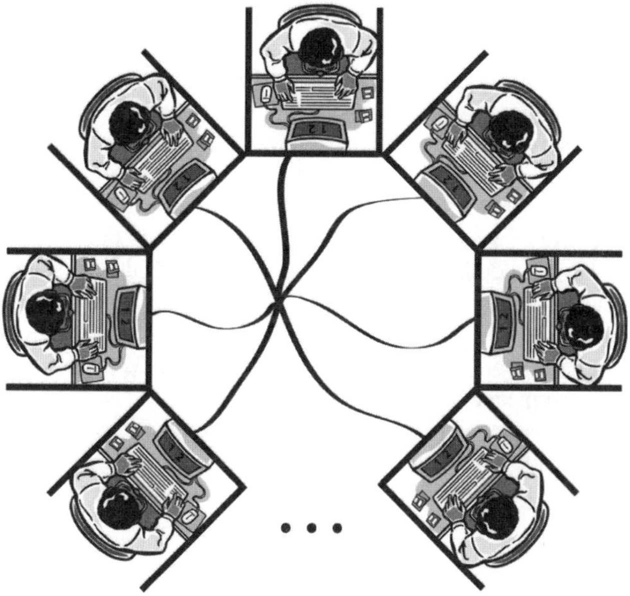

Fig. 13.2 Schematic illustration of the decision experiment (from [21]). Several test persons take decisions based on the aggregate information their computer displays. The computers are connected and can, therefore, exchange information. However, a direct communication among players is suppressed

$$f_1^{\text{eq}} = \frac{n_1}{N} = \frac{P_2^1}{P_1^1 + P_2^1} + \frac{1}{N} \frac{P_1^0 - P_2^0}{P_1^1 + P_2^1} \tag{13.4}$$

of persons choosing alternative 1. The *system optimum* corresponds to the maximum of the total payoff $n_1 P_1(n_1) + n_2 P_2(n_2)$, which lies by an amount of

$$\frac{1}{2N} \frac{P_1^0 - P_2^0}{P_1^1 + P_2^1} \tag{13.5}$$

below the user optimum. Therefore, only experiments with a few players allow to find out, whether the test persons adapt to the user or the system optimum. Small groups are also more suitable for the experimental investigation of the fluctuations in the system and of the long-term adaptation behavior. Schreckenberg, Selten et al. found that, on average, the test groups adapted relatively well to the user equilibrium. However, although it appears reasonable to stick to the same decision once the equilibrium is reached, the standard deviation stayed at a finite level. This was not only observed in *treatment 1*, where all players knew only their own (previously experienced) payoff, but also in *treatment 2*, where the payoffs $P_1(n_1)$ and $P_2(n_2)$ for both, 1- *and* 2-decisions, were transmitted to all players (analogous to radio news). Nevertheless, treatment 2 could decrease the changing rate and increase the average payoffs (cf. Fig. 13.3). For details regarding the statistical analysis see [43].

To explain the mysterious persistence in the changing behavior and explore possibilities to suppress it, we have repeated these experiments with more iterations and tested additional treatments. In the beginning, all treatments were consecutively applied to the same players in order to determine the response to different kinds of information (see Fig. 13.3). Afterwards, single treatments and variants of them have been repeatedly tested with different players to check our conclusions. Apart from this, we have generalized the experimental setup in the sense that it was not anymore restricted to route choice decisions: The test persons did not have any idea of the payoff functions in the beginning, but had to develop their own hypothesis about them. In particular, the players did not know that the payoff decreased with the number of persons deciding for the same alternative.

In *treatment 3*, every test person was informed about the own payoff $P_1(n_1)$ [or $P_2(n_2)$] and the *potential payoff*

$$P_2(N - n_1 + \epsilon N) = P_2(n_2) - \epsilon N P_2^1 \tag{13.6}$$

[or $P_1(N - n_2 + \epsilon N) = P_1(n_1) - \epsilon N P_1^1$] he or she would have obtained, if a fraction ϵ of persons had additionally chosen the other alternative (here: $\epsilon = 1/N$). Treatments 4 and 5 were variants of treatment 3, but some payoff parameters were changed in time to simulate varying environmental conditions. In *treatment 5*, each player additionally received an individual recommendation which alternative to choose.

The higher changing rate in treatment 1 compared to treatment 2 can be understood as effect of an exploration rate ν_1 required to find out which alternative performs better. It is also plausible that treatment 3 could further reduce the changing rate: In the user equilibrium with $P_1(n_1) = P_2(n_2)$, every player knew that he or she would *not* get *the same*, but a *reduced* payoff, if he or she would change the decision. That explains why the new treatment 3 could reach a great adaptation performance, reflected by a very low standard deviation and almost optimal average payoffs. The behavioral changes induced by the treatments were not only observed on average, but for all single individuals (see Fig. 13.4). Moreover, even the smallest individual cumulative payoff exceeded the highest one in treatment 1. Therefore, treatment 3's way of information presentation is much superior to the ones used today.

13.2 Is It Just an Unstable User Equilibrium?

In this section, we will investigate why players changed their decision in the user equilibrium at all. With $P_1(1, t) = \langle n_1(t) \rangle / N$ and $\langle n_i(t) \rangle = n_i(t)$ (as $n_i(t)$ are the *measured* numbers of i-decisions at time t), we find the following balance equation for the decision experiment:

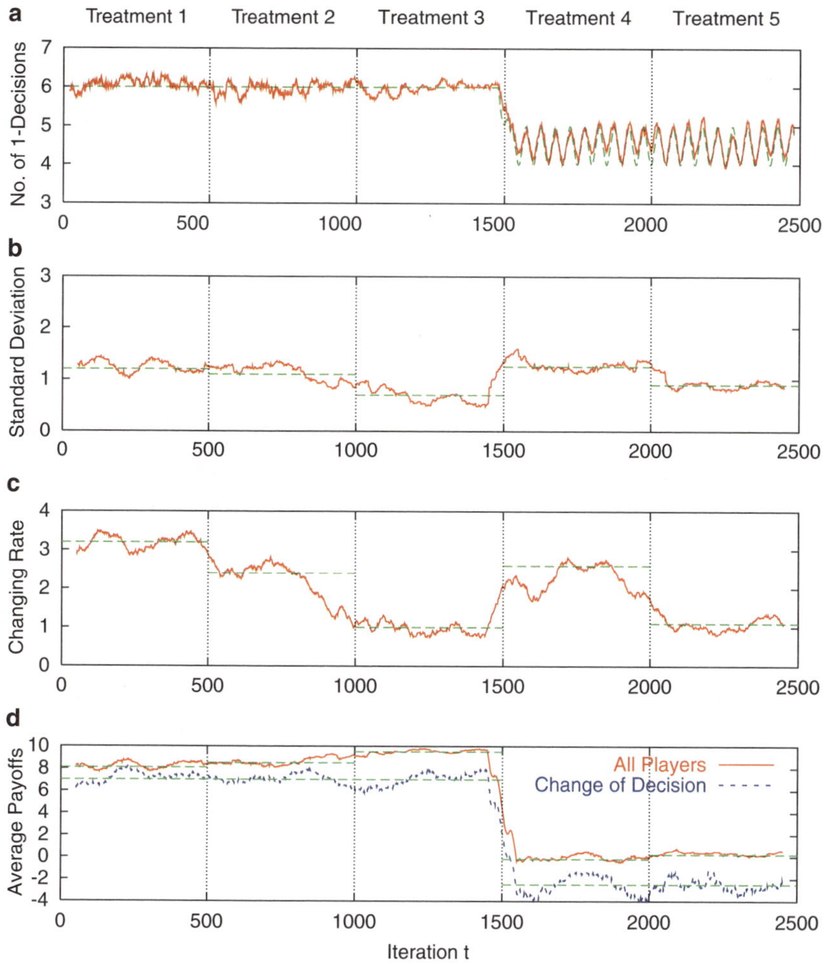

Fig. 13.3 Overview of treatments 1 to 5 [21] (with $N = 9$ and payoff parameters $P_2^0 = 28$, $P_1^1 = 4$, $P_2^1 = 6$, and $P_1^0 = 34$ for $0 \leq t \leq 1,500$, but a zick-zack-like variation between $P_1^0 = 44$ and $P_1^0 = -6$ with a period of 50 for $1,501 \leq t \leq 2,500$): **(a)** Average number of decisions for alternative 1 (*solid line*) compared to the user equilibrium (*broken line*), **(b)** standard deviation of the number of 1-decisions from the user equilibrium, **(c)** number of decision changes from one iteration to the next one, **(d)** average payoff per iteration for players who have changed their decision and for all players. The latter increased with a reduction in the changing rate, but normally stayed below the payoff in the user equilibrium (which is 1 on average in treatments 4 and 5, otherwise 10). The displayed moving time-averages [(**a**) over 40 iterations, (**b**)–(**d**) over 100 iterations] illustrate the systematic response to changes in the treatment every 500 iterations. *Dashed lines* in (**b**)–(**d**) show estimates of the stationary values after the transient period (to guide the eyes), while time periods around the dotted lines are not significant. Compared to treatment 1, treatment 3 managed to reduce the changing rate and to increase the average payoffs (three times more than treatment 2 did). These changes were systematic for *all* players (see Fig. 13.4). In treatment 4, the changing rate and the standard deviation went up, since the user equilibrium changed in time. The user-specific recommendations in treatment 5 could almost fully compensate for this. The above conclusions are also supported by additional experiments with single treatments

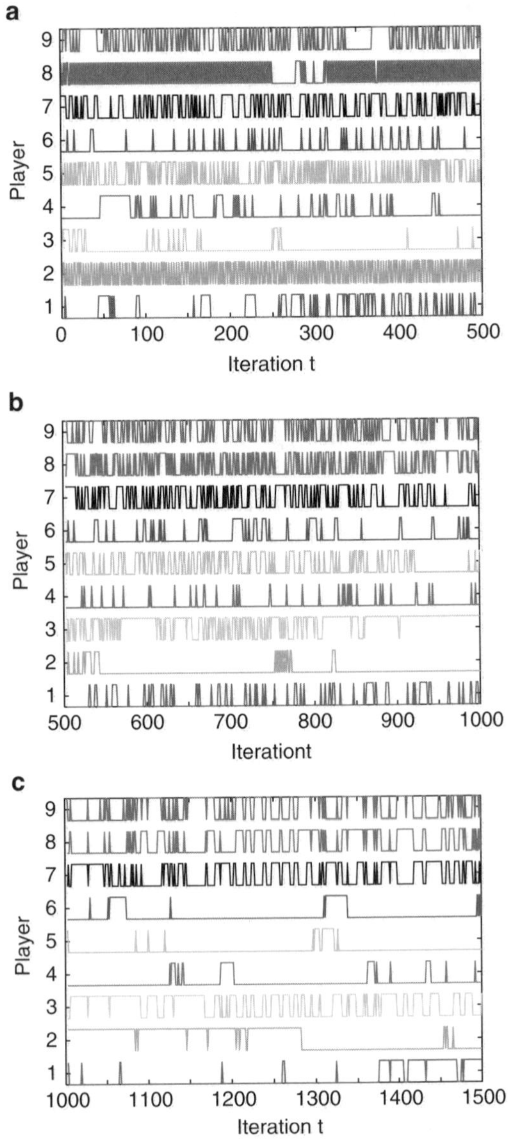

Fig. 13.4 Comparison of the individual decision behaviors in **(a)** treatment 1, **(b)** treatment 2, and **(c)** treatment 3 (from [21]). The *upper values* correspond to a decision for alternative 2, the *lower ones* for alternative 1. Note that some test persons showed similar behaviors (either more or less the same or almost opposite ones), although they could not talk to each other. This shows that there are some typical strategies how to react to a certain information configuration, i.e. to a certain decision distribution. The group has, in fact, to develop complementary strategies in order to reach a good adaptation performance. Identical strategies would perform poorly (as in the minority game [12–14]). Despite the mentioned complementary behavior, there is a characteristic reaction to changes in the treatment. For example, compared to treatment 2 all players reduce their changing rate in treatment 3

$$\langle n_1(t+1)\rangle - n_1(t) = p(1|2, n_1; t)n_2(t) - p(2|1, n_1; t)n_1(t). \tag{13.7}$$

Assuming stationary transition probabilities $p(2|1, n_1)$ (after a transient phase), the equilibrium distribution corresponds to

$$\langle n_1(t+1)\rangle = \langle n_1(t)\rangle = n_1(t). \tag{13.8}$$

Consequently, the equilibrium condition

$$p(2|1, n_1)n_1(t) = p(1|2, n_1)n_2(t) \tag{13.9}$$

should be fulfilled for the user equilibrium $n_1(t) = f_1^{\text{eq}}N$ and $n_2(t) = (1 - f_1^{\text{eq}})N$. This, however, is generally *not* compatible with the assumption

$$p(2|1, n_1) \propto \exp[P_2(N - n_1 + 1) - P_1(n_1)] \tag{13.10}$$

or similar specifications of the transition probability that increase monotonically with the payoff P_2 or the payoff difference $P_2 - P_1$! Since normally

$$\frac{p(2|1, f_1^{\text{eq}}N)}{p(1|2, f_1^{\text{eq}}N)} \neq \frac{1 - f_1^{\text{eq}}}{f_1^{\text{eq}}}, \tag{13.11}$$

the test persons would have serious problems reaching the user equilibrium. The decision distribution would possibly tend to oscillate around it, corresponding to an unstable user equilibrium. We have tested this potential interpretation of the ongoing tendency to change the decision. Figure 13.5 compares the changing rates and the standard deviations for a case where the equilibrium condition (13.9) should be valid and another case where it should be violated. However, the changing rate and standard deviation were higher in the first case, so that the hypothesis of an unstable equilibrium must be wrong. In the user equilibrium with $n_1(t) = f_1^{\text{eq}}N = N - n_2(t)$, the inflow $p(1|2, n_1)n_2(t)$ is, in fact, well balanced by the outflow

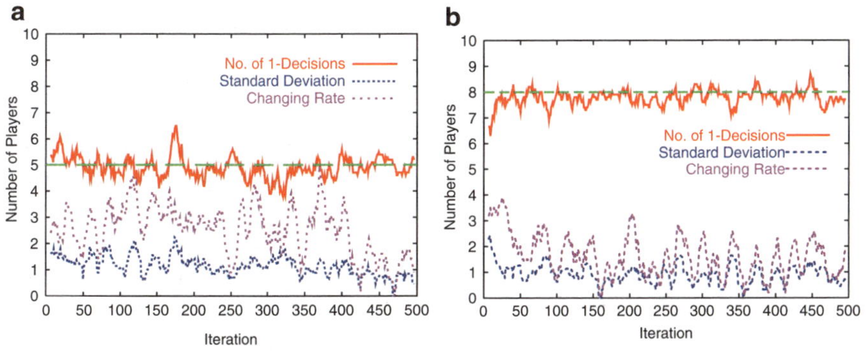

Fig. 13.5 Comparison of the group performance for treatment 2, when the user equilibrium corresponds to (**a**) $f_1^{\text{eq}} = 50\%$ 1-decisions (for $P_1^0 = 32$, $P_1^1 = 5$, $P_2^0 = 32$, $P_2^1 = 5$) or (**b**) $f_1^{\text{eq}} = 80\%$ 1-decisions (for $P_1^0 = 42$, $P_1^1 = 4$, $P_2^0 = 22$, $P_2^1 = 6$). If the user equilibrium were unstable for $f_1^{\text{eq}} \neq 1/2$, the changing rate and standard deviation should be lower in (**a**) than in (**b**). The observation contradicts this assumption. The persistent changing rate is also not caused by a difference between the system and the user optimum, since this is zero in (**a**) but one in (**b**). Instead, the higher changing rate for the symmetrical case $f_1^{\text{eq}} = 1/2$ is for statistical reasons (Remember that the variance of a binomial distribution $B(N, p)$ is $Np(1 - p)$ and becomes maximal for $p = 1/2$)

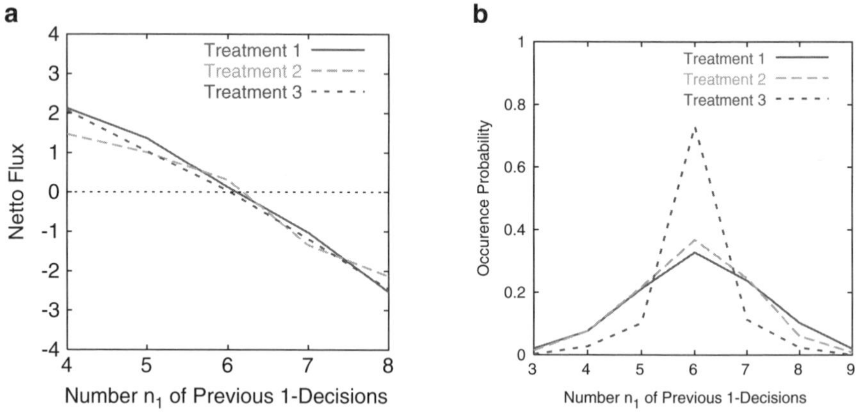

Fig. 13.6 **(a)** The netto flux $p(1|2,n_1)n_2(t) - p(1|2,n_1)n_1(t)$, which reflects the systematic part of decision changes, does not significantly depend on the treatment. As expected, it is zero in the user equilibrium, positive below it, and negative above it. **(b)** The treatment can influence the occurrence probability of decision distributions. Compared to treatments 1 and 2, the probability distribution is much more sharply peaked for treatment 3, implying a significantly smaller level of randomness during decision changes (a smaller "diffusion coefficient")

$p(1|2,n_1)n_1(t)$, as Fig. 13.6 shows. By the way, the results displayed in Fig. 13.5 also disprove the idea that a difference between the user and the system optimum may be the reason for the continuing changing behavior.

13.3 Explaining the Volatile Decision Dynamics

The reason for the pertaining changing behavior can be revealed by a more detailed analysis of the individual decisions in treatment 3. Figure 13.7 shows some kind of intermittent behavior, i.e. quiescent periods without changes, followed by turbulent periods with many changes. This is reminiscent of volatility clustering in stock market indices [19, 33, 38], where individuals also react to aggregate information reflecting all decisions (the trading transactions). Single players seem to change their decision to reach above-average payoffs. In fact, although the cumulative individual payoff is anticorrelated with the average changing rate, some players receive higher payoffs with larger changing rates than others. They profit from the overreaction in the system. Once the system is out of equilibrium, all players respond in one way or another. Typically, there are too many decision changes (see Figs. 13.7 and 13.9). The corresponding overcompensation, which had also been predicted by computer simulations [2, 5, 6, 20, 36, 46], gives rise to "turbulent" periods.

Finally, we note that the calm periods without decision changes tend to become longer in the course of time. That is, after a very long time period the individuals seem to learn not to change their behavior when the user equilibrium is reached. This is not only found in Fig. 13.7, but also visible in Fig. 13.3c after about 800

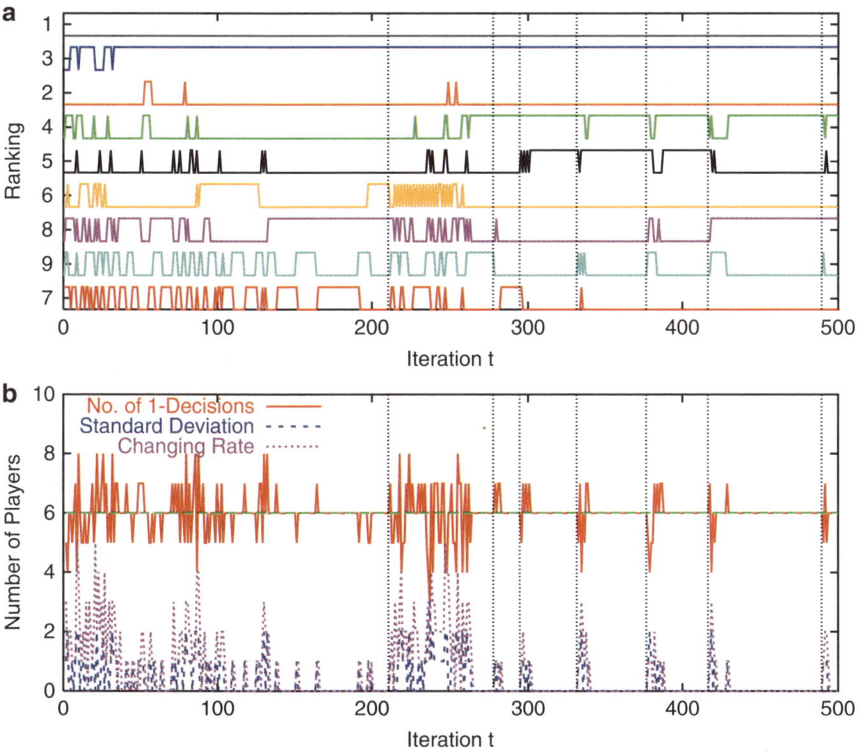

Fig. 13.7 Illustration of typical results for treatment 3 [21] (which was here the only treatment applied to the test persons, in contrast to Fig. 13.3). **(a)** Decisions of all nine players. Players are displayed from the *top* to the *bottom* in the order of increasing changing rate. Although the ranking of the cumulative payoff and the changing rate are anticorrelated, the relation is not monotonic. Note that turbulent or volatile periods characterized by many decision changes are usually triggered by individual changes after quiescent periods (*dotted lines*). **(b)** The changing rate is mostly larger than the (standard) deviation from the user equilibrium $n_1 = f_1^{\text{eq}} N = 6$, indicating an overreaction in the system

iterations. In larger systems (with more test persons) this transient period would take even longer, so that this stabilization effect could not be observed by Schreckenberg, Selten et al. [43].

13.4 Simulation of Reinforcement Learning and Emergence of Individual Response Patterns

A close look at Fig. 13.8a reveals additional details of decision behavior:

- Some players change their decision more frequently than others.
- Some test persons show similar behaviors (e.g., players 8 and 9 or 1 and 7 for $t \geq 400$), while some display almost opposite behaviors (e.g., players 7 and 8).

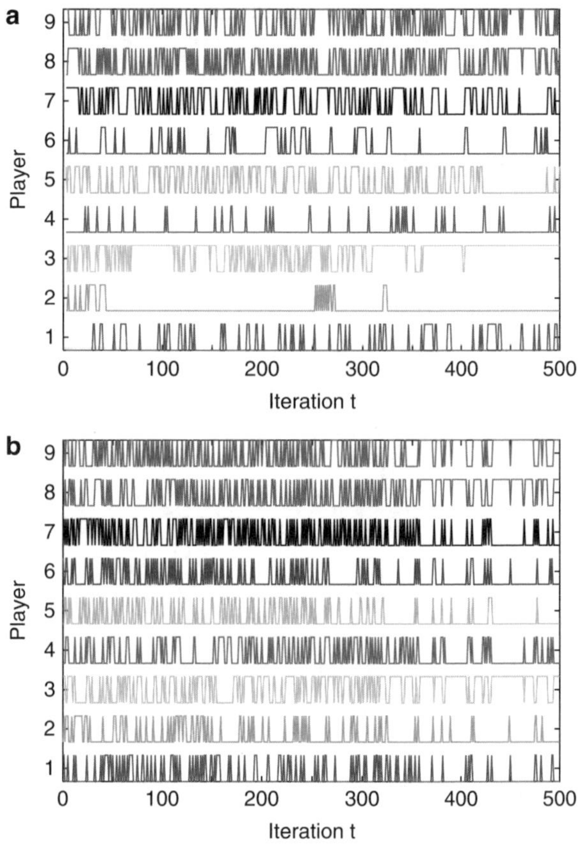

Fig. 13.8 (**a**) Typical individual decision changes of nine test persons exposed to treatment 2 with the parameters specified in Fig. 13.3. (**b**) Simulation of decision changes based on a model of reinforment learning (see main text) with parameter values $\delta = 0.01$, $q_0 = 0.4$, and $r = 0.995$

The second point is very surprising, as the players could not communicate with each other. However, both observations can be explained by the conjecture that the individuals develop different characteristic strategies how to react to specific information. "Movers" and "stayers" or direct and contrary strategies have, in fact, been observed by Schreckenberg, Selten et al. [43], and it is an interesting question, how they arise. The group has to develop complementary strategies in order to reach a good adaptation performance. As a consequence, if some players do not react to changing conditions, others will take the chance to earn additional payoff. This experimentally supports the behavior assumed in the theory of efficient markets. Note that identical strategies would perform poorly, as in the minority game [3, 12–14].

In order to reproduce the above described evolution of complementary strategies and other observed features, we have developed a simulation model based on reinforcement learning [18, 34]. At first, it appears reasonable to apply a learning strategy that reproduces the "law of relative effect", according to which the probability $p'_\alpha(i, t+1)$ of an individual α to choose alternative i at time $t + 1$

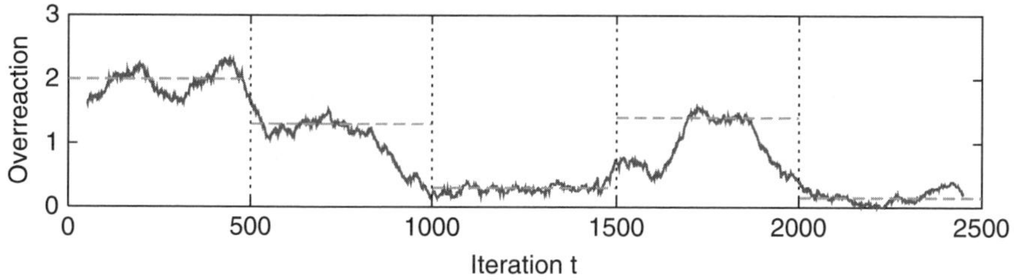

Fig. 13.9 Measured overreaction, i.e., difference between the actual number of decision changes (the changing rate) and the required one (the standard deviation) [21]. The overreaction can be significantly influenced by the treatment, i.e. the way of information presentation. The minimum overreaction was reached by treatment 5, i.e. user-specific recommendations

would reflect the relative frequency with which this alternative was successful in the past. This can, for example, be reached by means of the reinforcement rule

$$p'_\alpha(i,t+1) = \begin{cases} 1 - q_0[1 - p'_\alpha(i,t)] & \text{in case of a successful decision} \\ q_0 p'_\alpha(i,t) & \text{otherwise} \end{cases} \tag{13.12}$$

[34], where the way in which a successful decision is defined may vary from one situation to another. However, a probabilistic decision strategy, when applied by all individuals, produces a large amount of stochastic fluctuations, i.e. the user equilibrium is hard to maintain. More importantly, although the above learning strategy may explain a specialization in the individual behaviors (i.e. different decision probabilities, depending on the respective success history), it does not allow to understand the state-dependent probability of decision changes (see Fig. 13.12). We will, therefore, develop a model for the conditional (transition) probability $p_\alpha(i|i',n_1;t)$ of individual α to select alternative i, given that the previous decision was i' and n_1 individuals had taken decision 1. Furthermore, let us assume that each individual updates this transition probability according to the following scheme:

$$p_\alpha(i|i',n_1;t+1) = \begin{cases} \max[1-\delta, p_\alpha(i|i',n_1;t) + q(t+1)] & \text{for a successful decision,} \\ \min[\delta, p_\alpha(i|i',n_1;t) - q(t+1)] & \text{otherwise.} \end{cases} \tag{13.13}$$

Due to the normalization of transition probabilities, we have the additional relation

$$p_\alpha(3-i|i',n_1;t+1) = 1 - p_\alpha(i|i',n_1;t+1), \tag{13.14}$$

as $3 - i$ is the alternative of decision $i \in \{1,2\}$. The parameter $\delta \approx 0$ reflects a minimum changing probability, which ensures that there is always a certain readiness to adapt to a potentially changing environment. It is responsible for the stochastic termination of quiescent phases, in which nobody changes the decision.

Our simulations were run with $\delta = 0.01$, i.e. the minimum changing probability was assumed to be 1%.

The parameter $q(t)$ denotes the size of the adaptation step, by which the transition probability is increased in case of success or otherwise decreased, while the minimum and maximum functions guarantee that the transition probabilities $p_\alpha(i|i', n_1; t+1)$ stay between the minimum value δ and the maximum value $1-\delta$. A time-dependent choice such as

$$q(t) = q_0 r^t \tag{13.15}$$

with an initial value q_0 of q ($0 < q_0 < 1$) and a value of r slightly smaller than 1 allow one to describe that the learning rate is large in the beginning, when the different possible strategies are explored, but it eventually goes down, as the optimum strategy becomes more and more clear. In the course of time, this leads to the stabilization of a particular, history-dependent response pattern that characterizes the individual decision strategy. The resulting response pattern shows either a high likelihood to stay with the previous decision (with $p_\alpha \approx \delta$) or a high likelihood to change it (with $p_\alpha \approx 1 - \delta$), depending on the respective system state n_1 and previous decision i'. That is, the resulting strategy tends to be approximately deterministic, reflecting that the individual believes to know what is the "right" decision. This is markedly different from other decision models with reinforcement learning [18, 34]. Nevertheless, when averaging over all occurring system states, the individuals appear to play mixed strategies, i.e. they seem to show probabilistic (rather than almost deterministic) decision behavior (see Fig. 13.12). Therefore, our approach is expected to be consistent with the law of relative effect, but only in the statistical sense. Altogether, formula (13.15) reflects the observed trial-and-error behavior in the beginning (the "experimentation phase"), but a tendency to follow learned strategies later on without significant changes. The parameters δ, q_0, and r may, of course, be individual, but for reasons of simplicity we have assumed identical values in our simulations.

The way, in which a successful decision is defined, may depend on the respective situation or experiment. In our simulations of treatment 2, we have assumed that the decision is valued as successful, when

$$P_i(n_i(t+1)) \geq P_{3-i}(N - n_i(t+1)) = P_{3-i}(n_{3-i}(t+1)) \tag{13.16}$$

and

$$P_i(n_i(t+1)) \geq P_{i'}(n_{i'}(t)), \tag{13.17}$$

i.e. when the payoff was at least as large as for the other alternative $3 - i$ and not smaller than in the previous time step. The first decision was made randomly with probability 1/2. The following decisions were also randomly chosen, but in accordance with the respective transition probabilities, which were updated according to the above scheme.

The simulation results are in good qualitative agreement with the features observed in our experiments. We find an adaptation of the group to the user equilibrium with an average individual payoff of approximately 8.5, as in our experiments. Moreover, the changing rate is high in the beginning and decreases in the course of time (see Fig. 13.8b). As experimentally observed, some players change their decision more frequently than others, and we find almost similar or opposite behaviors after some time. That is, our simulations allow to reproduce that players develop individual strategies (i.e. response patterns, "roles", or "characters") in favour of a good group performance.

By means of our simulations, we can not only reproduce the main experimental observations. One can also optimize the group sizes and number of iterations of decision experiments. The above simulation concept is now used to design new experiments, which try to improve the system performance or even to establish the social optimum by particular information strategies. In the following section, we will, for example, introduce a possible concept for decision guidance.

13.4.1 Potentials and Limitations of "Decision Control"

To avoid overreaction, in treatment 5 we have recommended a number $f_1^{\mathrm{eq}}(t+1)N - n_1(t)$ of players to change their decision and the other ones to keep it. These user-specific recommendations helped the players to reach the smallest overreaction of all treatments (see Fig. 13.9) and a very low standard deviation, although the payoffs were changing in time (see Fig. 13.10). Treatment 4 shows how the group performance was affected by the time-dependent user equilibrium: Even without recommendations, the group managed to adapt to the changing conditions surprisingly well, but the standard deviation and changing rate were approximately as high as in treatment 2 (see Fig. 13.3). This adaptability (the collective "group intelligence") is based on complementary responses (direct and contrary ones [43], "movers" and "stayers", cf. Fig. 13.4). That is, if some players do not react to the changing conditions, others will take the chance to earn additional payoff. This experimentally supports the behavior assumed in the theory of efficient markets, but here the efficiency is limited by overreaction.

In most experiments, we found a constant and high compliance $C_S(t) \approx 0.92$ with recommendations to stay, but the compliance $C_M(t)$ with recommendations to change (to "move") [15, 31, 32, 44] turned out to vary in time. It decreased with the reliability of the recommendations (see Fig. 13.11a), which again dropped with the compliance.

Based on this knowledge, we have developed a model, how the competition for limited resources (such as road capacity) could be *optimally* guided by means of information services. Let us assume we had $n_1(t)$ 1-decisions at time t, but the optimal number of 1-decision at time $t+1$ is calculated to be $f_1^{\mathrm{eq}}(t+1)N \geq n_1(t)$. Our aim is to balance the deviation $f_1^{\mathrm{eq}}(t+1)N - n_1(t) \geq 0$ by the expected net number

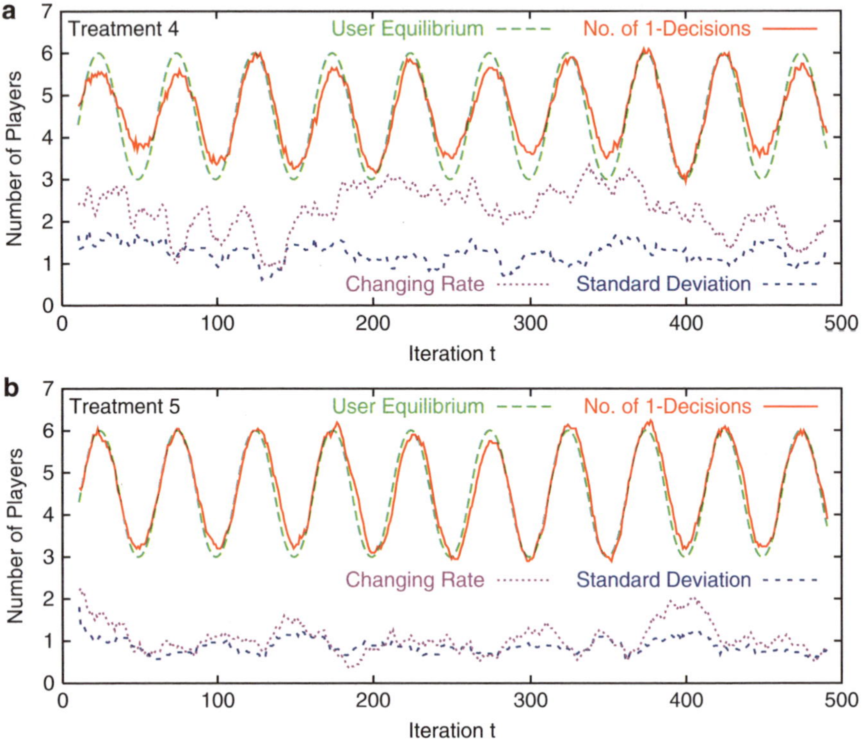

Fig. 13.10 Representative examples for **(a)** treatment 4 and **(b)** treatment 5 (from [21]). The displayed curves are moving time-averages over 20 iterations. Compared to treatment 4, the user-specific recommendations in treatment 5 (assuming $C_M = C_S = 1$, $R_1 = 0$, $R_2 = \max([f_1^{\text{eq}}(t+1)N - n_1(t) + B(t+1)]/n_2(t), 1)$, $I_1 = I_2 = 1$) could increase the group adaptability to the user equilibrium a lot, even if they had a systematic or random bias B (see Fig. 13.11a). The standard deviation was reduced considerably and the changing rate even more

$$\langle \Delta n_1(t+1)\rangle = \langle n_1(t+1) - n_1(t)\rangle = \langle n_1(t+1)\rangle - n_1(t) \tag{13.18}$$

of transitions from decision 2 to decision 1, i.e. $f_1^{\text{eq}}(t+1)N - n_1(t) = \langle \Delta n_1(t+1)\rangle$. In the case $f_1^{\text{eq}}(t+1)N - n_1(t) < 0$, indices 1 and 2 have to be interchanged.

Let us assume we give recommendations to fractions $I_1(t)$ and $I_2(t)$ of players who had chosen decision 1 and 2, respectively. The fraction of changing recommendations to previous 1-choosers shall be denoted by $R_1(t)$, and for previous 2-choosers by $R_2(t)$. Correspondingly, fractions of $[1 - R_1(t)]$ and $[1 - R_2(t)]$ receive a recommendation to stick to the previous decision. Moreover, $[1 - C_M(t)]$ is the *refusal probability* of recommendations to change, while $[1 - C_S(t)]$ is the refusal probability of recommendations to stay. Finally, we denote the spontaneous transition probability from decision 1 to 2 by $p_a(2|1, n_1; t)$ and the inverse transition probability by $p_a(1|2, n_1; t)$, in case a player does not receive any recommendation.

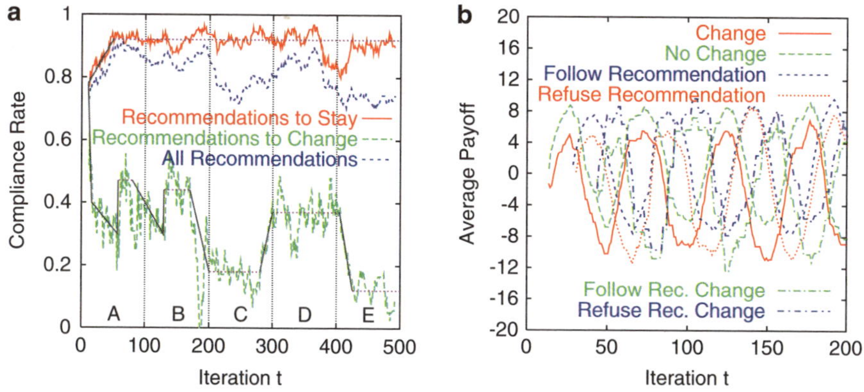

Fig. 13.11 **(a)** In treatment 5, the compliance to recommendations to change dropped considerably below the compliance to recommendations to stay. The compliance to changing recommendations was very sensitive to the degree of their reliability, i.e. participants followed recommendations just as much as they helped them to reach the user equilibrium (so that the bias B did not affect the small deviation from it, see Fig. 13.10b). While during time interval A, the recommendations would have been perfect, if all players had followed them, in time interval B the user equilibrium was overestimated by $B = +1$, in C it was underestimated by $B = -2$, in D it was randomly over- or underestimated by $B = \pm 1$, and in E by $B = \pm 2$. Obviously, a random error is more serious than a systematic one of the same amplitude. *Dotted non-vertical lines* illustrate the estimated compliance levels during the transient periods and afterwards (*horizontal dotted lines*). **(b)** The average payoffs varied largely with the decision behavior. Players who changed their decision got significantly lower payoffs on average than those who kept their previous decision. Even recommendations could not overcome this difference: It stayed profitable not to change, although it was generally better to follow recommendations than to refuse them. For illustrative reasons, the third and fourth line were shifted by 15, while the fifth and sixth line were shifted by 30 iterations (From [21])

This happens with probabilities $[1 - I_1(t)]$ and $[1 - I_2(t)]$, respectively. Both transition probabilities $p_a(2|1, n_1; t)$ and $p_a(1|2, n_1; t)$ are functions of the number $n_1(t) = N - n_2(t)$ of previous 1-decisions. The index a allows us to reflect different strategies or characters of players. The fraction of players pursuing strategy a is then denoted by $F_a(t)$. Applying methods summarized in [24, 25], the expected change $\langle \Delta n_1(t+1) \rangle$ of n_1 is given by the balance equation

$$
\begin{aligned}
\langle \Delta n_1(t+1) \rangle = & \sum_a p_a(1|2, n_1; t) F_a(t) [1 - I_2(t)] n_2(t) \\
& - \sum_a p_a(2|1, n_1; t) F_a(t) [1 - I_1(t)] n_1(t) \\
& + \sum_a \{ C^a_M(t) R_2(t) + [1 - C^a_S(t)][1 - R_2(t)] \} F_a(t) I_2(t) n_2(t) \\
& - \sum_a \{ C^a_M(t) R_1(t) + [1 - C^a_S(t)][1 - R_1(t)] \} F_a(t) I_1(t) n_1(t),
\end{aligned}
\tag{13.19}
$$

which should agree with $f_1^{\text{eq}}(t+1)N - n_1(t)$. We have evaluated the overall transition probabilities

$$p(1|2, n_1; t) = \sum_a p_a(1|2, n_1; t) F_a(t) \quad \text{and}$$

$$p(2|1, n_1; t) = \sum_a p_a(2|1, n_1; t) F_a(t). \tag{13.20}$$

According to classical decision theories [7, 9, 24, 25, 41], we would expect that the transition probabilities $p_a(2|1, n_1; t)$ and $p(2|1, n_1; t)$ should be monotonically increasing functions of the payoff $P_2(N - n_1(t))$, the payoff difference $P_2(N - n_1(t)) - P_1(n_1(t))$, the potential payoff $P_2(N - n_1(t) + \epsilon N)$, or the potential payoff gain $P_2(N - n_1(t) + \epsilon N) - P_1(n_1(t))$. All these quantities vary linearly with n_1, so that $p(2|1, n_1; t)$ should be a monotonic function of $n_1(t)$. A similar thing should apply to $p(1|2, n_1; t)$. Instead, the experimental data point to transition probabilities with a *minimum* at the user equilibrium (see Fig. 13.12a). That is, the players stick to a certain alternative for a longer time, when the system is close to the user equilibrium. This is a result of learning [16, 17, 26, 27, 39, 40]. In fact, we find a gradual change of the transition probabilities in time (see Fig. 13.12b). The corresponding "learning curves" reflect the players' adaptation to the user equilibrium. After the experimental determination of the transition probabilities $p(2|1, n_1; t)$, $p(1|2, n_1; t)$ and specification of the overall compliance probabilities

$$C_M(t) = \sum_a C_M^a(t) F_a(t), \qquad C_S(t) = \sum_a C_S^a(t) F_a(t), \tag{13.21}$$

we can guide the decision behavior in the system via the levels $I_i(t)$ of information dissemination and the fractions $R_i(t)$ of recommendations to change ($i \in \{1, 2\}$). These four degrees of freedom allow us to apply a variety of guidance strategies depending on the respective information medium. For example, a guidance by radio news is limited by the fact that $I_1(t) = I_2(t)$ is given by the average percentage of radio users. Therefore, (13.19) cannot always be solved by variation of the fractions of changing recommendations $R_i(t)$. User-specific services have much higher guidance potentials and could, for example, be transmitted via SMS. Among the different guidance strategies fulfilling (13.19), the one with the minimal statistical variance will be the best. However, it would already improve the present situation to inform everyone about the *fractions* $R_i(t)$ of participants who should change their decision, as users can learn to respond with varying frequencies (see Fig. 13.12). Some actually respond more sensitively than others (see Fig. 13.4), so that a group of users can reach a good overall performance based on individual strategies.

The outlined guidance strategy could, of course, also be applied to reach the system optimum rather than the user optimum. The values of $\Delta n_1(t+1)$ would just be different. Note, however, that the users would soon recognize that this guidance is

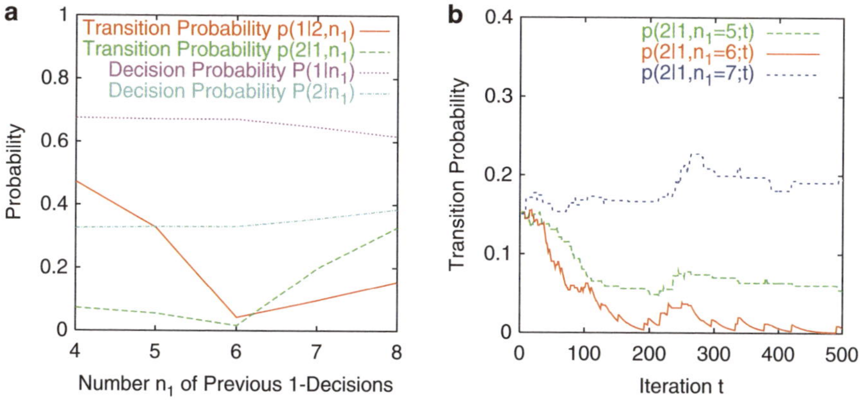

Fig. 13.12 Illustration of group-averaged decision distributions $P(i|n_1)$ and transition probabilities $p(i'|i,n_1;t)$ measured in treatment 3 (from [21]). **(a)** The probability $P(1|n_1)$ to choose alternative 1 was approximately 2/3, independently of the number n_1 of players who had previously chosen alternative 1. The probability $P(2|n_1)$ to choose alternative 2, given that n_1 players had chosen alternative 1, was always about 1/3. In contrast, the group-averaged transition probability $p(1|2,n_1)$ describing decision changes from alternative 2 to 1 did depend on the number n_1 of players who had chosen decision 1. The same was true for the inverse transition probability $p(2|1,n_1)$ from decision 1 to decision 2. Remarkably enough, these transition probabilities are not monotonically increasing with the payoff or the expected payoff gain, as they do not monotonically increase with n_1. Instead, the probability to change the decision shows a minimum at the user equilibrium $n_1 = f_1^{\rm eq} N = 6$. Figures 13.4 and 13.8 suggest that this transition probability does not reflect the individual transition probabilities. There rather seem to be typical response patterns (see Sect. 13.4), i.e. some individuals react only to large deviations from the user equilibrium, while others already react to small ones, so that the overall response of the group reaches a good adaptation performance. **(b)** The reason for the different transition probabilities is an adaptation process in which the participants learn to take fewer changing decisions, when the user equilibrium is reached or close by, but more, when the user equilibrium is far away (The curves were exponentially smoothed with $\alpha = 0.05$)

not suitable to reach the user optimum. Consequently, the compliance probabilities would gradually go down, which would affect the potentials and reliability of the guidance system.

In practical applications, we would determine the compliance probabilities $C_j(t)$ with $j \in \{M, S\}$ (and the transition probabilities) on-line with an exponential smoothing procedure according to

$$C_j(t+1) = \alpha C_j'(t) + (1-\alpha)C_j(t) \quad \text{with} \quad \alpha \approx 0.1, \tag{13.22}$$

where $C_j'(t)$ is the percentage of participants who have followed their recommendation at time t. As the average payoff for decision changes is normally lower than for staying with the previous decision (see Figs. 13.11 and 13.3d), a high compliance probability C_M is hard to achieve. That is, individuals who follow recommendations to change normally pay for reaching the user equilibrium (because of the overreaction in the system). Hence, there are no good preconditions to charge the players for recommendations, as we did in another treatment. Consequently, only a few players requested recommendations, which reduced their reliability, so that the overall performance of the system went down.

13.4.2 Master Equation Description of Iterated Decisions

The description of decisions that are taken at discrete time steps (e.g. on a day-to-day basis) is different from decisions in continuous time. We can, however, apply the time-discrete master equation [24] with $\Delta t = 1$, if there is no need to distinguish several characters a. As the number of individuals changing to the other alternative is given by a *binomial distribution*, we obtain the following expression for the configurational transition probability:

$$
\begin{aligned}
&P\big((n_1,n_2),t+1 \,\big|\, (n_1-\Delta n_1, n_2+\Delta n_1),t\big) \\
&= \sum_{k=0}^{\min(n_1-\Delta n_1, n_2)} \binom{n_2+\Delta n_1}{\Delta n_1 + k} p(1|2,n_1-\Delta n_1;t)^{\Delta n_1 + k} \\
&\times\big[1-p(1|2,n_1-\Delta n_1;t)\big]^{n_2-k} \binom{n_1-\Delta n_1}{k} p(2|1,n_1-\Delta n_1;t)^k \\
&\times\big[1-p(2|1,n_1-\Delta n_1;t)\big]^{n_1-\Delta n_1-k}.
\end{aligned}
\tag{13.23}
$$

This formula sums up the probabilities that Δn_1+k of $n_2+\Delta n_1$ previous 2-choosers change independently to alternative 1 with probability $p(1|2,n_1-\Delta n_1;t)$, while k of the $n_1-\Delta n_1$ previous 1-choosers change to alternative 2 with probability $p(2|1,n_1-\Delta n_1;t)$, so that the net number of changes is Δn_1. If $\Delta n_1 < 0$, the roles of alternatives 1 and 2 have to be interchanged. Only in the limits $p(1|2,n_1-\Delta n_1;t) \approx 0$ and $p(2|1,n_1-\Delta n_1;t) \approx 0$ corresponding to $\Delta t \approx 0$ do we get the approximation

$$
\begin{aligned}
&P\big((n_1,n_2),t+1 \,\big|\, (n_1-\Delta n_1, n_2+\Delta n_1),t\big) \\
&\approx \begin{cases} p(1|2,n_1-1;t)(n_2+1) & \text{if } \Delta n_1 = +1 \\ p(2|1,n_1+1;t)(n_1+1) & \text{if } \Delta n_1 = -1 \\ 0 & \text{otherwise,} \end{cases}
\end{aligned}
\tag{13.24}
$$

which is relevant for the time-continuous master equation.

The potential use of (13.23) is the calculation of the statistical variation of the decision distribution or, equivalently, the number n_1 of 1-choosers. It also allows one to determine the variance, which the optimal guidance strategy should minimize in favour of reliable recommendations.

13.5 Summary and Outlook

With the described decision experiments, we have explored different and identified superior ways of information presentation that facilitate to guide user decisions in the spirit of higher payoffs. By far the least standard deviations from the user equilibrium could be reached by presenting the own payoff and the potential payoff, if the respective participant (or a certain fraction of players) had additionally chosen the other alternative. Interestingly, the decision dynamics was found to be intermittent similar to the volatility clustering in stock markets, where individuals also react to aggregate information. This results from the desire to reach above-average payoffs, combined with the immanent overreaction in the system. We have also demonstrated that payoff losses due to a volatile decision dynamics (e.g., excess travel times) can be reduced via user-specific recommendations by a factor of three or more. Such kinds of results will be applied to the route guidance on German highways (see, for example, the project SURVIVE conducted by Nobel prize winner Reinhard Selten and Michael Schreckenberg). Optimal recommendations to reach the user equilibrium follow directly from the derived balance equation (13.19) for decision changes based on empirical transition and compliance probabilities. The quantification of the transition probabilities requires a novel stochastic description of the decision behavior, which is not just driven by the potential (gains in) payoffs, in contrast to intuition and established models. To understand these findings, one has to take into account reinforcement learning, which can also explain the emergence of individual response patterns (see Sect. 13.4).

Obviously, it requires both, theoretical and experimental efforts to get ahead in decision theory. In a decade from now, the microscopic theory of human interactions will probably have been developed to a degree that allows one to systematically derive social patterns and economic dynamics on this ground. This will not only yield a deeper understanding of socio-economic systems, but also help to more efficiently distribute scarce resources such as road capacities, time, space, money, energy, goods, or our natural environment. One day, similar guidance strategies as the ones suggested above may help politicians and managers to stabilize economic markets, to increase average *and* individual profits, and to decrease the unemployment rate.

Acknowledgements This study was partially supported by the ALTANA-Quandt foundation. The author wants to thank Prof. Aruka, Prof. Selten, and Prof. Schreckenberg for their invitations and fruitful discussions, Prof. Kondor and Dr. Schadschneider for inspiring comments, Tilo Grigat for preparing some of the illustrations, Martin Schönhof and Daniel Kern for their help in setting up and carrying out the decision experiments, and the test persons for their patience and ambitious playing until the end of our experiments. Hints regarding manuscript-related references are very much appreciated.

References

1. J. Adler, V. Blue, Towards the design of intelligent traveler information systems. Transport. Res. C **6**, 157–172 (1998)
2. R. Arnott, A. de Palma, R. Lindsey, Does providing information to drivers reduce traffic congestion? Transport. Res. A **25**, 309–318 (1991)
3. W.B. Arthur, Inductive reasoning and bounded rationality. Am. Econ. Rev. **84**, 406–411 (1994)
4. Articles in *Route Guidance and Driver Information*, IEE Conference Publications, Vol. **472** (IEE, London, 2000)
5. W. Barfield, T. Dingus, *Human Factors in Intelligent Transportation Systems* (Erlbaum, Mahwah, NJ, 1998)
6. M. Ben-Akiva, A. de Palma, I. Kaysi, Dynamic network models and driver information systems. Transport. Res. A **25**, 251–266 (1991)
7. M. Ben-Akiva, D.M. McFadden, et al., Extended framework for modeling choice behavior. Market. Lett. **10**, 187–203 (1999)
8. M. Ben-Akiva, J. Bottom, M.S. Ramming, Route guidance and information systems. Int. J. Syst. Contr. Engin. **215**, 317–324 (2001)
9. M. Ben-Akiva, S.R. Lerman, *Discrete Choice Analysis: Theory and Application to Travel Demand* (MIT Press, Cambridge, MA, 1997)
10. P. Bonsall, P. Firmin, M. Anderson, I. Palmer, P. Balmforth, Validating the results of a route choice simulator. Transport. Res. C **5**, 371–387 (1997)
11. P. Bonsall, The influence of route guidance advice on route choice in urban networks. Transportation **19**, 1–23 (1992)
12. D. Challet, M. Marsili, Y.-C. Zhang, Modeling market mechanism with minority game. Physica A **276**, 284–315 (2000)
13. D. Challet, Y.-C. Zhang, Emergence of cooperation and organization in an evolutionary game. Physica A **246**, 407ff (1997)
14. D. Challet, Y.-C. Zhang, On the minority game: Analytical and numerical studies. Physica A **256**, 514–532 (1998)
15. P.S.-T. Chen, K.K. Srinivasan, H.S. Mahmassani, Effect of information quality on compliance behavior of commuters under real-time traffic information. Transport. Res. Record **1676**, 53–60 (1999)
16. Y.-W. Cheung, D. Friedman, Individual learning in normal form games: Some laboratory results. Games Econ. Behav. **19**(1), 46–76 (1997)
17. I. Erev, A.E. Roth, Predicting how people play games: Reinforcement learning in experimental games with unique, mixed strategy equilibria. Am. Econ. Rev. **88**(4), 848–881 (1998)
18. D. Fudenberg, D. Levine, *The Theory of Learning in Games* (MIT Press, Cambridge, MA, 1998)
19. S. Ghashghaie, W. Breymann, J. Peinke, P. Talkner, Y. Dodge, Turbulent cascades in foreign exchange markets. Nature **381**, 767–770 (1996)
20. R. Hall, Route choice and advanced traveler information systems on a capacitated and dynamic network. Transport. Res. C **4**, 289–306 (1996)
21. D. Helbing, M. Schönhof, D. Kern, Volatile decision dynamics: Experiments, stochastic description, intermittency control, and traffic optimization. New J. Phys. **4**, 33.1–33.16 (2002)
22. D. Helbing, A section-based queueing-theoretical traffic model for congestion and travel time analysis, J. Phys. A: Math. Gen. **36**(46), L593-L598 (2003)
23. D. Helbing, Traffic and related self-driven many-particle systems. Rev. Mod. Phys. **73**, 1067–1141 (2001)
24. D. Helbing, *Quantitative Sociodynamics* (and references therein) (Kluwer Academic, Dordrecht, 1995)
25. D. Helbing, *Stochastische Methoden, nichtlineare Dynamik und quantitative Modelle sozialer Prozesse* (Shaker, Aachen, 1996)

26. J.B. van Huyck, J.P. Cook, R.C. Battlio, Selection dynamics, asymptotic stability, and adaptive behavior. J. Pol. Econ. **102**(5), 975–1005 (1994)
27. J.B. van Huyck, R.C. Battlio, R.O. Beil, Tacit coordination games, strategic uncertainty, and coordination failure. Am. Econ. Rev. **80**(1), 234–252 (1990)
28. Y. Iida, T. Akiyama, T. Uchida, Experimental analysis of dynamic route choice behavior. Transport. Res. B **26**, 17–32 (1992)
29. A. Khattak, A. Polydoropoulou, M. Ben-Akiva, Modeling revealed and stated pretrip travel response to advanced traveler information systems. Transport. Res. Record **1537**, 46–54 (1996)
30. H.N. Koutsopoulos, A. Polydoropoulou, M. Ben-Akiva, Travel simulators for data collection on driver behavior in the presence of information. Transport. Res. C **3**, 143–159 (1995)
31. M. Kraan, H.S. Mahmassani, N. Huynh, Traveler Responses to Advanced Traveler Information Systems for Shopping Trips: Interactive Survey Approach. Transport. Res. Record **1725**, 116 (2000)
32. R.D. Kühne, K. Langbein-Euchner, M. Hilliges, N. Koch, Evaluation of compliance rates and travel time calculation for automatic alternative route guidance systems on freeways. Transport. Res. Record **1554**, 153–161 (1996)
33. T. Lux, M. Marchesi, Scaling and criticality in a stochastic multi-agent model of a financial market. Nature **397**, 498–500 (1999)
34. M.W. Macy, A. Flache, Learning dynamics in social dilemmas. Proc. Natl. Acad. Sci. USA **99**(Suppl. 3), 7229–7236 (2002)
35. H.S. Mahmassani, D.-G. Stephan, Experimental investigation of route and departure time choice dynamics of urban commuters. Transport. Res. Records **1203**, 69–84 (1988)
36. H.S. Mahmassani, R. Jayakrishnan, System performance and user response under real-time information in a congested traffic corridor. Transport. Res. A **25**, 293–307 (1991)
37. H.S. Mahmassani, R.C. Jou, Transferring insights into commuter behavior dynamics from laboratory experiments to field surveys. Transport. Res. A **34**, 243–260 (2000)
38. R.N. Mantegna, H.E. Stanley, *Introduction to Econophysics: Correlations and Complexity in Finance* (Cambridge University, Cambridge, England, 1999)
39. J. Nachabar, Prediction, optimization, and learning in repeated games. Econometrica **65**, 275–309 (1997)
40. S. Nakayama, R. Kitamura, Route Choice Model with Inductive Learning. Transport. Res. Record **1725**, 63–70 (2000)
41. J. de D. Ortúzar, L.G. Willumsen, *Modelling Transport*, Chap. 7: Discrete-Choice Models (Wiley, Chichester, 1990)
42. M. Schreckenberg, R. Selten (eds.), *Human Behaviour and Traffic Networks* (Springer, Berlin, 2004)
43. M. Schreckenberg, R. Selten, T. Chmura, T. Pitz, J. Wahle, Experiments on day-to-day route choice (and references therein), e-print http://vwitme011.vkw.tu-dresden.de/TrafficForum/journalArticles/tf01080701.pdf, last accessed on March 8, 2012
44. K.K. Srinivasan, H.S. Mahmassani, Modeling Inertia and Compliance Mechanisms in Route Choice Behavior Under Real-Time Information. Transport. Res. Record **1725**, 45–53 (2000)
45. J. Wahle, A. Bazzan, F. Klügl, M. Schreckenberg, Decision dynamics in a traffic scenario. Physica A **287**, 669–681 (2000)
46. J. Wahle, A.L.C. Bazzan, F. Klügl, M. Schreckenberg, Anticipatory traffic forecast using multi-agent techniques, in *Traffic and Granular Flow '99*, ed. by D. Helbing, H.J. Herrmann, M. Schreckenberg, D.E. Wolf (Springer, Berlin, 2000), pp. 87–92

Chapter 14
Systemic Risks in Society and Economics*

14.1 Introduction

When studying systemic risks, i.e. risks that can trigger unexpected large-scale changes of a system or imply uncontrollable large-scale threats to it, scientific research has often focused on natural disasters such as earthquakes, tsunamis, hurricanes, vulcano outbreaks, or on failures of engineered systems such as blackouts of the electric power grids or nuclear accidents (as in Chernobyl). However, many major disasters hitting human societies relate to social problems [1–4]: This includes famines and other shortages of resources, wars, climate change, and epidemics, some of which are related to population density and population growth. Financial instabilities and economic crises are further examples of systemic risks.

Let us illustrate these risks by some numbers: World War I caused more than 15,000,000 victims, and World War II even 60,000,000 fatalities. The latter generated costs of 1,000 billion 1944 US$ and destroyed 1,710 cities, 70,000 villages, 31,850 industrial establishments, 40,000 miles of railroad, 40,000 hospitals, and 84,000 schools. Moreover, the world has seen many lossful wars ever since. The current financial and economic crises triggered an estimated loss of 4-20 trillion US$.

Climate change is expected to cause natural disasters, conflicts for water, food, land, migration, social and political instability. The related reduction of the world gross domestic product is expected to amount to 0.6 trillion US$ per year or more. Turning our attention to epidemics, the Spanish flu has caused 20-40 million deaths, and SARS has triggered losses of 100 billion US$.

Considering these examples, one could in fact say "The major risks are social", but they are still poorly understood. In fact, we know much more about the origin

*This chapter reprints a Case Study to be cited as D. Helbing (2010) Systemic Risks in Society and Economics. International Risk Governance Council (irgc), see http://irgc.org/IMG/pdf/Systemic_Risks_Helbing2.pdf.

DOI 10.1007/978-3-642-24004-1_14,

of the universe and about elementary particles than about the working of our socio-economic system. This situation must be urgently changed (see Sect. 14.5).

It is obvious that mankind must be better prepared for the crises to come. A variety of factors is currently driving the world out of equilibrium: Population growth, climate change, globalization, changes in the composition of populations, and the exploitation of natural resources are just some examples. As president of New York's Columbia University, Lee C. Bollinger formulated the problem as follows: "The forces affecting societies around the world ... are powerful and novel. The spread of global market systems ... are ... reshaping our world ..., raising profound questions. These questions call for the kinds of analyses and understandings that academic institutions are uniquely capable of providing. Too many policy failures are fundamentally failures of knowledge" [5].

We certainly need to increase our capacity to gain a better understanding of socio-economic systems, conditions triggering instabilities, alternative system designs, ways to avoid or mitigate crises, and side effects of policy measures. This contribution will shortly summarize the current knowledge of how systemic risks emerge in society, and give a variety of relevant examples.

14.2 Socio-Economic Systems as Complex Systems

An important aspect of social and economic systems is that they are *complex* systems (see Fig. 14.1) [6–38]. Other examples of complex systems are turbulent fluids, traffic flows, large supply chains, or ecological systems. The commonality of complex systems is that they are characterized by a large number of interacting (mutually coupled) system elements (such as individuals, companies, countries, cars, etc.) [7, 39–49]. These interactions are usually non-linear (see Sect. 14.2.1). Typically, this implies a rich system behavior [7]. In particular, such systems tend to behave *dynamic* rather than static, and *probabilistic* rather than deterministic. As a consequence, complex systems can show surprising or even *paradoxical behaviors*. The slower-is-faster effect [50, 51], according to which delays can sometimes speed up the efficiency of transport systems, may serve as an example.

Moreover, complex systems are often hardly predictable and uncontrollable. While we are part of many complex systems (such as traffic flows, groups or crowds, financial markets, and other socio-economic systems), our perception of them is mostly oversimplified [52, 53] or biased [54–56]. In fact, they challenge our established ways of thinking and are currently a nightmare for decision-makers [52]. The following subsections will explain these points in more detail.

Note that there are at least three different ways in which the term "complexity" is used:

1. *Structural complexity* applies, for example, to a car, which is a complicated system made up of many parts. These parts, however, are constructed in a way that makes them behave in a deterministic and predictable way. Therefore, a car is relatively easy to control.

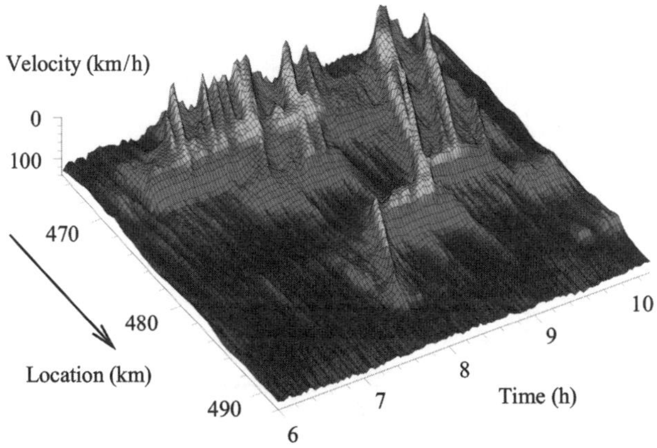

Fig. 14.1 Freeway traffic constitutes a dynamically complex system, as it involves the interaction of many independent driver-vehicle units with a largely autonomous behavior. Their interactions can lead to the self-organization of different kinds of traffic jams, the occurence of which is hard to predict (after [57])

2. *Dynamic complexity* may be illustrated by freeway traffic. Here, the interaction of many independent driver-vehicle units with a largely autonomous behavior can cause the self-organization of different kinds of traffic jams, the occurence of which is hard to predict (see Fig. 14.1).
3. *Algorithmic complexity* measures how the computer resources needed to simulate or optimize a system scale with system size.

This chapter mainly focuses on *dynamic* complexity.

14.2.1 Non-Linear Interactions and Power Laws

Systems with a complex system dynamics are mostly characterized by non-linear interactions among the elements or entities constituting the system (be it particles, objects, or individuals). Non-linear interactions are typical for systems in which elements mutually adapt to each other. That is, the elements are *influenced* by their environment, but at the same time, they also have an *impact* on their environment.

Non-linearity means that causes and effects are not proportional to each other. A typical case is a system that is hardly responsive to control attempts, or which shows sudden regime shifts when a "tipping point" is crossed [58–63] (see Fig. 14.2). Examples for this are sudden changes in public opinion (e.g. from smoking-tolerance to smoking bans, from pro- to anti-war mood, from a strict banking secret to transparency, or from buying pickup trucks to buying environment-friendly cars).

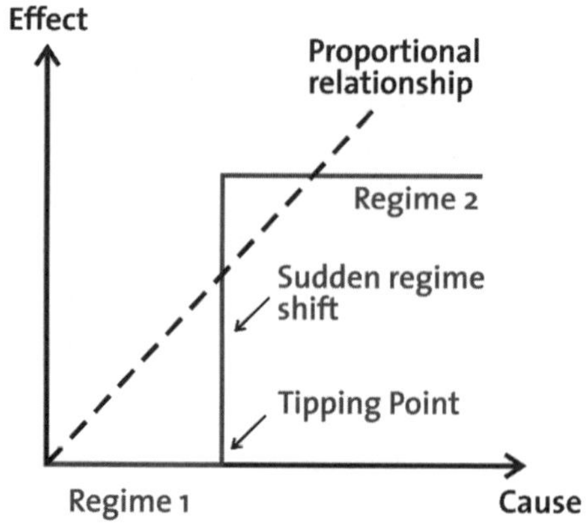

Fig. 14.2 Schematic illustration of one of the typical behaviors of complex systems: In regimes 1 and 2, a "cause" (such as a *control attempt*) has essentially no effect on the system, while at the "tipping point", an abrupt (and *often unexpected*) transition to a different system behavior occurs. A recent example is the sudden large-scale *erosion of the Swiss banking secret*, after UBS had handed over about 300 names of clients to an US authority

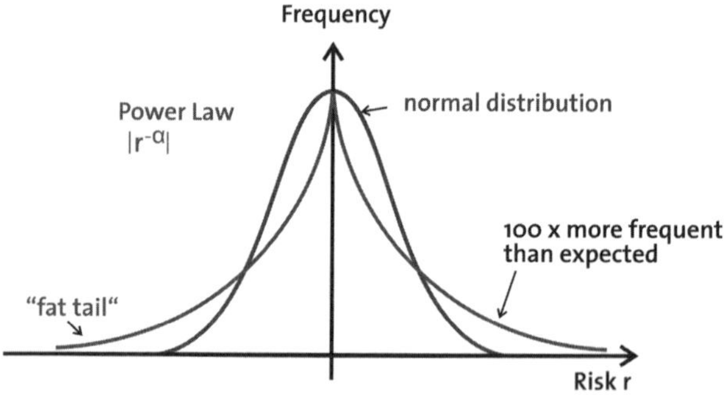

Fig. 14.3 When system components interact strongly, the normally distributed behavior of *separated* system elements often becomes (approximately) power-law distributed. As a consequence, fluctuations of any size can occur in the system, and extreme events are much more frequent than expected. Note that power laws are typical for a system at a critical point, also known as "tipping point"

14.2.2 Power Laws and Heavy-Tail Distributions

It is important to note that strong interactions among the system elements often change the statistical distributions characterizing their behavior. Rather than normal distributions, one typically finds (truncated) "power laws" or, more generally, so-called heavy-tail distributions [48, 49, 58] (see Fig. 14.3 and Sect. 14.2.4). These imply that extreme events occur much more frequently than expected. For example, the crash of the stock market on Black Monday was a 35 sigma event (where sigma stands for the standard deviation of the Dow Jones Index on a logarithmic

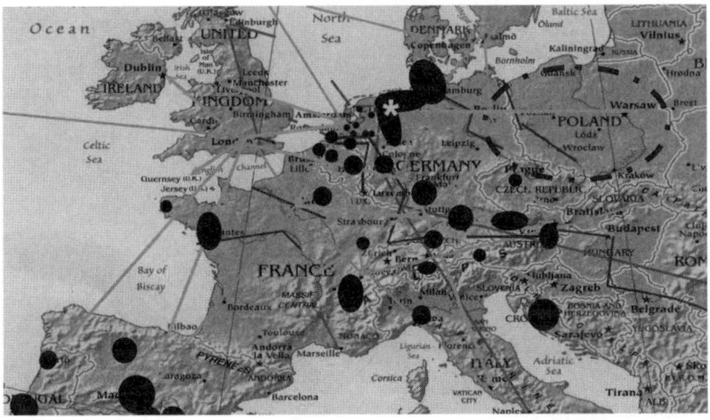

Fig. 14.4 Example of a blackout of the electrical power grid in Europe (from: EU project IRRIIS. E. Liuf (2007) Critical Infrastructure protection, R&D view). To allow for the transfer of a ship, one power line had to be temporarily disconnected in Northern Germany. This triggered an overload-related cascading effect [80], during which many power lines went out of operation. As a consequence, there were blackouts all over Europe (see black areas). The pattern illustrates how counter-intuitive and hardly predictable the behavior of complex systems with network interactions can be

scale). Other examples are the size distributions of floods, storms, earth quakes, or wars [1–4]. Obviously, the occurence of the respective heavy-tail distributions is highly important for the insurance business and for the risk assessment of financial derivatives.

14.2.3 Network Interactions and Systemic Risks Through Failure Cascades

A typical case of non-linear interactions are network interactions, which are ubiquitous in socio-economic systems [64–79]. These imply *feedback loops* and *vitious circles* or induce (often undesired) *side effects* [32]. (For example, the introduction of cigarette taxes has promoted smuggling and other criminal activities.) Moreover, network interactions are often the reason for a *cascading of failure events*. Examples for this are *epidemic spreading*, the failure of the interbank market during a financial crisis, the spreading of *traffic congestion*, or the *blackout of an electrical power system* (see Fig. 14.4).

Failure cascades (which are also called *chain reactions, avalanche* or *domino effects*) are the most common mechanism by which *local* risks can become *systemic* [81–84] (see Fig. 14.5). Systemic failures are usually triggered by one of the following reasons:

1. The parameters determining system stability are driven towards a so-called "critical point" or "tipping point", beyond which the system behavior becomes

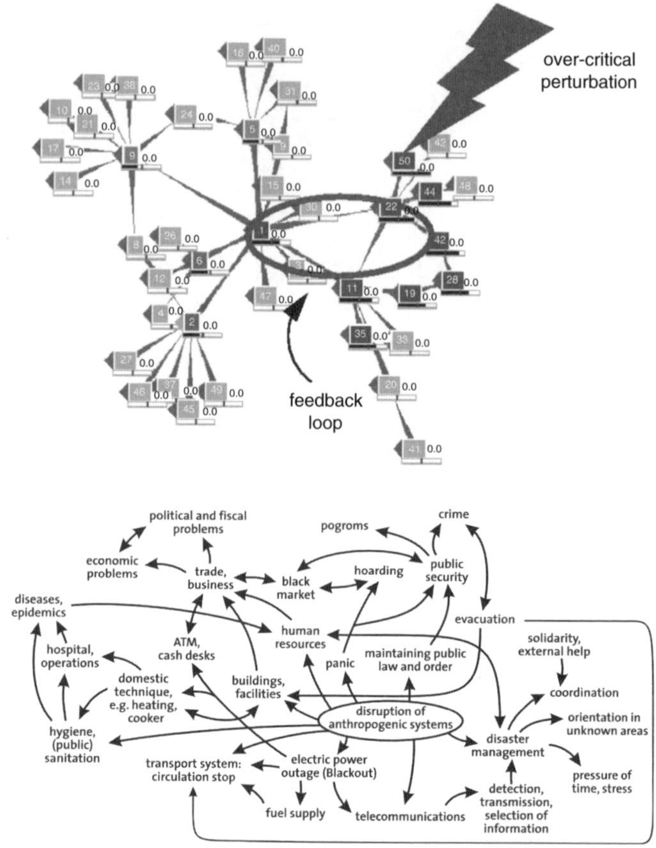

Fig. 14.5 Top: Schematic illustration of a networked system which is hit by an over-critical perturbation (e.g. a natural disaster). The problem of feedback cycles is highlighted. They can have "autocatalytic" (*escalation*) effects and act like vitious circles. Bottom: Illustration of cascading effects in socio-economic systems, which may be triggered by the *disruption* (*over-critical perturbation*) *of an anthropogenic system*. A more detailed picture can be given for specific disasters. Note that the largest financial damage of most disasters is caused by such cascading effects, i.e. the systemic impact of an over-critical perturbation (after [85])

unstable (see Sect. 14.2.1). For example, the *destabilization* of the former German Democratic Republic (GDR) triggered off spontaneous demonstrations in Leipzig, Germany, in 1989, which eventually caused the re-unification of Germany. This "peaceful revolution" shows that systemic instability does not necessarily imply systemic malfunctions. It can also induce a transition to a better and more robust system state after a transient transformation period. Further examples of spontaneous transitions by systemic destabilization are discussed in Sects. 14.2.4, 14.3, and 14.4.1.

2. The system is metastable (i.e. robust to small perturbations, which quickly disappear over time), but there occurs an *over-critical perturbation* (such as a

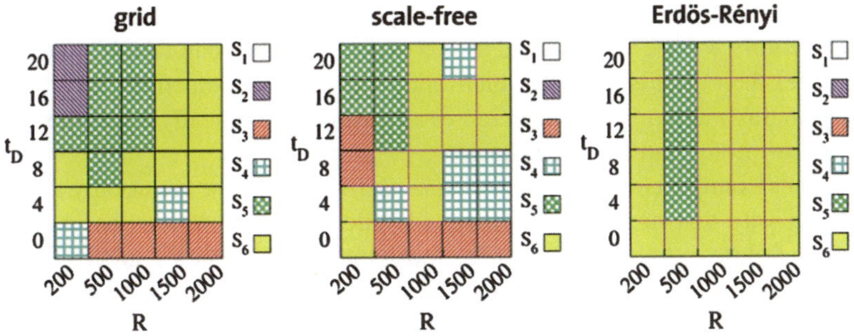

Fig. 14.6 The most efficient disaster response strategy depends on many factors such as the network type (after [84]). Here, we have studied six different disaster response strategies for regular grids, scale-free networks, and Erdös-Rényi random networks. The best strategy is a function of the resources R available for disaster response management and the time delay t_D before practical measures are taken. Obviously, there is no single strategy, which always performs well. This makes disaster response challenging, calling for scientific support

natural disaster), which harms the system functionality so much that this has damaging effects on other parts of the system [84] (see Fig. 14.6).

3. The system is metastable, but there is a *coincidence of several perturbations* in the network nodes or links such that their interaction happens to be overcritical and triggers off additional failures in other parts of the system [83]. In fact, disasters caused by *human error* [86, 87] are often based on a combination of several errors. In networked systems, the occurence of this case is just a matter of time.

14.2.4 Self-Organized or Self-Induced Criticality

A system may get into a critical state not only by external influences that are affecting system stability. It is known that some endogeneous processes can *automatically* drive the system towards a critical state, where avalanche or cascading effects of arbitrary size appear (reflecting the characteristic heavy-tail statistics at critical points, see Sect. 14.2.2). In such cases, the occurence of extreme events is expected, and we speak of "self-induced" or "self-organized criticality" (SOC) [88, 89].

It is likely that *bankruptcy cascades* can be understood in this way. The underlying mechanism is that a company or bank tries to make a better offer to customers or clients than the competing companies or banks do. This forces the competitors to make better offers as well. Eventually, the profit margins in a free market become so small that variations in the consumption rate can drive some companies or banks out of business, which creates economic problems for other companies or banks. Considering the interconnections between different companies or banks, this mechanism can cause bankruptcy cascades. Eventually, the number of competitors will be smaller, and as a consequence, they can take higher prices. Therefore, their profits go up, which encourages new competitors to enter the market. In this way, competition increases again and automatically drives the system back to low profits and bankruptcies.

Another example concerns *safety standards* [86, 87]. These are usually specified in such a way that normal perturbations would not cause serious harm or even systemic failures. As a consequence, most man-made systems are constructed in a way that makes them robust to small and moderate perturbations (in other words: meta-stable). However, the requirement of cost efficiency excerts pressure on decision-makers to restrict safety standards to what really appears to be needed, and not more. Consequently, if a large-scale failure has not occurred in a long time, decision-makers often conclude that the existing safety standards are higher than necessary and that there is some potential to reduce costs by decreasing them somewhat. Eventually, the standards are lowered so much that an over-critical perturbation occurs sooner or later, which causes a systemic failure. As a consequence, the safety-standards will be increased again, and the process will start from the beginning.

As a third example, let us discuss man-made systems with capacity limits such as *traffic or logistic systems*. These systems are often driven towards maximum efficiency, i.e. full usage of their capacity. However, when reaching this point of maximum efficiency, they also reach a tipping point, at which the system becomes dynamically unstable [90]. This is known, for example, from freeway and railway traffic. As a consequence, the system suffers an unexpected *capacity drop due to optimization efforts*, shortly after the maximum performance was reached.

Similarly to freeway traffic, engineers also try to avoid the occurence of congestion in *urban traffic*, which can be reached by re-routing strategies. A closer analysis shows that this optimization leads again to a sudden breakdown of the flow, once the maximum throughput is reached [91]. One may, therefore, conclude that *optimizing for the full usage of available system capacity implies the danger of an abrupt breakdown of the system performance with potentially very harmful consequences. To avoid this problem, one must know the capacity of the system and avoid to reach it. This can be done by requiring to respect sufficient safety margins.*

14.2.5 Limits of Predictability, Randomness, Turbulence and Chaos

The large number of non-linearly coupled system components can lead to a *complex dynamics* (see Fig. 14.7). Well-known examples for this are the phenomena of *turbulence* [92] and *chaos* [42, 93], which make the dynamics of the system unpredictable after a certain time period. A typical example are weather forecasts.

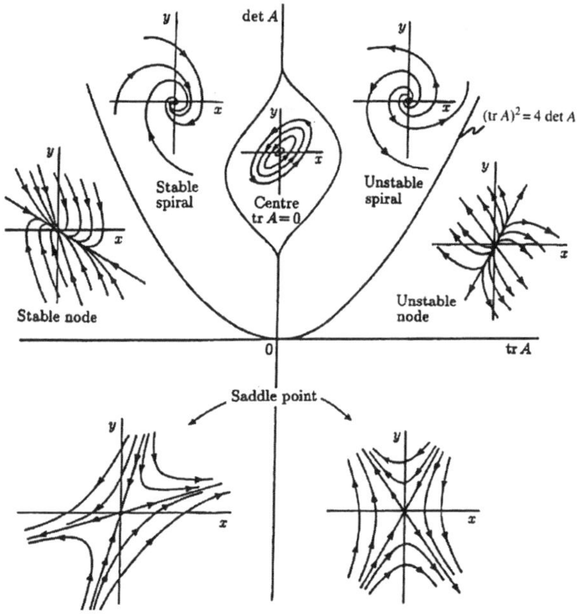

Fig. 14.7 Illustration of various cases of non-linear dynamics that can occur in complex systems (from [98], p. 504). Deterministic chaos and turbulence constitute further and even more complicated cases of non-linear system dynamics

The large sensitivity to small perturbations is sometimes called the "*butterfly effect*", suggesting that (in a chaotically behaving system) the flight of a butterfly could significantly change the system behavior after a sufficiently long time. A further obstacle for predicting the behavior of many complex systems is a *probabilistic* or stochastic *dynamics* [94, 95], i.e. the importance of *randomness*.

In socio-economic systems, there is furthermore a tendency of *self-fulfilling or self-destroying prophecy effects* [96] (and it is hard to say which effect will finally dominate, see the current response of the population to the swine flu campaign). *Stock markets* show both effects: On the one hand, the self-fulfilling prophecy effect leads to *herding behavior*, which creates *bubbles* [97]. On the other hand, the competition for the highest possible returns eventually destroys any predictable gains (otherwise everybody could become rich without having to work, thereby creating a "financial perpetuum mobile"). Altogether, this competition creates a (more or less) "efficient" and unpredictable stock market. A generalization of this principle is known as *Goodhart's law*.

14.2.6 The Illusion of Control

Besides the difficulties to predict the future behavior of complex systems, there are other effects which make them difficult to control:

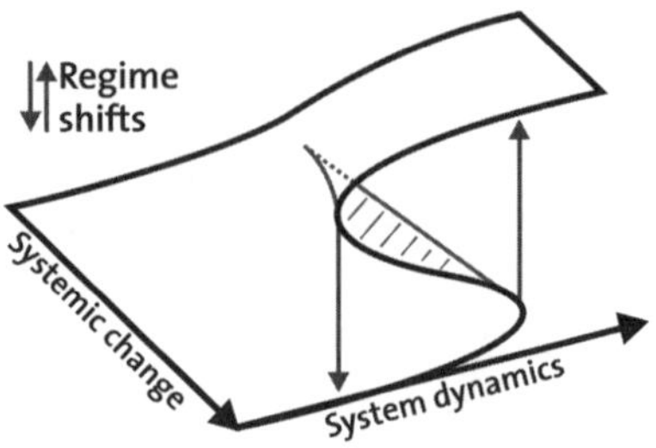

Fig. 14.8 When a complex system is changed (e.g. by *external control attempts*), its system parameters, stability, and dynamics may be affected. This figure illustrates the occurence of a so-called "cusp catastrophe". It implies discontinuous transitions (*regime shifts*) in the system dynamics

1. On the one hand, big changes may have *small or no effects* (see Fig. 14.2) and, when considering network interactions (see Sect. 14.2.3), even adverse effects. This reflects the *principle of Le Chatelier*,according to which a system tends to counteract external control attempts.
2. On the other hand, if the system is close to a "critical" or "tipping point", even small changes may cause a sudden "regime shift", also known as "phase transition" or "catastrophe" (see Figs. 14.2 and Sect. 14.8). In other words, small changes can sometimes have a big impact, and often very unexpectedly so. However, there are typically some *early warning signals* for such critical transitions [99]. This includes the phenomenon of "*slow relaxation*", which means that it takes a long time to dampen out perturbations in the system, i.e. to drive the system back to equilibrium.

   Another warning signal of potential regime shifts are "*critical fluctuations*", which normally obey a heavy-tail distribution (see Sect. 14.2.2). In other words, perturbations in the system tend to be larger than usual – a phenonenon which is also known as "flickering".
3. Control attempts may also be obstructed by "*irreducible randomness*", i.e. a degree of uncertainty or perturbation which cannot be eliminated (see Sect. 14.2.5).
4. *Delays* are another typical problem that often cause a failure of control [100]. The underlying reason is that delays may create an unstable system behavior (also when people attempt to compensate delays by anticipation). Typical examples are the *breakdown of traffic flows* and the occurence of *stop-and-go traffic*, which result from delayed speed adjustments of drivers to variations in the vehicle speeds ahead.

   Since many control attempts these days are based on the use of statistics, but compiling such statistics is time-consuming, delays may cause instabilities also in other areas of society. *Business cycles*, for example, may result from such delays as well (or may at least be intensified by them).
5. Finally, there is the problem of "*unknown unknowns*" [101], i.e. hidden factors which influence the system behavior, but have not been noticed before. By

definition, they appear unexpectedly. "*Structural instabilities*" [39] may create such effects. The appearance of a new species in an ecosystem is a typical example. In economics, this role is played by *innovations* or new products, which happen to change the social or economic world. Well-known examples for this are the invention of contraceptives, computers, or mobile phones.

14.2.7 The Logic of Failure

As a consequence of the above, complex systems cannot be controlled in the conventional way (like pressing a button or steering a car). Such control attempts will usually fail, as Doerner's book "The Logic of Failure" has impressively shown [52].

A typical failure scenario is as follows: A decision-maker tries to change the social system. It turns out that the measure taken does not have any effect (see Fig. 14.2). Therefore, he or she decides to intensify the measure. The effect may still not be as expected. Hence, an even more forceful control attempt is made. As a consequence, the system undergoes a sudden regime shift (see Figs. 14.2–14.8) and the system organizes itself in a different way (but not necessarily in the desired way). The decision-maker now tries to re-gain control and counteracts the unexpected change. If the attempts to stabilize the system are delayed, this can even lead to an oscillatory or chaotic system dynamics.

The right approach to influence complex systems is to support and strengthen the *self-organization* and *self-control* of the system by *mechanism design* (see Sect. 14.4.1). This basically means that coordination and cooperation in a complex system will appear by itself, if the interactions among the system elements are well chosen. That is, regulations should not specify *what exactly* the system elements should do, but set bounds to actions (define "rules of the game"), which give the system elements enough degrees of freedom to self-organize good solutions. If the interaction rules are suitable, such an approach will usually lead to a much more flexible and adaptive system behavior. Another advantage is "systemic robustness", i.e. the ability cope with challenges by external perturbations. Note however, that everything depends on the interactions of the system elements. Unsuitable interactions can, for example, cause that the system behaves dynamically *unstable* or that it gets trapped in a *suboptimal ("frustrated") state*. Hence, finding the right interaction rules is a great challenge for decision-makers, and complex systems scientists are needed to address them properly.

14.3 The Example of Financial Market Instability

One example of systemic risks that deserves more attention here is financial market instability [102–108]. The recent financial crises shows very clearly how cascading effects can lead to an uncontrollable dynamics and a relatively sudden systemic crises. What started with local problems concerning subprime mortgages eventually affected the mortgage companies, the home building industry, the financial markets, the US economy, and the world economy. This crisis has been explained in many ways. Widely discussed reasons include:

- The deregulation of financial markets.
- The explosive spread of derivatives (which reached a value of 15 times the gross product of the world).
- The apparently "riskless" securization of risky deals by credit default swaps, lowering lending standards.
- The opaqueness (intransparency) of financial derivatives.
- The failure of rating agencies due to the complexity of the financial products.
- Bad risk models (neglecting, for example, correlations and the heavy-tail character of the fluctuations).
- Calibration of risk models with historical data not reflecting the actual situation.
- Insufficient net assets of banks.
- Low interest rates to fight previous crises.
- The growth of over-capacities and other developments with pro-cylical effects.
- Short-term incentive structures ("bonus schemes") and "greed" of investment bankers and managers.

Less debated, but not less relevant reasons are [109–111]:

- The complexity of the financial system is larger than what is knowable. For example, many portfolios appear to contain too many different assets to support a reliable optimization with the amount of data available [112].
- In the "*arms race*" between banks (and other agents) with the regulators, regulators are sometimes in the weaker position. Therefore, financial market instability may result from the fact that instability is *beneficial* for some interest groups: It *requires* an unstable market to allow some people to become very rich in a short time: Instability implies opportunities for good investments. When GDP grows slowly, good returns mainly result from financial bubbles.
- The financial architecture has created a *complex system*, with a hard-to-predict and hard-to-control dynamics. Financial products ("derivatives") were constructed in a multi-level way, very much like a house of cards.
- The world-wide *network interdependencies* of all major banks have spread local risks all over the system to an extent that produced a *systemic* risk. It created a "global village" without any "firewalls" (security breaks).
- Delays in the adaptation of some markets build up *disequilibria in the system* with the potential of earthquake-like stress reliefs. As examples for this, one may take historical crashes in currency markets or recent drops in the values of certain AAA-rated stocks.
- The financial and economic system are organized in a way that allows for the occurrence of *strong correlations*. For example, when the strategies of companies all over the world become more and more similar (due to "group think" [113] or asking the same consultancy companies), a *lack of variety* (heterogeneity) results

in the system. This can cause (more or less) that either no company fails or many companies fail at the same time.

- An important factor producing *herding effects* [114, 115] and bubbles is the continuous information feedback regarding the investment decisions of others [116]. In this connection, it is important to underline that repeated interactions between decision-makers supports consensus, but creates *over-confidence* (i.e. a false feeling of safety, despite misjudgements of reality). Therefore, it undermines the "wisdom of crowds" [117, 118]. This problem may be further intensified by the *public media* which, in the worst case, may even create a *mass hysteria*.
- The price formation mechanism mixes material values and *psychology* in a single, one-dimensional quantity, the price. Therefore, the price dynamics is sensitive to factors such as trust, risk aversion, greed, and herding effects (the imitation of the behavior of others) [54–56, 119].
- A stability of single banks does not imply that the banking system cannot enter a state of *systemic instability*. (Monetary value is a matter of trust, and therefore a single event such as the failure of Lehmann Brothers could induce that banks were not anymore willing to lend money to each other. This triggered a liquidity crises so big that it would have caused the failure of the world financial system, if the central banks would not have quickly provided huge amounts of liquidity.)
- Lack of trust also reduces lending of cheap money to troubled companies, which may drive them into bankruptcy, thereby increasing a bank's problems.
- More generally, the economic system seems to have a tendency towards *self-organized critical behavior* (see Sect. 14.2.4).

Many of the above factors have contributed to strong non-linear couplings in the system. Furthermore, strong network interdependencies have been created through the interbank markets and complex financial derivatives. These features are already expected to imply cascade-like effects and a heavy-tail statistics (see Sect. 14.2.2). This tendency is expected to be *further* amplified by anticipation attempts in fluctuating markets. However, even more dangerous than the occurrence of fluctuations in the markets is the occurence of *strong correlations*. These can be promoted by economic cycles, herding effects, and the coupling of policies or regulation attempts to global risk indicators.

The worldwide crisis in the automobile sector in 2009 and the *quant meltdown* in August 2007 are good examples for the occurence of strong correlations. The latter may be understood as follows [120]: Returns of hedge fonds largely depend on their leverage. Therefore, there is an "evolutionary pressure" towards *high leverage*, which can increase volatility. In case of huge price jumps, however, banks tend to demand their loans back. This decreases the leverage of the affected hedge funds and thereby their chances to perform well in the future. Hence, large system-wide leverage levels are pre-requisites for collapses, and crises can emerge virtually "out of nothing", just through fluctuations. This example illustrates well how unsuitable risk-averse policies can create pro-cyclical effects, through which banks may harm their own interests.

14.4 Managing Complexity

Having discussed the particular challenges of complex systems, one may be left with the impression that such systems are just too difficult for us to handle. However, in the past decades, a variety of scientific techniques have been developed to address these challenges. These include:

- Large-scale data mining.
- Network analysis.
- Systems dynamics.
- Scenario modeling.
- Sensitivity analysis.
- Non-equilibrium statistical physics.
- Non-linear dyamics and chaos theory.
- Systems theory and cybernetics.
- Catastrophe theory.
- The statistics of extreme events.
- The theory of critical phenomena and, maybe most prominently these days:
- Agent-based modeling [129–133].

The methods developed by these fields allow us to better assess the sensitivity or robustness of systems and their dynamics, as will be shortly discussed in the following. They have also revealed that complex systems are not our "enemies". In fact, they possess a number of favorable properties, which can be used to our benefit.

14.4.1 How to Profit from Complex Systems

Understanding complex systems facilitates to utilize their interesting properties, which however requires one to work *with* the system rather than *against* it [121–128]. For example, complex systems tend to show *emergent (collective) properties*, i.e. properties that the single system components do not have. This is, for example, relevant for the possibility of *collective intelligence* [134–136]. One may also benefit from the fact that complex systems tend to *self-organize* in a way, which is adaptive to the enviroment and often *robust* and *resource-efficient* as well. This approach has, for example, been successfully applied to develop improved design principles for pedestrian facilities and other systems.

Technical control approaches based on self-organization principles become more and more available now. While previous *traffic control* on highways and in cities was based on a centralized optimization by supercomputers with expensive measurement and control infrastructures, currently developed approaches are based on *decentralized coordination strategies* (such as driver assistant systems or traffic lights that are flexibly controlled by local traffic flows).

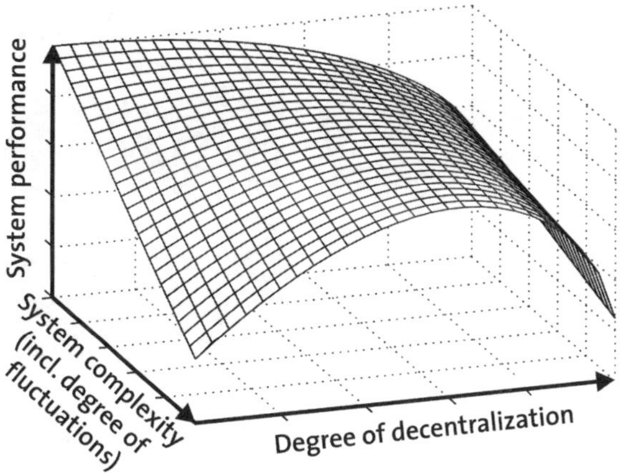

Fig. 14.9 One advantage of centralized control is quick large-scale coordination. However, disadvantages result from the vulnerability of the network, a tendency of information overload, the risk of selecting the wrong control parameters, and delays in adaptive feedback control. Because of greater flexibility to local conditions and greater robustness to perturbations, decentralized control approaches can perform better in complex systems with heterogeneous elements, large degree of fluctuations, and short-term predictability (after [139])

Centralized structures can reach a quick information exchange among remote parts of a system, but they become unstable beyond a certain critical size (as the *collapse of political states* and many unsuccessful *mergers of companies* show). In comparison, decentralized approaches are particularly suited to reach a flexible adjustment to local conditions and local coordination [137]. Some decentralized concepts for real-time control already exceed the performance of centralized ones, particularly in complex, hardly controllable, fluctuating enviroments, which require a quick and flexible response to the actual situation [138] (see Fig. 14.9). In fact, in a strongly varying world, strict stability and control is not possible anymore or excessively expensive (as the *public spending deficits* show). Therefore, a paradigm shift towards more flexible, agile, adaptive systems is needed, possible, and overdue. The best solutions are probably based on suitable combinations of centralized and decentralized approaches.

In social systems, the principle of self-organization, which is also known as principle of the "invisible hand", is ubiquitous. However, self-organization does not automatically lead to optimal results, and it may fail under extreme conditions (as is known, for example, from financial and traffic systems as well as dense pedestrian crowds).

A particularly important example of self-control is the establishment of *social norms*, which are like social forces guiding the behavior of people. In this way, *social order* can be created and maintained even without centralized regulations such as enforced laws. Nevertheless, one must be aware that the principles on which social cooperation and norms are based (for example, repeated interaction, trust and reputation, or altruistic sanctioning of deviant behavior) are fragile. Simple

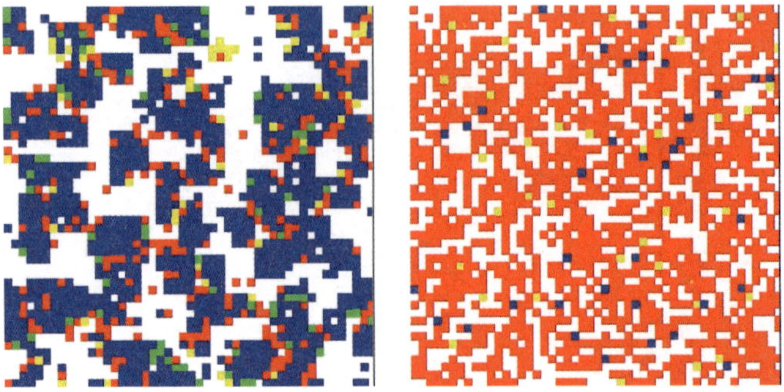

Fig. 14.10 Establishment of cooperation in a world with local interactions and local mobility (*left*) in comparison with the breakdown of cooperation in a world with global interactions and global mobility (*right*) (*blue = cooperators, red = defectors/cheaters/free-riders*) (after [140]). Note that the loss of solidarity results from a lack of neighborhood interactions, not from larger mobility

computer simulations suggest, for example, that a change from repeated local interactions (between family members, friends, colleagues, and neighbors) to non-recurring interactions with changing partners from all over the world may cause a breakdown of human cooperation [140]. Therefore, the on-going *globalization* could potentially destabilize our social systems [141–143] (see Fig. 14.10), which largely builds on norms and social cooperation. (Remember, for example, that the breakdown of the interbank market, which almost caused a collapse of the world financial system, was due to a *breakdown of the network of trust*.)

14.4.2 Reducing Network Vulnerability

In Sect. 14.2.3, we have seen that systemic risks are mostly based on cascade spreading effects in networks. However, the vulnerability of networks to such spreading events can be reduced. The following measures are often quite effective:

- The network structure can often been improved by *redundancy*, i.e. the provision of alternatives, so that an over-critical perturbation would only occur, if several nodes would fail or several links would break simultaneously.
- However, too much interconnectedness may be harmful, as it is provides the "infrastructure" for the system-wide spreading of an unexpected problem. Therefore, it makes sense to *limit the degree of connectedness* and the size of networks (in order to avoid a "too big to fail" problem).
- Alternatively, one can introduce "firewalls": Having several networks, each of them characterized by strong links, while the connections *between* the networks are weak, would allow to decouple the so defined supernetwork into several subnetworks (see Fig. 14.11). This principle of decompartementalization allows one to prevent the spreading of a problem over the whole system, if the disconnection strategy is well chosen. The principle of firewalls to protect

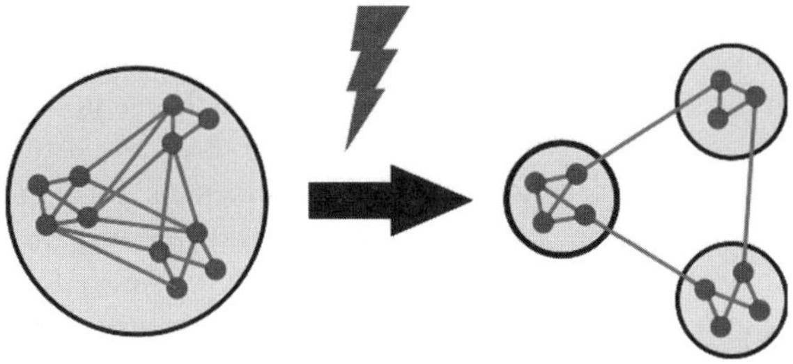

Fig. 14.11 A networked system should be constructed in a way that allows its quick decomposition or decompartementalization into weakly coupled (or, if necessary, *even uncoupled*) subnetworks. In such a way, failure cascades all over the system (or *large parts of it*) can be avoided, and most parts of it can be protected from damage

computer systems from malicious intrusion or the principle of electrical fuses to protect an electrical network from overload could certainly be transferred to other networked systems such as the financial system.

- For similar reasons, a *heterogeneity* (variety) among the nodes and/or links of a network (in terms of design principles and operation strategies) will normally increase its robustness.
- When fighting failure cascades in networks, a *quick response* to over-critical perturbations is absolutely decisive. If the time delay of disaster response management is small, its effectiveness depends in a complicated way on the network structure, the amount of resources, and the strategy of distributing them in the network (see Fig. 14.6). In case of significant delays, cascade spreading can hardly be mitigated, even when large resources are invested.
- A *moderate level of fluctuations* may be useful to destroy potentially harmful correlations (such as financial bubbles) in the system. Such fluctuations could be created by central banks (for the purpose of "bubble control") or by other regulators, depending on the system. Note, however, that a large degree of fluctuations can cause over-critical perturbations or coincidences of perturbations.
- An unhealthy degree of volatility can be lowered by *introducing conservation laws and/or frictional effects* in the system. This is expected to dampen fluctuations and, thereby, to reduce the likelihood of events that may trigger systemic risks.

Rather than applying these concepts permanently, it can make sense to use them *adaptively*, depending on the state of the system. When designing networked systems according to the above principles, one can certainly profit from the experience of physicists and engineers with other systems.

14.5 Summary, Discussion, and Outlook

In this contribution, we have summarized properties of complex systems and identified sources and drivers of systemic risks in socio-economic systems. Complex systems cannot be easily controlled. They rather tend to follow a self-organized eigendynamics, and conventional control attempts often have counter-intuitive and unintended effects.

As the example of ecosystems shows, a networked system can have an astonishing degree of *robustness* without any central control. Robustness just requires the right interaction rules, which may be implemented, for example, by social norms, laws, technological measures etc., depending on the system. Properly chosen rules will lead to a *self-regulation* or *self-control* of the system, but improper specifications can lead to *low performance* or *systemic instability*. For example, if the failure rate of system elements is reduced, this may lead to larger systemic failures later on. Moreover, it is probably good if the system is regularly exposed to stress, as this is expected to strengthen its immunity to perturbations.

It was particularly underlined that, in any larger networked system, it is essential to have *"firewalls" (security breaks)*, which facilitate its quick decomposition or decompartmentalization into disconnected or weakly connected subnetworks before a failure cascade has percolated through the whole system or large parts of it.

Among the success stories of complex systems research, one may mention the nobel prizes of Ilya Prigogine, Thomas Schelling, and Paul Krugmann. Some examples for *application areas* of complexity science are [144–148]

- The organization of the internet.
- Modern epidemiology.
- The prevention of crowd disasters.
- Innovative solutions to improve traffic flow.
- The understanding of global climate change.
- The enhancement of the reliability of energy supply.
- Modern disaster response management.
- Prediction markets and other methods using the wisdom of crowds.

However, many socio-economic crises still occur, because the system dynamics is not well enough understood, leading to serious management mistakes. In order to support decision-makers, scientists need to be put in a better position to address the increasing number of socio-economic problems. These mainly result from the fact that social and economic systems are rapidly changing, i.e. in a transformation process rather than in equilibrium.

We must close the gap between existing socio-economic problems and solutions, and create conditions allowing us to come up with solutions *before* a problem occurs. This requires to build up greater research capacities (a "socio-economic knowledge accelerator"). It will also be necessary to establish a new study direction ("integrative systems design") to provide decision-makers with solid knowledge regarding the behavior of complex systems, how to manage complexity in politics and economy, and how to cope with crises.

Finally, scientists need to have access to better and more detailed data. Special super-computing centers (as for climate research) would allow scientists to simulate model societies and study the impact of policy measures *before* their implementation. They would also support the development of contingency plans and the investigation of alternative ways of organization ("plan B"). Such centers will require a multi-disciplinary collaboration across the various relevant research areas, ranging from the socio-economic over the natural to the engineering sciences. For this, one needs to overcome the particular challenges of multidisciplinary research regarding organization, funding, and publication.

Considering that we know more about the origin of the universe than about the conditions for a stable society, a prospering economics, and enduring peace, we need nothing less than an "Apollo project for the socio-economic sciences". There is no time to lose, since there are already signs of critical fluctuations indicating possible regime shifts [149–154], which speak a clear language.

Acknowledgements This work was partially supported by the ETH Competence Center "Coping with Crises in Complex Socio-Economic Systems" (CCSS) through ETH Research Grant CH1-01 08-2. The author would like to thank Peter Felten for creating many of the illustrations shown in this contribution. Furthermore, the author is grateful for inspiring discussions with Kay Axhausen, Stefano Battiston, Lubos Buzna, Lars-Erik Cederman, Hans Herrmann, Imre Kondor, Matteo Marsili, Frank Schweitzer, Didier Sornette, Stefan Thurner, and many others.

References

1. D. Sornette, *Critical Phenomena in Natural Sciences: Chaos, Fractals, Selforganization and Disorder*. (Springer, Berlin, 2006)
2. S. Albeverio, V. Jentsch, H. Kantz, eds., *Extreme Events in Nature and Society*. (Springer, Berlin, 2006)
3. A. Bunde, J. Kropp, and H. J. Schellnhuber, eds., *The Science of Disasters. Climate Disruptions, Heart Attacks, and Market Crashes*. (Springer, Berlin, 2002)
4. M. Buchanan, *Ubiquity. Why Catastrophes Happen*. (Three Rivers, New York, 2000)
5. Office of the President of Columbia University, Lee C. Bollinger (2005) Announcing the Columbia Committee on Global Thought, see http://www.columbia.edu/cu/president/docs/communications/2005-2006/051214-committee-global-thought.html
6. W. Weidlich, *Sociodynamics: A Systematic Approach to Mathematical Modelling in the Social Sciences*. (Harwood Academic, Amsterdam, 2000)
7. H. Haken, *Synergetics: Introduction and Advanced Topics*. (Springer, Berlin, 2004)
8. D. Helbing, *Quantitative Sociodynamics*. (Kluwer Academic, Dordrecht, 1995)
9. R. Axelrod, *The Complexity of Cooperation: Agent-Based Models of Competition and Collaboration* (Princeton University, Princeton, NJ, 1997)
10. F. Schweitzer, ed. *Self-Organization of Complex Structures: From Individual to Collective Dynamics* (Gordon and Breach, Amsterdam, 1997)
11. D.S. Byrne, *Complexity Theory and the Social Sciences: An Introduction* (Routledge, New York, 1998)
12. W. Buckley, *Society–A Complex Adaptive System: Essays in Social Theory*. (Routledge, London, 1998)
13. S.Y. Auyang, *Foundations of Complex-System Theories in Economics, Evolutionary Biology, and Statistical Physics*. (Cambridge University, Cambridge, 1999)
14. F. Schweitzer, ed. *Modeling Complexity in Economic and Social Systems*. (World Scientific, Singapore, 2002)
15. R.K. Sawyer, *Social Emergence: Societies As Complex Systems* (Cambridge University, Cambridge, 2005)
16. A.S. Mikhailov, V. Calenbuhr, *From Cells to Societies*. (Springer, Berlin, 2006)

17. B. Blasius, J. Kurths, L. Stone *Complex Population Dynamics*. (World Scientific, Singapore, 2007)
18. M. Batty, *Cities and Complexity: Understanding Cities with Cellular Automata, Agent-Based Models, and Fractals*. (MIT Press, Cambridge, 2007)
19. S. Albeverio, D. Andrey, P. Giordano, A. Vancheri, *The Dynamics of Complex Urban Systems*. (Physica, Heidelberg, 2008)
20. K. Mainzer, *Thinking in Complexity: The Computational Dynamics of Matter, Mind, and Mankind*. (Springer, Berlin, 2007)
21. J.H. Miller, S.E. Page, *Complex Adaptive Systems: An Introduction to Computational Models of Social Life*. (Princeton University, Princeton, NJ, 2007)
22. J.M. Epstein, *Generative Social Science: Studies in Agent-Based Computational Modeling*. (Princeton University, Princeton, NJ, 2007)
23. B. Castellani, F. Hafferty, *Sociology and Complexity Science*. (Springer, Berlin, 2009)
24. D. Lane, D. Pumain, S.E. van der Leeuw, G. West, eds. *Complexity Perspectives in Innovation and Social Change*. (Springer, Berlin, 2009)
25. C. Castellano, S. Fortunato, V. Loreto, Statistical physics of social dynamics. *Rev. Mod. Phys.* **81**, 591–646 (2009)
26. P.W. Anderson, K. Arrow, D. Pines, eds. *The Economy as an Evolving Complex System*. (Westview, Boulder, Menzies, NK, 1988)
27. H.W. Lorenz, *Nonlinear Dynamical Equations and Chaotic Economy*. (Springer, Berlin, 1993)
28. P. Krugman, *The Self-Organizing Economy*. (Blackwell, Malden, MA, 1996)
29. W.B. Arthur, S.N. Durlauf, D.A. Lane, *The Economy as an Evolving Complex System II*. (Westview, Boulder, Menzies, NK, 1997)
30. R.N. Mantegna, H.E. Stanley, *An Introduction to Econophysics: Correlations and Complexity in Finance*. (Cambridge University, Cambridge, 1999)
31. D. Colander, ed. *The Complexity Vision and the Teaching of Economics*. (Elgar, Cheltenham, UK, 2000)
32. J.D. Sterman, *Business Dynamics: Systems Thinking and Modeling for a Complex World*. (Boston, McGraw Hill, 2000)
33. T. Puu, *Attractors, Bifurcations, & Chaos. Nonlinear Phenomena in Economics*. (Springer, Berlin, 2003)
34. L.E. Blume, S.N. Durlauf, eds. *The Economy as an Evolving Complex System III*. (Oxford University, Oxford, 2006)
35. B.K. Chakrabarti, A. Chakraborti, A. Chatterjee, *Econophysics and Sociophysics*. (Wiley, New York, 2006)
36. A. Chatterjee, B.K. Chakrabarti, eds. *Econophysics of Markets and Business Networks*. (Springer, Milan, 2007)
37. A.C.-L. Chian, *Complex Systems Approach to Economic Dynamics*. (Springer, Berlin, 2007)
38. V. Lacayo, What complexity science teaches us about social change, see http://www.communicationforsocialchange.org/mazi-articles.php?id=333
39. G. Nicolis and I. Prigogine, *Self-Organization in Nonequilibrium Systems*. (Wiley, New York, 1977)
40. A.S. Mikhailov, *Foundations of Synergetics I: Distributed Active Systems* (Springer, Berlin, 1991)
41. A.S. Mikhailov, A.Y. Loskutov, *Foundations of Synergetics II: Complex Patterns*. (Springer, Berlin, 1991)
42. S.H. Strogatz, *Nonlinear Dynamics and Chaos: With Applications To Physics, Biology, Chemistry, And Engineering*. (Westview Press, 2001)
43. N. Boccara, *Modeling Complex Systems*. (Springer, Berlin, 2003)
44. J. Jost, *Dynamical Systems: Examples of Complex Behaviour*. (Springer, Berlin, 2005)
45. G. Nicolis, *Foundations of Complex Systems*. (World Scientific, New York, 2007)
46. P. Érdi, *Complexity Explained*. (Springer, Berlin, 2007)
47. C. Gros, *Complex and Adaptive Dynamical Systems*. (Springer, Berlin, 2008)

48. M. Schroeder, *Fractals, Chaos, Power Laws*. (Dover, 2009)
49. A. Saichev, Y. Malevergne, D. Sornette, *Theory of Zipf's Law and Beyond*. (Springer, Berlin, 2010)
50. D. Helbing, I. Farkas, T. Vicsek, Simulating dynamical features of escape panic. Nature **407**, 487–490 (2000)
51. D. Helbing, A. Mazloumian, Operation regimes and slower-is-faster effect in the control of traffic intersections. Eur. Phys. J. B **70**(2), 257–274 (2009)
52. D. Dorner, *The Logic Of Failure: Recognizing and Avoiding Error in Complex Situations*. (Basic, New York, 1997)
53. D. Helbing, Managing complexity in socio-economic systems. Eur. Rev. **17**(2), 423–438 (2009)
54. R.H. Thaler, ed. *Advances in Behavioral Finance*. (Russell Sage Foundation, New York, 1993)
55. C.F. Camerer, G. Loewenstein, M. Rabin, eds. *Advances in Behavioral Economics*. (Princeton University, Princeton, NJ, 2004)
56. R.H. Thaler, ed. *Advances in Behavioral Finance*, Vol. II. (Princeton University, Princeton, NJ, 2005)
57. M. Schönhof, D. Helbing, Empirical features of congested traffic states and their implications for traffic modeling. Transport. Sci. **41**(2), 135–166 (2007)
58. H.E. Stanley, *Introduction to Phase Transitions and Critical Phenomena*. (Oxford University, 1987)
59. N. Goldenfeld, *Lectures On Phase Transitions and the Renormalization Group*. (Westview, Boulder, CO, 1992)
60. E.C. Zeeman, ed. *Catastrophe Theory*. (Addison-Wesley, London, 1977)
61. T. Poston, I. Stewart, *Catastrophe Theory and Its Applications*. (Dover, Mineola, 1996)
62. M. Gladwell, *The Tipping Point: How Little Things Can Make a Big Difference*. (Back Bay, Boston, 2002)
63. V.I. Arnol'd, G.S. Wassermann, R.K. Thomas, *Catastrophe Theory*. (Springer, Berlin, 2004)
64. P. Doreian, F.N. Stokman, eds. *Evolution of Social Networks*. (Routledge, Amsterdam, 1997)
65. D.J. Watts, *Small Worlds*. (Princeton University, Princeton, NJ, 1999)
66. B.A. Huberman, *The Laws of the Web: Patterns in the Ecology of Information*. (MIT Press, Cambridge, 2001)
67. R. Albert, L.-A. Barabasi, Statistical mechanics of complex networks. Rev. Mod. Phys. **74**, 47–97 (2002)
68. S. Bornholdt, H.G. Schuster, *Handbook of Graphs and Networks: From the Genome to the Internet*. (Wiley-VCH, Germany 2003)
69. S.N. Dorogovtsev, J.F.F. Mendes, *Evolution of Networks*. (Oxford University, Oxford 2003)
70. P.J. Carrington, J. Scott, S. Wassermann, *Models and Methods in Social Network Analysis*. (Cambridge University, New York, 2005)
71. M. Newman, A.-L. Barabasi, D.J. Watts, eds. *The Structure and Dynamics of Networks*. (Princeton University, Princeton, NJ, 2006)
72. F. Vega-Redondo, *Complex Social Networks*. (Cambridge University, Cambridge, 2007)
73. M.O. Jackson, *Social and Economic Networks*. (Princeton University, Princeton, NJ, 2008)
74. J. Bruggeman, *Social Networks*. (Routledge, New York, 2008)
75. A. Barrat, M. Barthélemy, A. Vespignani, *Dynamical Processes on Complex Networks*. (Cambridge University, Cambridge, 2008)
76. J. Reichardt, *Structure in Complex Networks*. (Springer, Berlin, 2009)
77. A.K. Naimzada, S. Stefani, A. Torriero, eds. *Networks, Topology, and Dynamics*. (Springer, Berlin, 2009)
78. A. Pyka, A. Scharnhorst, eds. *Innovation Networks*. (Springer, Berlin, 2009)
79. T. Gross, H. Sayama, eds. *Adaptive Networks*. (Springer, Berlin, 2009)
80. I. Simonsen, L. Buzna, K. Peters, S. Bornholdt, D. Helbing, Transient dynamics increasing network vulnerability to cascading failures. *Phys. Rev. Lett.* **100**, 218701 (2008)
81. J. Lorenz, S. Battiston, F. Schweitzer, Systemic risk in a unifying framework for cascading processes on networks. *Eur. Phys. J. B* **71**(4), 441–460 (2009)

82. D. Helbing, C. Khnert, Assessing interaction networks with applications to catastrophe dynamics and disaster management. Physica A **328**, 584–606 (2003)
83. L. Buzna, K. Peters, D. Helbing, Modelling the dynamics of disaster spreading in networks. Physica A **363**, 132–140 (2006)
84. L. Buzna, K. Peters, H. Ammoser, C. Kühnert, D. Helbing, Efficient response to cascading disaster spreading. Phys. Rev. E **75**, 056107 (2007)
85. D. Helbing, H. Ammoser, C. Khnert, Disasters as extreme events and the importance of network interactions for disaster response management. In [19], pp. 319–348 (2005)
86. J. Reason, *Human Error*. (Cambridge University, Cambridge, 1990)
87. J.R. Chiles, *Inviting Disaster: Lessons From the Edge of Technology*. (Harper, New York, 2002)
88. P. Bak, *How nature works: the science of self-organized criticality*. (Springer, Berlin, 1999)
89. H.J. Jensen, *Self-Organized Criticality: Emergent Complex Behavior in Physical and Biological Systems*. (Cambridge University, Cambridge, 1998)
90. D. Helbing, Traffic and related self-driven many-particle systems. Rev. Mod. Phys. **73**, 1067–1141 (2001)
91. D. de Martino, L. Dall'Asta, G. Bianconi, M. Marsili, Congestion phenomena on complex networks. Phys. Rev. E **79**, 015101 (2009)
92. P.A. Davidson, *Turbulence*. (Cambridge University, Cambridge, 2004)
93. H.G. Schuster, W. Just, *Deterministic Chaos*. (Wiley-VCH, Weinheim, 2005)
94. N.G. van Kampen, *Stochastic Processes in Physics and Chemistry*. (North-Holland, Amsterdam, 2007)
95. G. Deco, B. Schürmann, *Information Dynamics*. (Springer, Berlin, 2001)
96. R.A. Jones, *Self-fulfilling Prophecies: Social, Psychological, and Physiological Effects of Expectancies*. (Lawrence Erlbaum, 1981)
97. R.E.A. Farmer, *Macroeconomics of Self-fulfilling Prophecies*. (MIT Press, Cambridge, MA, 1999)
98. J.D. Murray, *Mathematical Biology*, Vol. I. (Springer, Berlin, 2003)
99. M. Scheffer, J. Bascompte, W.A. Brock, V. Brovkin, S.R. Carpenter, V. Dakos, H. Held, E.H. van Nes, M. Rietkerk, G. Sugihara, Early-warning signals for critical transitions. Nature **461**, 53–59 (2009)
100. W. Michiels, S.-I. Niculescu, *Stability and Stabilization of Time-Delay Systems*. siam – Society for Industrial and Applied Mathematics, (Philadelphia, USA, 2007)
101. N.N. Taleb, *The Black Swan: The Impact of the Highly Improbable*. (Random House, New York, 2007)
102. E.E. Peters, *Chaos and Order in the Capital Markets*. (Wiley, New York, 1996)
103. S. Claessens, K.J. Forbes, eds. *International Financial Contagion*. (Kluwer Academic, Dordrecht, 2001)
104. K. Ilinsiki, *Physics of Finance. Gauge Modelling in Non-Equilibrium Pricing*. (Wiley, Chichester, 2001)
105. B.M. Roehner, *Patterns of Speculation*. (Cambridge University, Cambridge, 2002)
106. D. Sornette, *Why Stock Markets Crash: Critical Events in Complex Financial Systems* (Princeton University, Princeton, NJ, 2004)
107. B. Mandelbrot, R.L. Hudson, *The Misbehavior of Markets: A Fractal View of Financial Turbulence*. (Basic, New York, 2006)
108. M. Faggini, T. Lux, eds. *Coping with the Complexity of Economics*. (Springer, Milan, 2009)
109. R.J. Breiding, M. Christen, D. Helbing Lost robustness. Naissance Newsletter, 8–14 (April 2009)
110. R.M. May, S.A. Levin, G. Sugihara, Ecology for bankers. Nature **451**, 893–895 (2008)
111. S. Battiston, D. Delli Gatti, M. Gallegati, B. C.N. Greenwald, J.E. Stiglitz, Liaisons dangereuses: Increasing connectivity, risk sharing and systemic risk, (2009) see http://www3.unicatt.it/unicattolica/CentriRicerca/CSCC/allegati/delligatti.pdf
112. I. Kondor, I. Varga-Haszonits, Divergent estimation error in portfolio optimization and in linear regression. Eur. Phys. J. B **64**, 601–605 (2008)

113. G.A. Akerlof, R.J. Shiller, *Animal Spirits. How Human Psychology Drives the Economy, and Why It Matters for Global Capitalism*. (Princeton University, Princeton, NJ, 2009)
114. G. Le Bon, *The Crowd: A Study of the Popular Mind*. (Dover, New York, 2002)
115. W. Trotter, *Instincts of the Herd in Peace and War*. (Cosimo Classics, New York, 2005)
116. C. Mackay, *Extraordinary popular delusions and the madness of crowds* (2003)
117. J. Surowiecki, *The Wisdom of Crowds*. (Anchor, New York, 2005)
118. P. Ball, *Critical Mass*. (Arrow, London, 2004)
119. R. Falcone, K.S. Barber, J. Sabater-Mir, M.P. Singh, eds. *Trust in Agent Societies*. (Springer, Berlin, 2009)
120. S. Thurner, J.D. Farmer, J. Geanakoplos, Leverage causes fat tails and clustered volatility. (2009) E-print http://arxiv.org/abs/0908.1555.
121. R. Axelrod, M.D. Cohen, *Harnessing Complexity: Organizational Implications of a Scientific Frontier*. (Basic, New York, 2001)
122. H. Eisner, *Managing Complex Systems: Thinking Outside the Box* (Wiley, New York 2005)
123. L. Hurwicz, S. Reiter, *Designing Economic Mechanisms*. (Cambridge University, New York, 2006)
124. M. Salzano, D. Colander, eds. *Complexity Hints for Economic Policy*. (Springer, Berlin, 2007)
125. E. Schöll, H.G. Schuster, eds. *Handbook of Chaos Control*. (Wiley-VCH, Weinheim, 2008)
126. D. Grass, J.P. Caulkins, G. Feichtinger, G. Tragler, D.A. Behrens, *Optimal Control of Nonlinear Processes: With Applications in Drugs, Corruption, and Terror*. (Springer, Berlin, 2008)
127. D. Helbing, eds. *Managing Complexity: Insights, Concepts, Applications*. (Springer, Berlin, 2008)
128. L.A. Cox Jr., *Risk Analysis of Complex and Uncertain Systems*. (Springer, New York, 2009)
129. M. Gallegati, A. Kirman, eds. *Beyond the Representative Agent*. (Edward Elgar, Cheltenham, UK 1999)
130. A. Consiglio, eds. *Artificial Markets Modeling*. (Springer, Berlin, 2007)
131. M. Aoki, H. Yoshikawa, *Reconstructing Macroeconomics*. (Cambridge University Press, Cambridge, 2007)
132. K. Schredelseker, F. Hauser, eds. *Complexity and Artificial Markets*. (Springer, Berlin, 2008)
133. D. Delli Gatti, E. Gaffeo, M. Gallegati, G. Giulioni, A. Palestrini, *Emergent Macroeconomics. An Agent-Based Approach to Business Fluctuations*. (Springer, Milan, 2008)
134. J. Surowiecki, *The Wisdom of Crowds*. (Anchor, New York, 2005)
135. H. Rheingold, *Smart Mobs*. (Perseus, Cambridge, MA, 2003)
136. D. Floreano, C. Mattiussi, *Bio-Inspired Artificial Intelligence: Theories, Methods, and Technologies*. (MIT Press, Cambridge, MA, 2008)
137. D. Helbing, A. Deutsch, S. Diez, K. Peters, Y. Kalaidzikis, K. Padberg, S. Lämmer, A. Johansson, G. Breier, F. Schulze, M. Zerial, BioLogistics and the struggle for efficiency: Concepts and perspectives. *Advances in Complex Systems*, (2009); in print, see e-print http://www.santafe.edu/research/publications/wpabstract/200910041
138. S. Lämmer, D. Helbing, Self-control of traffic lights and vehicle flows in urban road networks. J. Stat. Phys. (JSTAT), P04019 (2008)
139. K. Windt, T. Philipp, F. Böse, Complexity cube for the characterization of complex production systems. Int. J. Comput. Integrated Manuf. **21**(2), 195–200 (2007)
140. D. Helbing, W. Yu, H. Rauhut, Self-organization and emergence in social systems: Modeling the coevolution of social environments and cooperative behavior, (2009); see e-print http://www.santafe.edu/research/publications/wpabstract/200907026
141. E. Ostrom, *Governing the Commons. The Evolution of Institutions for Collective Action*. (Cambridge University, New York, 1990)
142. G. Hardin, *Living within Limits: Ecology, Economics, and Population Taboos*. (Oxford University, New York, 1995)
143. J.A. Baden, D.S. Noonan, eds. *Managing the Commons*. (Indiana University, Bloomington, Indiana, 1998)
144. GIACS Complexity Roadmap, see http://users.isi.it/giacs/roadmap

145. The Complex Systems Society Roadmap, see http://www.soms.ethz.ch/research/complexityscience/roadmap
146. ComplexityNET report on European Research Landscape, see http://www.soms.ethz.ch/research/complexityscience/European_Complexity_Landscape_D2.2_short_report.pdf
147. EU report on Tackling Complexity in Science, see http://www.soms.ethz.ch/research/complexityscience/EU_complexity_report.pdf
148. OECD Global Science Forum report on Applications of Complexity Science for Public Policy, see http://www.oecd.org/dataoecd/44/41/43891980.pdf
149. J.A. Tainter, *The Collapse of Complex Societies.* (Cambridge University, Cambridge, 1988)
150. J.E. Stiglitz, *Globalization and Its Discontents*. (Norton, New York, 2003)
151. D. Meadows, J. Randers, D. Meadows, *Limits to Growth. The 30-Year Update*. (Chelsea Green Publishing, Withe River Junction, Vermont, 2004)
152. J. Diamond, *Collapse. How Societies Choose to Fall or Succeed.* (Penguin, New York, 2005)
153. R. Costanza, L.J. Graumlich, W. Steffen, eds. *Sustainability of Collapse?* (MIT Press, Cambridge, MA, 2007)
154. P. Krugman, *The Return of Depression Economics and the Crisis of 2008*. (Norton, New York, 2009)

Chapter 15
Managing Complexity*

15.1 What Is Special About Complex Systems?

Many of us have been raised with the idea of cause and effect, i.e. some stimulus-response theory of the world. Particularly, small causes would have small effects and large causes would have large effects. This is, in fact, true for "linear systems", where cause and effect are proportional to each other. Such behavior is often found close to the equilibrium state of a system. However, when complex systems are driven far from equilibrium, non-linearities dominate, which can cause many kinds of "strange" and counter-intuitive behaviors. In the following, we will mention a few. We all have been surprised by these behaviors many times.

While linear system have no more than *one* stationary state (equilibrium) or *one* optimal solution, the situation for non-linear systems is different. They can have *multiple* stationary solutions or optima (see Fig. 15.1), which has several important implications:

- The resulting state is history-dependent. Different initial conditions will not automatically end up in the same state [1]. This is sometimes called "hysteresis".
- It may be hard to find the best, i.e. the "global" optimum in the potentially very large set of local optima. Many non-linear optimization problems are "NP hard", i.e. the computational time needed to determine the best state tends to explode with the size of the system [2]. In fact, many optimization problems are "combinatorially complex".

*This chapter reprints a previous publication with kind permission of the copyright owner, Springer Publishers. It is requested to cite this work as follows: D. Helbing and *S. Lämmer*, Managing complexity: An introduction. Pages 1–16 in D. Helbing (ed.) *Managing Complexity* (Springer, Berlin, 2008).

D. Helbing (ed.), *Social Self-Organization*, Understanding Complex Systems, DOI 10.1007/978-3-642-24004-1_15, © Springer-Verlag Berlin Heidelberg 2012

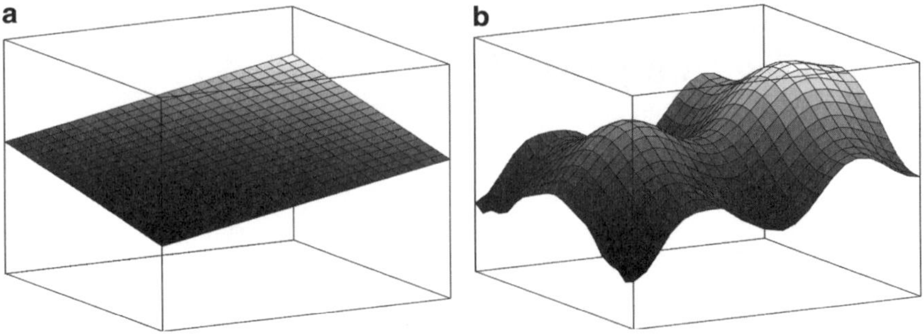

Fig. 15.1 Illustration of linear and non-linear functions. While linear functions have one maximum in a limited area (*left*), non-linear functions may have many (local) maxima (*right*)

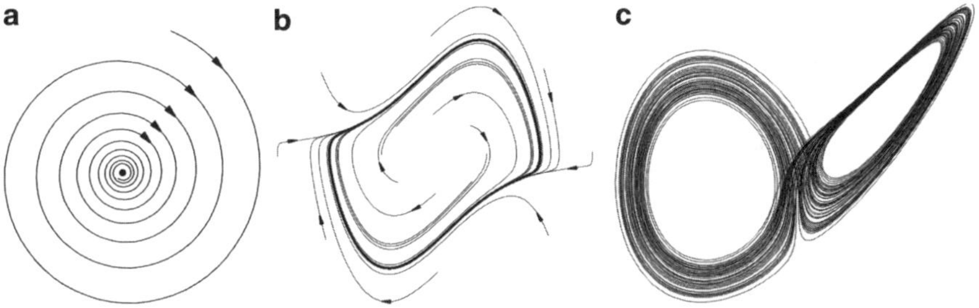

Fig. 15.2 Illustration of trajectories that converge towards (**a**) a stable stationary point, (**b**) a limit cycle, and (**c**) a strange attractor

15.1.1 Chaotic Dynamics and Butterfly Effect

It may also happen that the stationary solutions are unstable, i.e. any small perturbation will drive the system away from the stationary state until it is attracted by another state (a so-called "attractor"). Such attractors may be other stationary solutions, but in many cases, they can be of oscillatory nature (e.g. "limit cycles"). Chaotically behaving systems [3] are characterized by "strange attractors", which are non-periodic (see Fig. 15.2). Furthermore, the slightest change in the trajectory of a chaotic system ("the beat of a butterfly's wing") will eventually lead to a completely different dynamics. This is often called the "butterfly effect" and makes the behavior of chaotic systems unpredictable (beyond a certain time horizon), see Fig. 15.3.

15.1.2 Self-organization, Competition, and Cooperation

Systems with non-linear interactions do not *necessarily* behave chaotically. Often, they are characterized by "emergent", i.e. spontaneous coordination or

Fig. 15.3 Illustration of the "butterfly effect", i.e. the separation of neighboring trajectories in the course of time

synchronization [4–6]. Even coordinated states, however, may sometimes be undesired. A typical example for this are stop-and-go waves in freeway traffic [7], which are a results of an instability of the traffic flow due to the delayed velocity adjustments of vehicles.

Self-organization is typical in driven many-component systems [7] such as traffic, crowds, organizations, companies, or production plants. Such systems have been successfully modeled as many-particle or multi-agent systems. Depending on the respective system, the components are vehicles, individuals, workers, or products (or their parts). In these systems, the energy input is absorbed by frictional effects. However, the frictional effect is not homogeneous, i.e. it is not the same everywhere. It rather depends on the local interactions among the different components of the system, which leads to spatio-temporal pattern formation.

The example of social insects like ants, bees, or termites shows that simple interactions can lead to complex structures and impressive functions. This is often called "swarm intelligence" [8]. Swarm intelligence is based on local (i.e. decentralized) interactions and can be used for the self-organization and self-steering of complex systems. Some recent examples are traffic assistance [9] systems or self-organized traffic light control [9, 10]. However, if the interactions are not appropriate, the system may be characterized by unstable dynamics, breakdowns and jamming, or it may be trapped in a local optimum (a "frustrated state").

Many systems are characterized by a competition for scarce resources. Then, the question whether and how a system optimum is reached is often studied with methods from "game theory" [11–13]. Instead of reaching the state that maximizes the overall success, the system may instead converge to a user equilibrium, where the success ("payoff") of every system component is the *same*, but lower than it *could* be. This happens, for example, in traffic systems with the consequence of excess travel times [14]. In conclusion, if everybody tries to reach the best outcome for him- or herself, this may lead to overall bad results and social dilemmas [15] (the "tragedy of the commons" [16]). Sometimes, however, the system optimum can only be reached by complicated coordination in space and/or time, e.g. by suitable turn-taking behavior (see Fig. 15.4). We will return to this issue in Sect. 15.2.4, when we discuss the "faster-is-slower" effect.

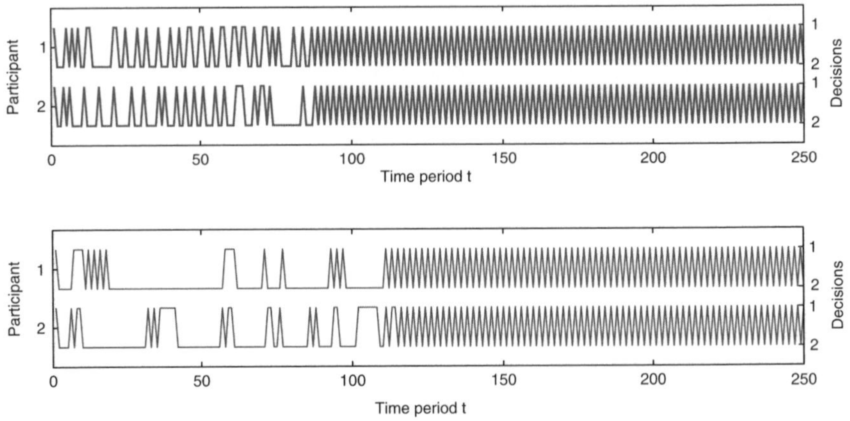

Fig. 15.4 Emergence of turn-taking behavior: After some time, individuals may learn to improve their average success by choosing both possible options in an alternating and coordinated way

15.1.3 Phase Transitions and Catastrophe Theory

One typical feature of complex systems is their robustness with respect to perturbations, because the system tends to get back to its "natural state", the attractor. However, as mentioned above, many complex system can assume different states. For this reason, we may have transitions from one system state ("phase" or attractor) to another one. These phase transitions occur at so-called "critical points" that are reached by changes of the system parameters (which are often slowly changing variables of the system). When system parameters come close to critical points, small fluctuations may become a dominating influence and determine the future fate of the system. Therefore, one speaks of "critical fluctuations" [1].

In other words, large fluctuations are a sign of a system entering an unstable regime, indicating its potential transition to another system state, which may be hard to anticipate. Another indicator of potential instability is "critical slowing down". However, once the critical point is passed, the system state may change quite rapidly. The relatively abrupt change from one system state to an often completely different one is studied by "catastrophe theory" [17]. One can distinguish a variety of different types of catastrophes, but we cannot go into all these details, here.

15.1.4 Self-organized Criticality, Power Laws, and Cascading Effects

At the critical point itself, fluctuations are not only dominating, they may even become *arbitrarily* large. Therefore, one often speaks of "scale-free" behavior,

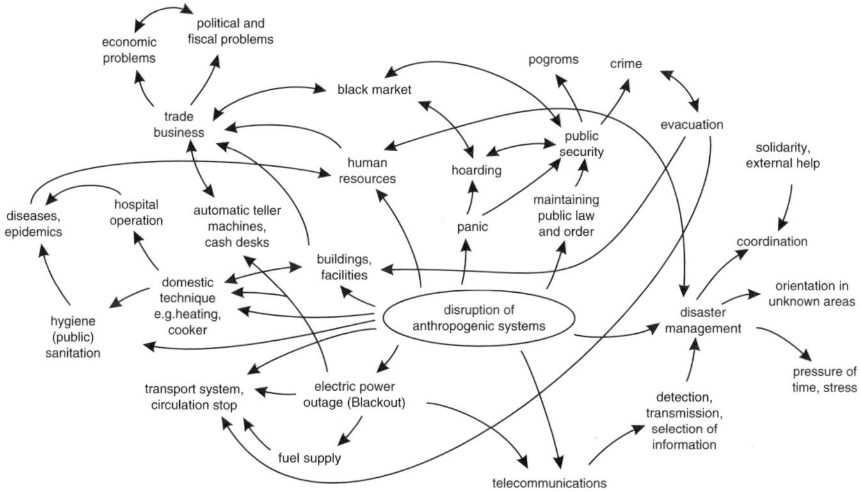

Fig. 15.5 Illustration of the interaction network in athropogenic systems. When the system is seriously challenged, this is likely to cause cascading failures along the arrows of this network (after [20])

which is typically characterized by power laws [18, 19]. Note that, for power laws, the variance and the expected value (the average) of a variable may be undefined!

One possible implication of power laws are cascading effects. The classical example is a sand pile, where more and more grains are added on top [21]. Eventually, when the critical "angle of repose" is reached, one observes avalanches of sand grains of all possible sizes, and the avalanche size distribution is given by a power law. The angle of repose, by the way, even determines the stability of the famous pyramids in Egypt.

Cascading effects are the underlying reason for many disasters, where the failure of one element of a system causes the failure of another one (see Fig. 15.5). Typical examples for this dynamics are blackouts of electrical power grids and the spreading of epidemics, rumors, bankruptcies or congestion patterns. This spreading is often along the links of the underlying causality or interaction networks [20].

"Self-organized criticality" [21, 22] is a particularly interesting phenomenon, where a system is driven towards a critical point. This is not uncommon for economic systems or critical infrastructures: Due to the need to minimize costs, safety margins will not be chosen higher than necessary. For example, they will be adjusted to the largest system perturbation that has occurred in the last so-and-so many years. As a consequence, there will be no failures in a long time. But then, controllers start to argue that one could safe money by reducing the standards. Eventually, the safety margins will be low enough to be exceeded by *some* perturbation, which may finally trigger a disaster.

Waves of bankruptcies [23, 24] are not much different. The competition for customers forces companies to make better and better offers, until the profits have reached a critical value and some companies will die. This will reduce the competitive pressure among the remaining companies and increase the profits again. As a consequence, new competitors will enter the market, which eventually drives the system back to the critical point.

15.2 Some Common Mistakes in the Management of Complex Systems

The particular features of complex systems have important implications for organizations, companies, and societies, which are complex multi-component systems themselves. Their counter-intuitive behaviors result from often very complicated feedback loops in the system, which cause many management mistakes and undesired side effects. Such effects are particularly well-known from failing political attempts to improve the social or economic conditions.

15.2.1 The System Does Not Do What You Want It to Do

One of the consequences of the non-linear interactions between the components of a complex system is that the internal interactions often dominate the external control attempts (or boundary conditions). This is particularly obvious for group dynamics [25, 26].

It is quite typical for complex systems that, many times, large efforts have no significant effect, while sometimes, the slightest change (even a "wrong word") has a "revolutionary" impact. This all depends on whether a system is close to a critical state (which will lead to the latter situation) or not (then, many efforts to change the system will be in vain). In fact, complex systems often counteract the action. In chemical systems, this is known as Le Chatelier's principle.[1] Therefore, if it is necessary to change a system, the right strategy is to drive it to a critical point first. Then, it will be easy to drive it into a new state, but the potential problem is that the resulting state is often hard to predict.

Regarding such predictions, classical time series analysis will normally provide bad forecasts. The problem of opinion polls to anticipate election results when the mood in the population is changing, is well-known. In many cases, the expectations of a large number of individuals, as expressed by the stock prices at real or virtual stock markets, is more indicative than results of classical extrapolation. Therefore, auction-based mechanisms have been proposed as a new prediction tool. Recently,

---

[1] Specifically, Le Chatelier's principle says: "If a chemical system at equilibrium experiences a change in concentration, temperature, or total pressure, the equilibrium will shift in order to minimize that change."

there are even techniques to forecast the future with small groups [27]. This, however, requires to correct for individual biases by fitting certain personality parameters. These reflect, for example, the degree of risk aversion.

15.2.2 Guided Self-organization Is Better Than Control

The previous section questions the classical control approach, which is, for example, used to control machines. But it is also frequently applied to business and societies, when decision-makers attempt to regulate all details by legislation, administrative procedures, project definitions, etc. These procedures are very complicated and time-consuming, sensitive to gaps, prone to failures, and they often go along with unanticipated side effects and costs. However, a complex system cannot be controlled like a bus, i.e. steering it somewhere may drive it to some unexpected state.

Biological systems are very differently designed. They do not specify all procedures in detail. Otherwise cells would be much too small to contain all construction plans in their genetic code, and the brain would be too small to perform its incredible tasks. Rather than trying to control all details of the system behavior, biology makes use of the self-organization of complex systems rather than "fighting" it. It *guides* self-organization, while forceful control would destroy it [28].

Detailed control would require a large amount of energy, and would need further resources to put and keep the components of an artificial system together. That means, overriding the self-organization in the system is costly and inefficient. Instead, one could use self-organization principles as part of the management plan. But this requires a better understanding of the natural behavior of complex systems like companies and societies.

15.2.3 Self-organized Networks and Hierarchies

Hierarchies are a classical way to control systems. However, strict hierarchies are only optimal under certain conditions. Particularly, they require a high reliability of the nodes (the staff members) and the links (their exchange).

Experimental results on the problem solving performance of groups [29] show that small groups can find solutions to difficult problems faster than any of their constituting individuals, because groups profit from complementary knowledge and ideas. The actual performance, however, sensitively depends on the organization of information flows, i.e. on who can communicate with whom. If communication is unidirectional, for example, this can reduce performance. However, it may also be inefficient if everybody can talk to everyone else. This is, because the number of potential (bidirectional) communicative links grows like $N(N-1)/2$, where N denotes the number of group members. The number of communicative or

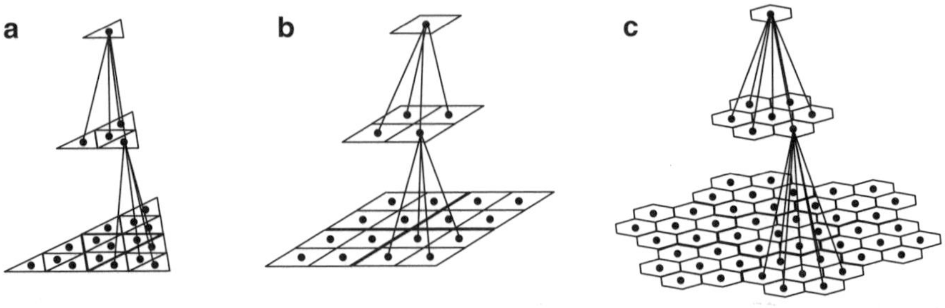

Fig. 15.6 Illustration of different kinds of hierarchical organization. As there are no alternative communication links, strict hierarchies are vulnerable to the failure of nodes or links

group-dynamical constellations even grows as $(3^N - 2^{N+1} + 1)/2$. Consequently, the number of possible information flows explodes with the group size, which may easily overwhelm the communication and information processing capacity of individuals. This explains the slow speed of group decision making, i.e. the inefficiency of large committees. It is also responsible for the fact that, after some transient time, (communication) activities in large (discussion) groups often concentrate on a few members only, which is due to a self-organized information bundling and differentiation (role formation) process. A similar effect is even observed in insect societies such as bee hives: When a critical colony size is exceeded, a few members develop hyperactivity, while most colony members become lazy [30].

This illustrates the tendency of bundling and compressing information flows, which is most pronounced in strict hierarchies. But the performance of strictly hierarchical organizations (see Fig. 15.6) is vulnerable for the following reasons:

- Hierarchical organizations are not robust with respect to failure of nodes (due to illness of staff members, holidays, quitting the job) or links (due to difficult personal relationships).
- They often do not connect interrelated activities in different departments well.
- Important information may get lost due to the filtering of information implied by the bundling process.
- Important information may arrive late, as it takes time to be communicated over various hierarchical levels.

Therefore, hierarchical networks with short-cuts are expected to be superior to strictly hierarchical networks [31–33]. They can profit from alternative information paths and "small-world" effects [34].

Note that the spontaneous formation of hierarchical structures is not untypical in social systems: Individuals form groups, which form companies, organizations, and parties, which make up a society or nation. A similar situation can be found in biology, where organelles form cells, cells form organs, and organs form bodies. Another example is well-known from physics, where elementary particles form nuclei, which combine to atoms with electrons. The atoms form chemical molecules, which organize themselves as solids. These make up cellestial bodies, which form solar systems, which again establish galaxies.

Obviously, the non-linear interactions between the different elements of the system give rise to a formation of different *levels*, which are *hierarchically* ordered one below another. While changes on the lowest hierarchical level are fastest, changes on the highest level are slow.

On the lowest level, we find the strongest interactions among its elements. This is obviously the reason for the fast changes on the lowest hierarchical level. If the interactions are attractive, *bonds* will arise. These cause the elements to behave no longer completely individually, but to form *units* representing the elements of the next level. Since the attractive interactions are more or less "saturated" by the bonds, the interactions *within* these units are stronger than the interactions *between* them. The relatively weak *residual interactions* between the formed units induce their relatively slow dynamics [35].

In summary, a general interdependence between the interaction strength, the changing rate, and the formation of hierarchical levels can be found, and the existence of different hierarchical levels implies a "separation of time scales".

The management of organizations, production processes, companies, and political changes seems to be quite different today: The highest hierarchy levels appear to take a *strong* influence on the system on a relatively *short* time scale. This does not only require a large amount of resources (administrative overhead). It also makes it difficult for the lower, less central levels of organization to adjust themselves to a changing environment. This complicates large-scale coordination in the system and makes it more costly. Strong interference in the system may even destroy self-organization in the system instead of using its potentials. Therefore, the restructuring of companies can easily fail, in particularly if it is applied too often. A good example is given in [36].

Governments would be advised to focus their activities on coordination functions, and on adaptations that are relevant for long time scales, i.e. applicable for 100 years or so. Otherwise the individuals will not be able to adjust to the boundary conditions set by the government. If the government tries to adjust to the population and the people try to adjust to the socio-economic conditions on the same time scale of months or years, the control attempts are expected to cause a potentially chaotic dynamics and a failure of control.

Anyway, detailed regulations hardly ever reach more fairness. They rather reduce flexibility, and make the anyway required processes inefficient, slow, complicated, and expensive. As a consequence, many people will not be able to utilize their rights without external help, while a specialized minority will be able to profit from the regulations or exploit them.

15.2.4 Faster Is Often Slower

Another common mistake is to push team members to their limits and have machines run at maximum speed. In many cases, this will not maximize productivity and throughput, but rather frustration. Most systems require some spare capacity to run smoothly. This is well illustrated by queuing systems: If the arrival rate reaches the service rate, the average waiting time will grow enormously. The same applies to the variation of the waiting time. Jamming and full buffers will be an unfavorable, but likely side effect. And there will be little reserves in case of additional demand.

The situation becomes even more difficult by dynamic interaction effects, when a system is driven to its limits. In traffic systems, for example, this leads to a "capacity drop". Such a capacity drop occurs often unexpectedly and is a sign of inefficiencies due to dynamical friction or obstruction effects. It results from increasing coordination problems when sufficient space or time are lacking. The consequence is often a "faster-is-slower effect" [38] (see Fig. 15.7). This effect has been observed in many traffic, production, and logistic systems. Consequently, it is often not good if everybody is doing his or her best. It is more important to adjust to the other activities and processes in order to reach a harmonic and well coordinated overall dynamics. Otherwise, more and more conflicts, inefficiencies and mistakes will ruin the overall performance.

15.2.5 The Role of Fluctuations and Heterogeneity

Let us finally discuss the role of fluctuations and heterogeneity. Fluctuations are often considered unfavorable, as they are thought to produce disorder. They can also trigger instabilities and breakdowns, as is known from traffic flows. But in some systems, fluctuations can also have positive effects.

While a large fluctuation strength, in fact, tends to destroy order, medium fluctuation levels may even cause a *noise-induced ordering* (see Fig. 15.8). An eventual increase in the degree of order in the system is particularly expected if the system tends to be trapped in local minima ("frustrated states"). Only by means of fluctuations, it is possible to escape these traps and to eventually find better solutions.

Fluctuations are also needed to develop different behavioral roles under initially identical conditions. This eventually leads to a differentiation and specialization (heterogeneity), which often helps to reach a better group performance [40] (see Fig. 15.9).

Furthermore, the speed of evolution also profits from variety and fluctuations ("mutations"). Uniformity, i.e. if everybody behaves and thinks the same, will lead to a poor adaptation to changing environmental or market conditions. In contrast, a large variety of different approaches (i.e. a heterogeneous population) will imply a large innovation rate [41]. The innovation rate is actually expected to be proportional to the variance of individual solutions. Therefore, strong norms, "monocultures", and the application of identical strategies all over the world due to the trend towards globalization implies dangers.

This trend is re-inforced by "herding effects" [7]. Whenever the future is hard to predict, people tend to orient at the behavior of others. This may easily lead to wrong collective decisions, even of highly intelligent people. This danger can be only reduced by supporting and maintaining a plurality of opinions and solutions.

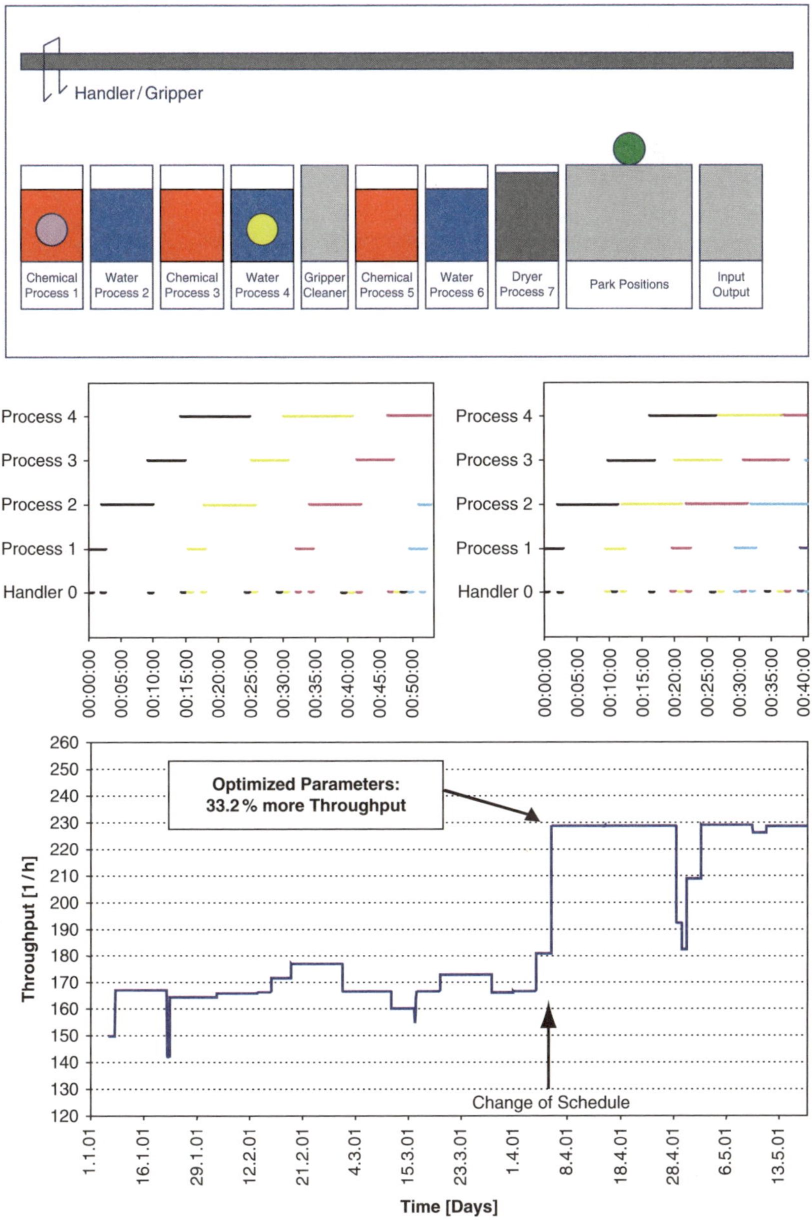

Fig. 15.7 *Top*: Schematic representation of the successive processes of a wet bench, i.e a particular supply chain in semiconductor production. *Middle*: The Gantt diagrams illustrate the treatment times of the first four of several more processes, where we have used the same colors for processes belonging to the same run, i.e. the same set of wafers. The *left diagram* shows the original schedule, while the *right* one shows an optimized schedule based on the "slower-is-faster effect". *Bottom*: The increase in the throughput of a wet bench by switching from the original production schedule to the optimized one was found to be 33%, in some cases even higher (after [37])

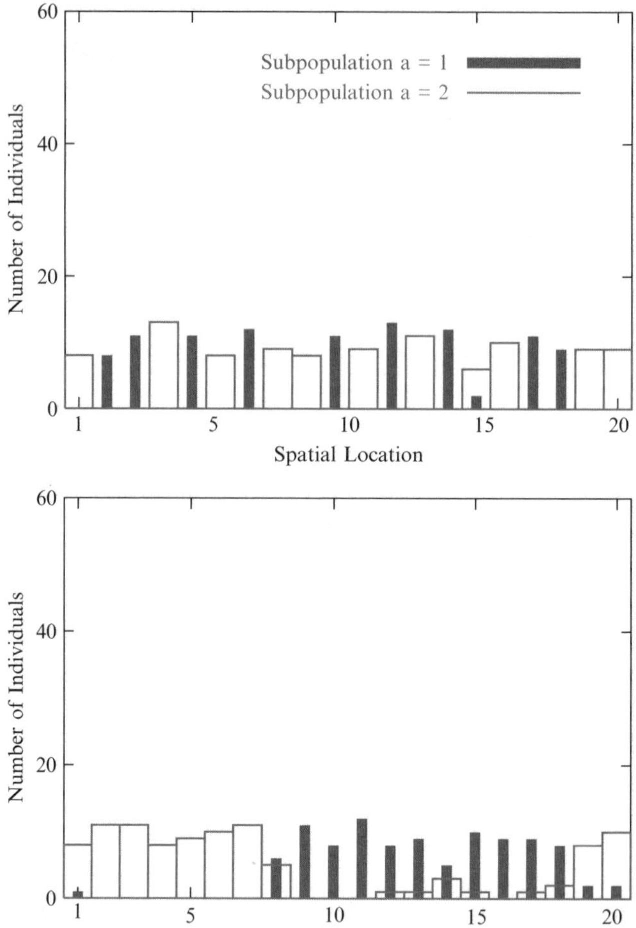

Fig. 15.8 Illustration of frequency distributions of behaviors in space (after [39]). *Top*: Separation of oppositely moving pedestrians perpendicularly to their walking direction for a low fluctuation strength. *Bottom*: Noise-induced ordering for medium fluctuation levels leads to a clear separation into two spatial areas. This reduces frictional effects and increases the efficiency of motion

15.3 Summary and Outlook

In this contribution, we have given a short overview of some properties and particularities of complex systems. Many of their behaviors may occur unexpectedly (due to "catastrophies" or phase transitions), and they are often counter-intuitive, e.g. due to feedback loops and side effects. Therefore, the response of complex systems to control attempts can be very different from the intended or predicted one.

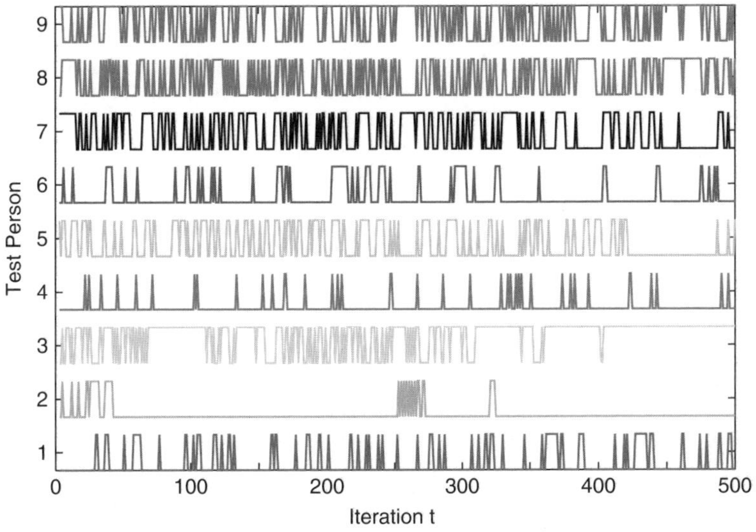

Fig. 15.9 Typical individual decision changes of nine test persons in a route choice experiment with two alternative routes. Note that we find almost similar or opposite behaviors after some time. The test persons develop a few kinds of complementary strategies ("roles") in favour of a good group performance (after [40])

Complex behavior in space and time is found for many multi-component systems with non-linear interactions. Typical examples are companies, organizations, administrations, or societies. This has serious implications regarding suitable control approaches. In fact, most control attempts are destined to fail. It would, however, be the wrong conclusion that one would just have to apply more force to get control over the system. This would destroy the self-organization in the system, on which social systems are based.

We need to obtain a better understanding of how to make use of the natural tendencies and behaviors at work. A management that supports and guides the natural self-organization in the system would perform much more efficiently than an artificially constructed system that requires continuous forcing. Companies and countries that manage to successfully apply the principle of self-organization will be the future winners of the on-going global competition.

In conclusion, we are currently facing a paradigm shift in the management of complex systems, and investments into complexity research will be of competitive advantage.

References

1. H. Haken, *Synergetics* (Springer, Berlin, 1977)
2. G. Ausiello, P. Crescenzi, G. Gambosi, et al., *Complexity and Approximation – Combinatorial optimization problems and their approximability properties* (Springer, Berlin, 1999)

3. S.H. Strogatz, *Nonlinear Dynamics and Chaos: With Applications to Physics, Biology, Chemistry and Engineering* (Perseus, New York, 2001)
4. Y. Kuramoto, *Chemical Oscillations, Waves, and Turbulence* (Springer, Berlin, 1984)
5. A. Pikovsky, M. Rosenblum, J. Kurths, *Synchronization: A Universal Concept in Nonlinear Sciences* (Cambridge University Press, Cambridge, 2003)
6. S.C. Manrubia, A.S. Mikhailov, D.H. Zanette, *Emergence of Dynamical Order. Synchronization Phenomena in Complex systems* (World Scientific, Singapore, 2004)
7. D. Helbing, Traffic and related self-driven many-particle systems. Rev. Mod. Phys. **73**, 1067 (2001)
8. E. Bonabeau, M. Dorigo, G. Theraulaz, Swarm Intelligence: From Natural to Artificial Systems. Santa Fe Institute Studies in the Sciences of Complexity Proceedings (1999)
9. A. Kesting, M. Schönhof, S. Lämmer, M. Treiber, D. Helbing, Decentralized approaches to adaptive traffic control. In *Managing Complexity: Insights, Concepts, Applications* ed. by D. Helbing (Springer, Berlin, 2008)
10. D. Helbing, S. Lämmer, Verfahren zur Koordination konkurrierender Prozesse oder zur Steuerung des Transports von mobilen Einheiten innerhalb eines Netzwerkes [Method to Coordinate Competing Processes or to Control the Transport of Mobile Units within a Network]. Pending patent DE 10 2005 023 742.8 (2005)
11. R. Axelrod, *The Evolution of Cooperation* (Basic Books, New York, 1985)
12. J. von Neumann, O. Morgenstern, A. Rubinstein, H.W. Kuhn, *Theory of Games and Economic Behavior* (Princeton University Press, Princeton, 2004)
13. T.C. Schelling, *The Strategy of Conflict* (Harvard University Press, Cambridge, 2006)
14. D. Helbing, M. Schönhof, H.-U. Stark, J.A. Holyst, How individuals learn to take turns: Emergence of alternating cooperation in a congestion game and the prisoner's dilemma. Adv. Complex Syst. **8**, 87 (2005)
15. N.S. Glance, B.A. Huberman, The dynamics of social dilemmas. Sci. Am. **270**, 76 (1994)
16. G. Hardin, The Tragedy of the Commons. Science **162**, 1243 (1968)
17. E.C. Zeeman, *Catastrophe Theory* (Addison-Wesley, London, 1977)
18. H.E. Stanley, *Introduction to Phase Transitions and Critical Phenomena* (Oxford University Press, Oxford, 1971)
19. M. Schroeder, *Fractals, Chaos, Power Laws: Minutes from an Infinite Paradise* (Freeman, New York, 1992)
20. D. Helbing, H. Ammoser, C. Kühnert, Disasters as extreme events and the importance of network interactions for disaster response management, in *The Unimaginable and Unpredictable: Extreme Events in Nature and Society*, ed. by S. Albeverio, V. Jentsch, H. Kantz (Springer, Berlin, 2005), pp. 319–348
21. P. Bak, C. Tang, K. Wiesenfeld, Self-organized criticality: An explanation of $1/f$ noise. Phys. Rev. Lett. **59**, 381 (1987)
22. P. Bak, *How Nature Works: The Science of Self-Organized Criticality* (Copernicus, New York, 1996)
23. A. Aleksiejuk, J.A. Hołyst, A simple model of bank bankruptcies. Physica A **299**(1-2), 198 (2001)
24. A. Aleksiejuk, J.A. Hołyst, G. Kossinets, Self-organized criticality in a model of collective bank bankruptcies. Int. J. Mod. Phys. C **13**, 333 (2002)
25. S.L. Tubbs, *A Systems Approach to Small Group Interaction* (McGraw-Hill, Boston, 2003)
26. H. Arrow, J.E. McGrath, J.L. Berdahl, *Small Groups as Complex Systems: Formation, Coordination, Development, and Adaptation* (Sage, CA, 2000)
27. K.-Y. Chen, L.R. Fine, B.A. Huberman, Predicting the Future. Inform. Syst. Front. **5**, 47 (2003)
28. A.S. Mikhailov, Artificial life: an engineering perspective, in *Evolution of Dynamical Structures in Complex Systems*, ed. by R. Friedrich, A. Wunderlin (Springer, Berlin, 1992), pp. 301–312
29. F-L. Ulschak, *Small Group Problem Solving: An Aid to Organizational Effectiveness* (Addison-Wesley Reading Mass., MA, 1981)

30. J. Gautrais, G. Theraulaz, J.-L. Deneubourg, C. Anderson, Emergent polyethism as a consequence of increased colony size in insect societies. J. Theor. Biol. **215**, 363 (2002)
31. D. Helbing, H. Ammoser, C. Kühnert, Information flows in hierarchical networks and the capability of organizations to successfully respond to failures, crises, and disasters. Physica A **363**, 141 (2006)
32. L.A. Adamic, E. Adar, Friends and neighbors on the web. Social Networks **25**(3), 211–230 (2003)
33. D. Stauffer, P.M.C. de Oliveira, Optimization of hierarchical structures of information flow. Int. J. Mod. Phys. C **17**, 1367 (2006)
34. D.J. Watts, S.H. Strogatz, Collective dynamics of smallworld networks. Nature **393**, 440 (1998)
35. D. Helbing, Quantitative Sociodynamics, in *Stochastic Methods and Models of Social Interaction Processes* (Kluwer Academic, Dordrecht, 1995)
36. M. Christen, G. Bongard, A. Pausits, N. Stoop, R. Stoop, Managing autonomy and control in economic systems. In *Managing Complexity: Insights, Concepts, Applications* ed. by D. Helbing (Springer, Berlin, 2008)
37. D. Fasold, Optimierung logistischer Prozessketten am Beispiel einer Nassätzanlage in der Halbleiterproduktion. MA thesis, TU Dresden (2001)
38. D. Helbing, T. Seidel, S. Lämmer, K. Peters, Self-organization principles in supply networks and production systems, in *Econophysics and Sociophysics - Trends and Perspectives*, ed. by B.K. Chakrabarti, A. Chakraborti, A. Chatterjee (Wiley, Weinheim, 2006), pp. 535–558
39. D. Helbing, T. Platkowski, Self-organization in space and induced by fluctuations. Int. J. Chaos Theor. Appl. **5**, 47–62 (2000)
40. D. Helbing, Dynamic decision behavior and optimal guidance through information services: Models and experiments, in *Human Behaviour and Traffic Networks*, ed. by M. Schreckenberg, R. Selten (Springer, Berlin, 2004), pp. 47–95
41. D. Helbing, M. Treiber, N.J. Saam, Analytical investigation of innovation dynamics considering stochasticity in the evaluation of fitness. Physical Review E **71**, 067101 (2005)

Chapter 16
Challenges in Economics*

16.1 Introduction

"How did economists get it so wrong?". Facing the financial crisis, this question was brilliantly articulated by the Nobel prize winner of 2008, Paul Krugman, in the New York Times [2]. A number of prominent economists even sees a failure of academic economics [3]. Remarkably, the following declaration has been signed by more than 2000 scientists [4]: "Few economists saw our current crisis coming, but this predictive failure was the least of the field's problems. More important was the profession's blindness to the very possibility of catastrophic failures in a market economy ... the economics profession went astray because economists, as a group, mistook beauty, clad in impressive-looking mathematics, for truth ... economists fell back in love with the old, idealized vision of an economy in which rational individuals interact in perfect markets, this time gussied up with fancy equations ... Unfortunately, this romanticized and sanitized vision of the economy led most economists to ignore all the things that can go wrong. They turned a blind eye to the limitations of human rationality that often lead to bubbles and busts; to the problems of institutions that run amok; to the imperfections of markets—especially financial markets—that can cause the economy's operating system to undergo sudden, unpredictable crashes; and to the dangers created when regulators don't believe in regulation. ... When it comes to the all-too-human problem of recessions and depressions, economists need to abandon the neat but wrong solution of assuming that everyone is rational and markets work perfectly."

Apparently, it has not always been like this. DeLisle Worrell writes: "Back in the sixties ... we were all too aware of the limitations of the discipline: it was

*This chapter reprints part of a previous publication to be cited as: D. Helbing and **S. Balietti**, Fundamental and Real-World Challenges in Economics. *Science and Culture* **76**(9/10), 399–417 (2010).

D. Helbing (ed.), *Social Self-Organization*, Understanding Complex Systems, DOI 10.1007/978-3-642-24004-1_16, static where the world was dynamic, it assumed competitive markets where few existed, it assumed rationality when we knew full well that economic agents were not rational … economics had no way of dealing with changing tastes and technology … Econometrics was equally plagued with intractable problems: economic observations are never randomly drawn and seldom independent, the number of excluded variables is always unmanageably large, the degrees of freedom unacceptably small, the stability of significance tests seldom unequivocably established, the errors in measurement too large to yield meaningful results … " [5].

In the following, we will try to identify the scientific challenges that must be addressed to come up with better theories in the near future. This comprises practical challenges, i.e. the real-life problems that must be faced (see Sect. 16.2), and fundamental challenges, i.e. the methodological advances that are required to solve these problems (see Sect. 16.3). After this, we will discuss, which contribution can be made by related scientific disciplines such as econophysics and the social sciences.

The intention of this contribution is constructive. It tries to stimulate a fruitful scientific exchange, in order to find the best way out of the crisis. According to our perception, the economic challenges we are currently facing can *only* be mastered by large-scale, multi-disciplinary efforts and by innovative approaches [6]. We fully recognize the large variety of non-mainstream approaches that has been developed by "heterodox economists". However, the research traditions in economics seem to be so powerful that these are not paid much attention to. Besides, there is no agreement on which of the alternative modeling approaches would be the most promising ones, i.e. the heterogeneity of alternatives is one of the problems, which slows down their success. This situation clearly implies institutional challenges as well, but these go beyond the scope of this contribution and will therefore be addressed in the future.

16.2 Real-World Challenges

Since decades, if not since hundreds of years, the world is facing a number of recurrent socio-economic problems, which are obviously hard to solve. Before addressing related fundamental scientific challenges in economics, we will therefore point out practical challenges one needs to pay attention to. This basically requires to classify the multitude of problems into packages of interrelated problems. Probably, such classification attempts are subjective to a certain extent. At least, the list presented below differs from the one elaborated by Lomborg *et al.* [7], who identified the following top ten problems: air pollution, security/conflict, disease control, education, climate change, hunger/malnutrition, water sanitation, barriers to migration and trade, transnational terrorism and, finally, women and development. The following (non-ranked) list, in contrast, is more focused on socio-economic factors rather than resource and engineering issues, and it is more oriented at the roots of problems rather than their symptoms:

1. *Demographic change* of the population structure (change of birth rate, migration, integration…)
2. *Financial and economic (in)stability* (government debts, taxation, and inflation/deflation; sustainability of social benefit systems; consumption and investment behavior…)
3. *Social, economic and political participation and inclusion* (of people of different gender, age, health, education, income, religion, culture, language, preferences; reduction of unemployment…)
4. *Balance of power* in a multi-polar world (between different countries and economic centers; also between individual and collective rights, political and company power; avoidance of monopolies; formation of coalitions; protection of pluralism, individual freedoms, minorities…)
5. *Collective social behavior* and opinion dynamics (abrupt changes in consumer behavior; social contagion, extremism, hooliganism, changing values; breakdown of cooperation, trust, compliance, solidarity…)
6. *Security and peace* (organized crime, terrorism, social unrest, independence movements, conflict, war…)
7. *Institutional design* (intellectual property rights; over-regulation; corruption; balance between global and local, central and decentral control…)
8. *Sustainable use of resources and environment* (consumption habits, travel behavior, sustainable and efficient use of energy and other resources, participation in recycling efforts, environmental protection…)
9. *Information management* (cyber risks, misuse of sensitive data, espionage, violation of privacy; data deluge, spam; education and inheritance of culture…)
10. *Public health* (food safety; spreading of epidemics [flu, SARS, H1N1, HIV], obesity, smoking, or unhealthy diets…)

Some of these challenges are interdependent.

16.3 Fundamental Challenges

In the following, we will try to identify the fundamental theoretical challenges that need to be addressed in order to understand the above practical problems and to draw conclusions regarding possible solutions.

The most difficult part of scientific research is often not to find the right answer. The problem is to ask the right questions. In this context it can be a problem that people are trained to think in certain ways. It is not easy to leave these ways and see the problem from a new angle, thereby revealing a previously unnoticed solution. Three factors contribute to this:

1. We may overlook the relevant facts because we have not learned to see them, i.e. we do not pay attention to them. The issue is known from internalized norms, which prevent people from considering possible alternatives.

2. We know the stylized facts, but may not have the right tools at hand to interpret them. It is often difficult to make sense of patterns detected in data. Turning data into knowledge is quite challenging.
3. We know the stylized facts and can interpret them, but may not take them seriously enough, as we underestimate their implications. This may result from misjudgements or from herding effects, i.e. from a tendency to follow traditions and majority opinions. In fact, most of the issues discussed below have been pointed out before, but it seems that this did not have an effect on mainstream economics so far or on what decision-makers know about economics. This is probably because mainstream theory has become a norm [8], and alternative approaches are sanctioned as norm-deviant behavior [9, 10].

As we will try to explain, the following fundamental issues are not just a matter of approximations (which often lead to the right understanding, but wrong numbers). Rather they concern *fundamental* errors in the sense that certain conclusions following from them are seriously misleading. As the recent financial crisis has demonstrated, such errors can be very costly. However, it is not trivial to see what dramatic consequences factors such as dynamics, spatial interactions, randomness, non-linearity, network effects, differentiation and heterogeneity, irreversibility or irrationality can have.

16.3.1 Homo Economicus

Despite of criticisms by several nobel prize winners such as Reinhard Selten (1994), Joseph Stiglitz and George Akerlof (2001), or Daniel Kahneman (2002), the paradigm of the *homo economicus*, i.e. of the "perfect egoist", is still the dominating approach in economics. It assumes that people would have quasi-infinite memory and processing capacities and would determine the best one among all possible alternative behaviors by strategic thinking (systematic utility optimization), and would implement it into practice without mistakes. The Nobel prize winner of 1976, Milton Friedman, supported the hypothesis of *homo economicus* by the following argument: "irrational agents will lose money and will be driven out of the market by rational agents" [11]. More recently, Robert E. Lucas Jr., the Nobel prize winner of 1995, used the *rationality* hypothesis to narrow down the class of empirically relevant equilibria [12].

The rational agent hypothesis is very charming, as its implications are clear and it is possible to derive beautiful and powerful economic theorems and theories from it. The best way to illustrate *homo economicus* is maybe a company that is run by using optimization methods from operation research, applying supercomputers. Another example are professional chess players, who are trying to anticipate the possible future moves of their opponents. Obviously, in both examples, the future course of actions can not be fully predicted, even if there are no random effects and mistakes.

It is, therefore, no wonder that people have repeatedly expressed doubts regarding the realism of the rational agent approach [13, 14]. Bertrand Russell, for example, claimed: "Most people would rather die than think". While this seems to be a rather extreme opinion, the following scientific arguments must be taken seriously:

1. Human cognitive capacities are bounded [16, 17]. Already phone calls or conversations can reduce people's attention to events in the environment a lot. Also, the abilities to memorize facts and to perform complicated logical analyses are clearly limited.
2. In case of NP-hard optimization problems, even supercomputers are facing limits, i.e. optimization jobs cannot be performed in real-time anymore. Therefore, approximations or simplifications such as the application of heuristics may be necessary. In fact, psychologists have identified a number of heuristics, which people use when making decisions [18].
3. People perform strategic thinking mainly in important new situations. In normal, everyday situation, however, they seem to pursue a satisficing rather than optimizing strategy [17]. Meeting a certain aspiration level rather than finding the optimal strategy can save time and energy spent on problem solving. In many situation, people even seem to perform routine choices [14], for example, when evading other pedestrians in counterflows.
4. There is a long list of cognitive biases which question rational behavior [19]. For example, individuals are favorable of taking small risks (which are preceived as "chances", as the participation in lotteries shows), but they avoid large risks [20]. Furthermore, non-exponential temporal discounting may lead to paradoxical behaviors [21] and requires one to rethink, how future expectations must be modeled.
5. Most individuals have a tendency towards other-regarding behavior and fairness [22, 23]. For example, the dictator game [24] and other experiments [25] show that people tend to share, even if there is no reason for this. Leaving a tip for the waiter in a restaurant that people visit only once is a typical example (particularly in countries where tipping is not common) [26]. Such behavior has often been interpreted as sign of social norms. While social norms can certainly change the payoff structure, it has been found that the overall payoffs resulting from them do not need to create a user or system optimum [27–29]. This suggests that behavioral choices may be irrational in the sense of non-optimal. A typical example is the existence of unfavorable norms, which are supported by people although nobody likes them [30].
6. Certain optimization problems can have an infinite number of local optima or Nash equilibria, which makes it impossible to decide what is the best strategy [31].
7. Convergence towards the optimal solution may require such a huge amount of time that the folk theorem becomes useless. This can make it practically impossible to play the best response strategy [32].

8. The optimal strategy may be deterministically chaotic, i.e. sensitive to arbitrarily small details of the initial condition, which makes the dynamic solution unpredictable on the long run ("butterfly effect") [33, 34]. This fundamental limit of predictability also implies a limit of control—two circumstances that are even more true for non-deterministic systems with a certain degree of randomness.

In conclusion, although the rational agent paradigm (the paradigm of *homo economicus*) is theoretically powerful and appealing, there are a number of empirical and theoretical facts, which suggest deficiencies. In fact, most methods used in financial trading (such as technical analysis) are not well compatible with the rational agent approach. Even if an optimal solution exists, it may be undecidable for practical reasons or for theoretical ones [35, 36]. This is also relevant for the following challenges, as *boundedly rational agents* may react inefficently and with delays, which questions the efficient market hypothesis, the equilibrium paradigm, and other fundamental concepts, calling for the consideration of spatial, network, and time-dependencies, heterogeneity and correlations etc. It will be shown that these points can have dramatic implications regarding the predictions of economic models.

16.3.2 The Efficient Market Hypothesis

The efficient market hypothesis (EMH) was first developed by Eugene Fama [37] in his Ph.D. thesis and rapidly spread among leading economists, who used it as an argument to promote laissez-faire policies. The EMH states that current prices reflect all publicly available information and (in its stronger formulation) that prices instantly change to reflect new public information.

The idea of self-regulating markets goes back to Adam Smith [38], who believed that "the free market, while appearing chaotic and unrestrained, is actually guided to produce the right amount and variety of goods by a so-called "invisible hand"." Furthermore, "by pursuing his own interest, [the individual] frequently promotes that of the society more effectually than when he intends to promote it" [39]. For this reason, Adam Smith is often considered to be the father of free market economics. Curiously enough, however, he also wrote a book on "The Theory of Moral Sentiments" [40]. "His goal in writing the work was to explain the source of mankind's ability to form moral judgements, in spite of man's natural inclinations towards self-interest. Smith proposes a theory of sympathy, in which the act of observing others makes people aware of themselves and the morality of their own behavior … [and] seek the approval of the "impartial spectator" as a result of a natural desire to have outside observers sympathize with them" [38]. Such a reputation-based concept would be considered today as indirect reciprocity [41].

Of course, there are criticisms of the efficient market hypothesis [42], and the Nobel prize winner of 2001, Joseph Stiglitz, even believes that "There is not invisible hand" [43]. The following list gives a number of empirical and theoretical arguments questioning the efficient market hypothesis:

1. Examples of market failures are well-known and can result, for example, in cases of monopolies or oligopolies, if there is not enough liquidity or if information symmetry is not given.
2. While the concept of the "invisible hand" assumes something like an optimal self-organization [44], it is well-known that this requires certain conditions, such as symmetrical interactions. In general, however, self-organization does not necessarily imply system-optimal solutions. Stop-and-go traffic [45] or crowd disasters [46] are two obvious examples for systems, in which individuals competitively try to reach individually optimal outcomes, but where the optimal solution is dynamically unstable.
3. The limited processing capacity of boundedly rational individuals implies potential delays in their responses to sensorial inputs, which can cause such instabilities [47]. For example, a delayed adaptation in production systems may contribute to the occurrence of business cycles [48]. The same applies to the labor market of specially skilled people, which cannot adjust on short time scales. Even without delayed reactions, however, the competitive optimization of individuals can lead to suboptimal individual results, as the "tragedy of the commons" in public goods dilemmas demonstrates [49, 50].
4. Bubbles and crashes, or more generally, extreme events in financial markets should not occur, if the efficient market hypothesis was correct (see next subsection).
5. Collective social behavior such as "herding effects" as well as deviations of human behavior from what is expected from rational agents can lead to such bubbles and crashes [51], or can further increase their size through feedback effects [52]. Cyclical feedbacks leading to oscillations are also known from the beer game [53] or from business cycles [48].

16.3.3 Equilibrium Paradigm

The efficient market paradigm implies the equilibrium paradigm. This becomes clear, if we split it up into its underlying hypotheses:

1. The market *can* be in equilibrium, i.e. there exists an equilibrium.
2. There is one and only one equilibrium.
3. The equilibrium is stable, i.e. any deviations from the equilibrium due to "fluctuations" or "perturbations" tend to disappear eventually.
4. The relaxation to the equilibrium occurs at an infinite rate.

Note that, in order to act like an "invisible hand", the stable equilibrium (Nash equilibrium) furthermore needs to be a system optimum, i.e. to maximize the average utility. This is true for coordination games, when interactions are well-mixed and exploration behavior as well as transaction costs can be neglected [54]. However, it is not fulfilled by so-called social dilemmas [49].

Let us discuss the evidence for the validity of the above hypotheses one by one:

1. A market is a system of extremely many dynamically coupled variables. Theoretically, it is not obvious that such a system would have a stationary solution. For example, the system could behave periodic, quasi-periodic, chaotic, or turbulent [81–83, 85–87, 94]. In all these cases, there would be no convergence to a stationary solution.
2. If a stationary solution exists, it is not clear that there are no further stationary solutions. If many variables are non-linearly coupled, the phenomenon of multistability can easily occur [55]. That is, the solution to which the system converges may not only depend on the model parameters, but also on the initial condition, history, or perturbation size. Such facts are known as path-dependencies or hysteresis effects and are usually visualized by so-called phase diagrams [56].
3. In systems of non-linearly interacting variables, the existence of a stationary solution does not necessarily imply that it is stable, i.e. that the system will converge to this solution. For example, the stationary solution could be a focal point with orbiting solutions (as for the classical Lotka-Volterra equations [57]), or it could be unstable and give rise to a limit cycle [58] or a chaotic solution [33], for example (see also item 1). In fact, experimental results suggest that volatility clusters in financial markets may be a result of over-reactions to deviations from the fundamental value [59].
4. An infinite relaxation rate is rather unusual, as most decisions and related implemenations take time [15, 60].

The points listed in the beginning of this subsection are also questioned by empirical evidence. In this connection, one may mention the existence of business cycles [48] or unstable orders and deliveries observed in the experimental beer game [53]. Moreover, bubbles and crashes have been found in financial market games [61]. Today, there seems to be more evidence against than for the equilibrium paradigm.

In the past, however, most economists assumed that bubbles and crashes would not exist (and many of them still do). The following quotes are quite typical for this kind of thinking (from [62]): In 2004, the Federal Reserve chairman of the U.S., Alan Greenspan, stated that the rise in house values was "not enough in our judgment to raise major concerns". In July 2005 when asked about the possibility of a housing bubble and the potential for this to lead to a recession in the future, the present U.S. Federal Reserve chairman Ben Bernanke (then Chairman of the Council of Economic Advisors) said: "It's a pretty unlikely possibility. We've never had a decline in housing prices on a nationwide basis. So, what I think is more likely is that house prices will slow, maybe stabilize, might slow consumption spending a bit. I don't think it's going to drive the economy too far from it's full path though." As late as May 2007 Bernanke stated that the Federal Reserve "do not expect significant spillovers from the subprime market to the rest of the economy".

According to the classical interpretation, sudden changes in stock prices result from new information, e.g. from innovations ("technological shocks"). The dynamics in such systems has, for example, been described by the method of comparative statics (i.e. a series of snapshots). Here, the system is assumed to be in equilibrium in each moment, but the equilibrium changes adiabatically (i.e. without delay), as the system parameters change (e.g. through new facts). Such a treatment of system dynamics, however, has certain deficiencies:

1. The approach cannot explain changes in or of the system, such as phase transitions ("systemic shifts"), when the system is at a critical point ("tipping point").
2. It does not allow one to understand innovations and other changes as results of an endogeneous system dynamics.
3. It cannot describe effects of delays or instabilities, such as overshooting, self-organization, emergence, systemic breakdowns or extreme events (see Sect. 16.3.4).
4. It does not allow one to study effects of different time scales. For example, when there are fast autocatalytic (self-reinfocing) effects and slow inhibitory effects, this may lead to pattern formation phenomena in space and time [63, 64]. The formation of settlements, where people agglomerate in space, may serve as an example [65, 66].
5. It ignores long-term correlations such as memory effects.
6. It neglects frictional effects, which are often proportional to change ("speed") and occur in most complex systems. Without friction, however, it is difficult to understand entropy and other path-dependent effects, in particular irreversibility (i.e. the fact that the system may not be able to get back to the previous state) [67]. For example, the unemployment rate has the property that it does not go back to the previous level in most countries after a business cycle [68].

16.3.4 Prevalence of Linear Models

Comparative statics is, of course, not the only method used in economics to describe the dynamics of the system under consideration. As in physics and other fields, one may use a linear approximation around a stationary solution to study the response of the system to fluctuations or perturbations [69]. Such a linear stability analysis allows one to study, whether the system will return to the stationary solution (which is the case for a stable [Nash] equilibrium) or not (which implies that the system will eventually be driven into a new state or regime).

In fact, the great majority of statistical analyses use *linear* models to fit empirical data (also when they do not involve time-dependencies). It is know, however, that linear models have special features, which are not representative for the rich variety of possible functional dependencies, dynamics, and outcomes. Therefore, the neglection of non-linearity has serious consequences:

1. As it was mentioned before, phenomena like multiple equilibria, chaos or turbulence cannot be understood by linear models. The same is true for self-organization phenomena or emergence. Additionally, in non-linearly coupled systems, usually "more is different", i.e. the system may change its behavior

fundamentally as it grows beyond a certain size. Furthermore, the system is often hard to predict and difficult to control (see Sect. 16.3.8).
2. Linear modeling tends to overlook that a strong coupling of variables, which would show a normally distributed behavior in separation, often leads to fat tail distributions (such as "power laws") [70, 71]. This implies that extreme events are *much* more frequent than expected according to a Gaussian distribution. For example, when additive noise is replaced by multiplicative noise, a number of surprising phenomena may result, including noise-induced transitions [72] or directed random walks ("ratchet effects") [73].
3. Phenomena such as catastrophes [74] or phase transition ("system shifts") [75] cannot be well understood within a linear modeling framework. The same applies to the phenomenon of "self-organized criticality" [79] (where the system drives itself to a critical state, typically with power-law characteristics) or cascading effects, which can result from network interactions (overcritically challenged network nodes or links) [77,78]. It should be added that the relevance of network effects resulting from the on-going globalization is often underestimated. For example, "the stock market crash of 1987, began with a small drop in prices which triggered an avalanche of sell orders in computerized trading programs, causing a further price decline that triggered more automatic sales." [80]

Therefore, while linear models have the advantage of being analytically solvable, they are often unrealistic. Studying non-linear behavior, in contrast, often requires numerical computational approaches. It is likely that most of today's unsolved economic puzzles cannot be well understood through linear models, no matter how complicated they may be (in terms of the number of variables and parameters) [81–94]. The following list mentions some areas, where the importance of non-linear interdependencies is most likely underestimated:

- Collective opinions, such as trends, fashions, or herding effects.
- The success of new (and old) technologies, products, etc.
- Cultural or opinion shifts, e.g. regarding nuclear power, genetically manipulated food, etc.
- The "fitness" or competitiveness of a product, value, quality perceptions, etc.
- The respect for copyrights.
- Social capital (trust, cooperation, compliance, solidarity, ...).
- Booms and recessions, bubbles and crashes.
- Bank panics.
- Community, cluster, or group formation.
- Relationships between different countries, including war (or trade war) and peace.

16.3.5 Representative Agent Approach

Another common simplification in economic modeling is the representative agent approach, which is known in physics as mean field approximation. Within this framework, time-dependencies and non-linear dependencies are often considered, but it is assumed that the interaction with other agents (e.g. of one company with all the other companies) can be treated as if this agent would interact with an average agent, the "representative agent".

Let us illustrate this with the example of the public goods dilemma. Here, everyone can decide whether to make an individual contribution to the public good or not. The sum of all contributions is multiplied by a synergy factor, reflecting the benefit of cooperation, and the resulting value is equally shared among all people. The prediction of the representative agent approach is that, due to the selfishness of agents, a "tragedy of the commons" would result [49]. According to this, everybody should free-ride, i.e. nobody should make a contribution to the public good and nobody would gain anything. However, if everybody would contribute, everybody could multiply his or her contribution by the synergy factor. This example is particularly relevant as society is facing a lot of public goods problems and would not work without cooperation. Everything from the creation of public infrastructures (streets, theaters, universities, libraries, schools, the World Wide Web, Wikipedia etc.) over the use of environmental resources (water, forests, air, etc.) or of social benefit systems (such as a public health insurance), maybe even the creation and maintainance of a commonly shared language and culture are public goods problems (although the last examples are often viewed as coordination problems). Even the *process* of creating public goods is a public good [95].

While it is a well-known problem that people tend to make unfair contributions to public goods or try to get a bigger share of them, individuals cooperate much more than one would expect according to the representative agent approach. If they would not, society could simply not exist. In economics, one tries to solve the problem by introducing taxes (i.e. another incentive structure) or a "shadow of the future" (i.e. a strategic optimization over infinite time horizons in accordance with the rational agent approach) [96,97]. Both comes down to changing the payoff structure in a way that transforms the public good problem into another one that does not constitute a social dilemma [98]. However, there are other solutions of the problem. When the realm of the mean-field approximation underlying the representative agent approach is left and spatial or network interactions or the heterogeneity among agents are considered, a miracle occurs: Cooperation can survive or even thrive through correlations and co-evolutionary effects [99–101].

A similar result is found for the public goods game with costly punishment. Here, the representative agent model predicts that individuals avoid to invest into punishment, so that punishment efforts eventually disappear (and, as a consequence, cooperation as well). However, this "second-order free-rider problem" is naturally resolved and cooperation can spread, if one discards the mean-field approximation and considers the fact that interactions take place in space or social networks [56]. Societies can overcome the tragedy of the commons even without transforming the incentive structure through taxes. For example, social norms as well as group dynamical and reputation effects can do so [102]. The representative agent approach implies just the opposite conclusion and cannot well explain the mechanisms on which society is built.

It is worth pointing out that the relevance of public goods dilemmas is probably underestimated in economics. Partially related to Adam Smith's belief in an "invisible hand", one often assumes underlying coordination games and that they would automatically create a harmony between an individually and system optimal state in the course of time [54]. However, running a stable financial system and economy is most likely a public goods problem. Considering unemployment, recessions always go along with a breakdown of solidarity and cooperation. Efficient production clearly requires mutual cooperation (as the counter-example of countries with many strikes illustrates). The failure of the interbank market when banks stop lending to each other, is a good example for the breakdown of both, trust and cooperation. We must be aware that there are many other systems that would not work anymore, if people would lose their trust: electronic banking, e-mail and internet use, Facebook, eBusiness and eGovernance, for example. Money itself would not work without trust, as bank panics and hyperinflation scenarios show. Similarly, cheating customers by selling low-quality products or selling products at overrated prices, or by manipulating their choices by advertisements rather than informing them objectively and when they want, may create profits on the short run, but it affects the trust of customers (and their willingness to invest). The failure of the immunization campaign during the swine flu pandemics may serve as an example. Furthermore, people would probably spend more money, if the products of competing companies were better compatible with each other. Therefore, on the long run, more cooperation among companies and with the customers would pay off and create additional value.

Besides providing a misleading picture of how cooperation comes about, there are a number of other deficiencies of the representative agent approach, which are listed below:

1. Correlations between variables are neglected, which is acceptable only for "well-mixing" systems. According to what is known from critical phenomena in physics, this approximation is valid only, when the interactions take place in high-dimensional spaces or if the system elements are well connected. (However, as the example of the public goods dilemma showed, this case does not necessarily have beneficial consequences. Well-mixed interactions could rather cause a breakdown of social or economic institutions, and it is conceivable that this played a role in the recent financial crisis.)
2. Percolation phenomena, describing how far an idea, innovation, technology, or (computer) virus spreads through a social or business network, are not well reproduced, as they depend on details of the network structure, not just on the average node degree [103].
3. The heterogeneity of agents is ignored. For this reason, factors underlying economic exchange, perturbations, or systemic robustness [104] cannot be well described. Moreover, as socio-economic differentiation and specialization imply heterogeneity, they cannot be understood as emergent phenomena within a representative agent approach. Finally, it is not possible to grasp innovation without the consideration of variability. In fact, according to evolutionary theory,

the innovation rate would be zero, if the variability was zero [105]. Furthermore, in order to explain innovation in modern societies, Schumpeter introduced the concept of the "political entrepreneur" [106], an extra-ordinarily gifted person capable of creating disruptive change and innovation. Such an extraordinary individual can, by definition, not be modeled by a "representative agent".

One of the most important drawbacks of the representative agent approach is that it cannot explain the fundamental fact of economic exchange, since it requires one to assume a heterogeneity in resources or production costs, or to consider a variation in the value of goods among individuals. Ken Arrow, Nobel prize winner in 1972, formulated this point as follows [107]: "One of the things that microeconomics teaches you is that individuals are not alike. There is heterogeneity, and probably the most important heterogeneity here is heterogeneity of expectations. If we didn't have heterogeneity, there would be no trade."

We close this section by mentioning that economic approaches, which go beyond the representative agent approach, can be found in Refs. [108, 109].

16.3.6 Lack of Micro-Macro Link and Ecological Systems Thinking

Another deficiency of economic theory that needs to be mentioned is the lack of a link between micro- and macroeconomics. Neoclassical economics implicitly assumes that individuals make their decisions in isolation, using only the information received from static market signals. Within this oversimplified framework, macro-aggregates are just projections of some representative agent behavior, instead of the outcome of complex interactions with asymmetric information among a myriad of heterogeneous agents.

In principle, it should be understandable how the macroeconomic dynamics results from the microscopic decisions and interactions on the level of producers and consumers [81, 110] (as it was possible in the past to derive micro-macro links for other systems with a complex dynamical behavior such as interactive vehicle traffic [111]). It should also be comprehensible how the macroscopic level (the aggregate econonomic situation) feeds back on the microscopic level (the behavior of consumers and producers), and to understand the economy as a complex, adaptive, self-organizing system [112, 113]. Concepts from evolutionary theory [114] and ecology [115] appear to be particularly promising [116]. This, however, requires a recognition of the importance of heterogeneity for the system (see the the previous subsection).

The lack of ecological thinking implies not only that the sensitive network interdependencies between the various agents in an economic system (as well as minority solutions) are not properly valued. It also causes deficiencies in the development and implementation of a sustainable economic approach based on recycling and renewable resources. Today, forestry science is probably the best developed scientific discipline concerning sustainability concepts [117]. Economic growth to maintain social welfare is a serious misconception. From other scientific disciplines, it is well known that stable pattern formation is also possible for a constant (and potentially sustainable) inflow of energy [69, 118].

16.3.7 Optimization of System Performance

One of the great achievements of economics is that it has developed a multitude of methods to use scarce resources efficiently. A conventional approach to this is optimization. In principle, there is nothing wrong about this approach. Nevertheless, there are a number of problems with the way it is usually applied:

1. One can only optimize for one goal at a time, while usually, one needs to meet several objectives. This is mostly addressed by weighting the different goals (objectives), by executing a hierarchy of optimization steps (through ranking and prioritization), or by applying a satisficing strategy (requiring a minimum performance for each goal) [119, 120]. However, when different optimization goals are in conflict with each other (such as maximizing the throughput and minimizing the queue length in a production system), a sophisticated time-dependent strategy may be needed [121].
2. There is no unique rule what optimization goal should be chosen. Low costs? High profit? Best customer satisfaction? Large throughput? Competitive advantage? Resilience? [122] In fact, the choice of the optimization function is arbitrary to a certain extent and, therefore, the result of optimization may vary largely. Goal selection requires strategic decisions, which may involve normative or moral factors (as in politics). In fact, one can often observe that, in the course of time, different goal functions are chosen. Moreover, note that the maximization of certain objectives such as resilience or "fitness" depends not only on factors that are under the control of a company. Resilience and "fitness" are functions of the whole system, in particularly, they also depend on the competitors and the strategies chosen by them.
3. The best solution may be the combination of two bad solutions and may, therefore, be overlooked. In other words, there are "evolutionary dead ends", so that gradual optimization may not work. (This problem can be partially overcome by the application of evolutionary mechanisms [120]).
4. In certain systems (such as many transport, logistic, or production systems), optimization tends to drive the system towards instability, since the point of maximum efficiency is often in the neighborhood or even identical with the point of breakdown of performance. Such breakdowns in capacity or performance can result from inefficiencies due to dynamic interaction effects. For example, when traffic flow reaches its maximum capacity, sooner or later it breaks down. As a consequence, the road capacity tends to drop during the time period where it is most urgently needed, namely during the rush hour [45, 123].

5. Optimization often eliminates reduncancies in the system and, thereby, increases the vulnerability to perturbations, i.e. it decreases robustness and resilience.
6. Optimization tends to eliminate heterogeneity in the system [80], while heterogeneity frequently supports adaptability and resilience.
7. Optimization is often performed with centralized concepts (e.g. by using supercomputers that process information collected all over the system). Such centralized systems are vulnerable to disturbances or failures of the central control unit. They are also sensitive to information overload, wrong selection of control parameters, and delays in adaptive feedback control. In contrast, decentralized control (with a certain degree of autonomy of local control units) may perform better, when the system is complex and composed of many heterogeneous elements, when the optimization problem is NP hard, the degree of fluctuations is large, and predictability is restricted to short time periods [77, 124]. Under such conditions, decentralized control strategies can perform well by adaptation to the actual local conditions, while being robust to perturbations. Urban traffic light control is a good example for this [121, 125].
8. Further on, today's concept of quality control appears to be awkward. It leads to a never-ending contest, requiring people and organizations to fulfil permanently increasing standards. This leads to over-emphasizing measured performance criteria, while non-measured success factors are neglected. The engagement into non-rewarded activities is discouraged, and innovation may be suppressed (e.g. when evaluating scientists by means of their h-index, which requires them to focus on a big research field that generates many citations in a short time).

   While so-called "beauty contests" are considered to produce the best results, they will eventually absorb more and more resources for this contest, while less and less time remains for the work that is actually to be performed, when the contest is won. Besides, a large number of competitors have to waste considerable resources for these contests which, of course, have to be paid by someone. In this way, private and public sectors (from physicians over hospitals, administrations, up to schools and universities) are aching under the evaluation-related administrative load, while little time remains to perform the work that the corresponding experts have been trained for. It seems naïve to believe that this would not waste resources. Rather than making use of individual strengths, which are highly heterogeneous, today's way of evaluating performance enforces a large degree of conformity.

There are also some problems with parameter fitting, a method based on optimization as well. In this case, the goal function is typically an error function or a likelihood function. Not only are calibration methods often "blindly" applied in practice (by people who are not experts in statistics), which can lead to overfitting (the fitting of meaningless "noise"), to the neglection of collinearities (implying largely variable parameter values), or to inaccurate and problematic parameter determinations (when the data set is insufficient in size, for example, when large portfolios are to be optimized [126]). As estimates for past data are not necessarily indicative for the future, making predictions with interpolation approaches can be quite problematic (see also Sect. 16.3.3 for the challenge of time dependence). Moreover, classical calibration methods do not reveal inappropriate model specifications (e.g. linear ones, when non-linear models would be needed, or unsuitable choices of model variables). Finally, they do not identify unknown unknowns (i.e. relevant explanatory variables, which have been overlooked in the modeling process).

16.3.8 Control Approach

Managing economic systems is a particular challenge, not only for the reasons discussed in the previous section. As large economic systems belong to the class of complex systems, they are hard or even impossible to manage with classical control approaches [76, 77].

Complex systems are characterized by a large number of system elements (e.g. individuals, companies, countries, ...), which have non-linear or network interactions causing mutual dependencies and responses. Such systems tend to behave dynamic rather than static and probabilistic rather than deterministic. They usually show a rich, hardly predictable, and sometimes paradoxical system behavior. Therefore, they challenge our way of thinking [127], and their controllability is often overestimated (which is sometimes paraphrased as "illusion of control") [80, 128, 129]. In particular, causes and effects are typically not proportional to each other, which makes it difficult to predict the impact of a control attempt.

A complex system may be unresponsive to a control attempt, or the latter may lead to unexpected, large changes in the system behavior (so-called "phase transitions", "regime shifts", or "catastrophes") [75]. The unresponsiveness is known as principle of Le Chatelier or Goodhart's law [130], according to which a complex system tends to counteract external control attempts. However, regime shifts can occur, when the system gets close to so-called "critical points" (also known as "tipping points"). Examples are sudden changes in public opinion (e.g. from pro to anti-war mood, from a smoking tolerance to a public smoking ban, or from buying energy-hungry sport utilities vehicles (SUVs) to buying environmentally-friendly cars).

Particularly in case of network interactions, big changes may have small, no, or unexpected effects. Feedback loops, unwanted side effects, and circuli vitiosi are quite typical. Delays may cause unstable system behavior (such as bull-whip effects) [53], and over-critical perturbations can create cascading failures [78]. Systemic breakdowns (such as large-scale blackouts, bankruptcy cascades, etc.) are often a result of such domino or avalanche effects [77], and their probability of occurrence as well as their resulting damage are usually underestimated. Further examples are epidemic spreading phenomena or disasters with an impact on the socio-economic system. A more detailed discussion is given in Refs. [76, 77].

Other factors contributing to the difficulty to manage economic systems are the large heterogeneity of system elements and the considerable level of randomness as well as the possibility of a chaotic or turbulent dynamics (see Sect. 16.3.4). Furthermore, the agents in economic systems are responsive to information, which can create self-fulfilling or self-destroying prophecy effects. Inflation may be viewed as example of such an effect. Interestingly, in some cases one even does not know in advance, which of these effects will occur.

It is also not obvious that the control mechanisms are well designed from a cybernetic perspective, i.e. that we have sufficient information about the system and suitable control variables to make control feasible. For example, central banks do not have terribly many options to influence the economic system. Among them are performing open-market operations (to control money supply), adjustments in fractional-reserve banking (keeping only a limited deposit, while lending a large part of the assets to others), or adaptations in the discount rate (the interest rate charged to banks for borrowing short-term funds directly from a central bank). Nevertheless, the central banks are asked to meet multiple goals such as:

- To guarantee well-functioning and robust financial markets.
- To support economic growth.
- To balance between inflation and unemployment.
- To keep exchange rates within reasonable limits.

Furthermore, the one-dimensional variable of "money" is also used to influence *individual* behavior via taxes (by changing behavioral incentives). It is questionable, whether money can optimally meet all these goals at the same time (see Sect. 16.3.7). We believe that a computer, good food, friendship, social status, love, fairness, and knowledge can only to a certain extent be replaced by and traded against each other. Probably for this reason, social exchange comprises more than just material exchange [131–133].

It is conceivable that financial markets as well are trying to meet too many goals at the same time. This includes:

- To match supply and demand.
- To discover a fair price.
- To raise the foreign direct investment (FDI).
- To couple local economies with the international system.
- To facilitate large-scale investments.
- To boost development.
- To share risk.
- To support a robust economy, and
- To create opportunities (to gamble, to become rich, etc.).

Therefore, it would be worth stuyding the system from a cybernetic control perspective. Maybe, it would work better to separate some of these functions from each other rather than mixing them.

16.3.9 Human Factors

Another aspect that tends to be overlooked in mainstream economics is the relevance of psychological and social factors such as emotions, creativity, social norms, herding effects, etc. It would probably be wrong to interpret these effects just as a result of perception biases (see Sect. 16.3.1). Most likely, these human factors serve certain functions such as supporting the creation of public goods [102] or collective intelligence [134, 135].

As Bruno Frey has pointed out, economics should be seen from a social science perspective [136]. In particular, research on happiness has revealed that there are more incentives than just financial ones that motivate people to work hard [133]. Interestingly, there are quite a number of factors which promote volunteering [132].

It would also be misleading to judge emotions from the perspective of irrational behavior. They are a quite universal and a relatively energy-consuming way of signalling. Therefore, they are probably more reliable than non-emotional signals. Moreover, they create empathy and, consequently, stimulate mutual support and a readiness for compromises. It is quite likely that this creates a higher degree of cooperativeness in social dilemma situations and, thereby, a higher payoff on average as compared to emotionless decisions, which often have drawbacks later on.

16.3.10 Information

Finally, there is no good theory that would allow one to assess the relevance of information in economic systems. Most economic models do not consider information as an explanatory variable, although information is actually a stronger driving force of urban growth and social dynamics than energy [137]. While we have an information theory to determine the number of bits required to encode a message, we are lacking a theory, which would allow us to assess what kind of information is relevant or important, or what kind of information will change the social or economic world, or history. This may actually be largely dependent on the perception of pieces of information, and on normative or moral issues filtering or weighting information. Moreover, we lack theories describing what will happen in cases of coincidence or contradiction of several pieces of information. When pieces of information interact, this can change their interpretation and, thereby, the decisions and behaviors resulting from them. That is one of the reasons why socio-economic systems are so hard to predict: "Unknown unknowns", structural instabilities, and innovations cause emergent results and create a dynamics of surprise [138].

16.4 Role of Other Scientific Fields

16.4.1 Econophysics, Ecology, Computer Science

The problems discussed in the previous two sections pose interesting practical and fundamental challenges for economists, but also other disciplines interested in understanding economic systems. Econophysics, for example, pursues a physical approach to economic systems, applying methods from statistical physics [81], network theory [139, 140], and the theory of complex systems [85, 87]. A contribution of physics appears quite natural, in fact, not only because of its tradition in detecting and modeling regularities in large data sets [141]. Physics also has a lot of experience how to theoretically deal with problems such as time-dependence, fluctuations, friction, entropy, non-linearity, strong interactions, correlations, heterogeneity, and many-particle simulations (which can be easily extended towards multi-agent simulations). In fact, physics has influenced economic modeling already in the past. Macroeconomic models, for example, were inspired by thermodynamics. More recent examples of relevant contributions by physicists concern models of self-organizing conventions [54], of geographic agglomeration [65], of innovation spreading [142], or of financial markets [143], to mention just a few examples.

One can probably say that physicists have been among the pioneers calling for new approaches in economics [81, 87, 143–147]. A particularly visionary book beside Wolfgang Weidlich's work was the "Introduction to Quantitative Aspects of Social Phenomena" by Elliott W. Montroll and Wade W. Badger, which addressed by mathematical *and* empirical analysis subjects as diverse as population dynamics, the arms race, speculation patterns in stock markets, congestion in vehicular traffic as well as the problems of atmospheric pollution, city growth and developing countries already in 1974 [148].

Unfortunately, it is impossible in our paper to reflect the numerous contributions of the field of econophysics in any adequate way. The richness of scientific contributions is probably reflected best by the Econophysics Forum run by Yi-Cheng Zhang [149]. Many econophysics solutions are interesting, but so far they are not broad and mighty enough to replace the rational agent paradigm with its large body of implications and applications. Nevertheless, considering the relatively small number of econophysicists, there have been many promising results. The probably largest fraction of publications in econophysics in the last years had a data-driven or computer modeling approach to financial markets [143]. But econophyics has more to offer than the analysis of financial data (such as fluctuations in stock and foreign currency exchange markets), the creation of interaction models for stock markets, or the development of risk management strategies. Other scientists have focused on statistical laws underlying income and wealth distributions, non-linear market dynamics, macroeconomic production functions and conditions for economic growth or agglomeration, sustainable economic systems, business cycles, microeconomic interaction models, network models, the growth of companies, supply and production systems, logistic and transport networks, or innovation dynamics and diffusion. An overview of subjects is given, for example, by Ref. [152] and the contributions to annual spring workshop of the Physics of Socio-Economic Systems Division of the DPG [153].

To the dissatisfaction of many econophysicists, the transfer of knowledge often did not work very well or, if so, has not been well recognized [150]. Besides scepticism on the side of many economists with regard to novel approaches introduced by "outsiders", the limited resonance and level of interdisciplinary exchange in the past was also caused in part by econophysicists. In many cases, questions have been answered, which no economist asked, rather than addressing puzzles economists are interested in. Apart from this, the econophysics work was not always presented in a way that linked it to the traditions of economics and pointed out deficiencies of existing models, highlighting the relevance of the new approach well. Typical responses are: Why has this model been proposed and not another one? Why has this simplification been used (e.g. an Ising model of interacting spins rather than a rational agent model)? Why are existing models not good enough to describe the same facts? What is the relevance of the work compared to previous publications? What practical implications does the finding have? What kind of paradigm shift does the approach imply? Can existing models be modified or extended in a way that solves the problem without requiring a paradigm shift? Correspondingly, there have been criticisms not only by mainstream economists, but also by colleagues, who are open to new approaches [151].

Therefore, we would like to suggest to study the various economic subjects from the perspective of the above-mentioned fundamental challenges, and to contrast econophysics models with traditional economic models, showing that the latter leave out important features. It is important to demonstrate what properties of economic systems cannot be understood for fundamental reasons within the mainstream framework (i.e. cannot be dealt with by additional terms within the modeling class that is conventionally used). In other words, one needs to show why a paradigm shift is unavoidable, and this requires careful argumentation. We are not claiming that this has not been done in the past, but it certainly takes an additional effort to explain the essence of the econophysics approach in the language of economics, particularly as mainstream economics may not always provide suitable terms and frameworks to do this. This is particularly important, as the number of econophysicists is small compared to the number of economists, i.e. a minority wants to convince an established majority. To be taken seriously, one must also demonstrate a solid knowledge of related previous work of economists, to prevent the stereotypical reaction that the subject of the paper has been studied already long time ago (tacitly implying that it does not require another paper or model to address what has already been looked at before).

A reasonable and promising strategy to address the above fundamental and practical challenges is to set up multi-disciplinary collaborations in order to combine the best of all relevant scientific methods and knowledge. It seems plausible that this will generate better models and higher impact than working in separation, and it will stimulate scientific innovation. Physicists can contribute with their experience in handling large data sets, in creating and simulating mathematical models, in developing useful approximations, in setting up laboratory experiments and measurement concepts. Current research activities in economics do not seem to put enough focus on:

- Modeling approaches for complex systems [154].
- Computational modeling of what is not analytically tractable anymore, e.g. by agent-based models [155–157].
- Testable predictions and their empirical or experimental validation [164].
- Managing complexity and systems engineering approaches to identify alternative ways of organizing financial markets and economic systems [91, 93, 165], and
- An advance testing of the effectiveness, efficiency, safety, and systemic impact (side effects) of innovations, before they are implemented in economic systems. This is in sharp contrast to mechanical, electrical, nuclear, chemical and medical drug engineering, for example.

Expanding the scope of economic thinking and paying more attention to these natural, computer and engineering science aspects will certainly help to address the theoretical and practical challenges posed by economic systems. Besides physics, we anticipate that also evolutionary biology, ecology, psychology, neuroscience, and artificial intelligence will be able to make significant contributions to the understanding of the roots of economic problems and how to solve them. In conclusion, there are interesting scientific times ahead.

16.4.2 Social Sciences

It is a good question, whether answering the above list of fundamental challenges will sooner or later solve the practical problems as well. We think, this is a precondition, but it takes more, namely the consideration of social factors. In particular, the following questions need to be answered:

1. How to understand human decision-making? How to explain deviations from rational choice theory and the decision-theoretical paradoxes? Why are people risk averse?
2. How does consciousness and self-consciousness come about?
3. How to understand creativity and innovation?
4. How to explain homophily, i.e. the fact that individuals tend to agglomerate, interact with and imitate *similar* others?
5. How to explain social influence, collective decision making, opinion dynamics and voting behavior?
6. Why do individuals often cooperate in social dilemma situations?
7. How do indirect reciprocity, trust and reputation evolve?

8. How do costly punishment, antisocial punishment, and discrimination come about?
9. How can the formation of social norms and conventions, social roles and socialization, conformity and integration be understood?
10. How do language and culture evolve?
11. How to comprehend the formation of group identity and group dynamics? What are the laws of coalition formation, crowd behavior, and social movements?
12. How to understand social networks, social structure, stratification, organizations and institutions?
13. How do social differentiation, specialization, inequality and segregation come about?
14. How to model deviance and crime, conflicts, violence, and wars?
15. How to understand social exchange, trading, and market dynamics?

We think that, despite the large amount of research performed on these subjects, they are still not fully understood. The ultimate goal would be to formulate mathematical models, which would allow one to understand these issues as emergent phenomena based on first principles, e.g. as a result of (co-)evolutionary processes. Such first principles would be the basic facts of human capabilities and the kinds of interactions resulting from them, namely:

1. Birth, death, and reproduction.
2. The need of and competition for resources (such as food and water).
3. The ability to observe their environment (with different senses).
4. The capability to memorize, learn, and imitate.
5. Empathy and emotions.
6. Signaling and communication abilities.
7. Constructive (e.g. tool-making) and destructive (e.g. fighting) abilities.
8. Mobility and (limited) carrying capacity.
9. The possibility of social and economic exchange.

Such features can, in principle, be implemented in agent-based models [158–163]. Computer simulations of many interacting agents would allow one to study the phenomena emerging in the resulting (artificial or) model societies, and to compare them with stylized facts [163, 168, 169]. The main challenge, however, is not to program a seemingly realistic computer game. We are looking for scientific models, i.e. the underlying assumptions need to be validated, and this requires to link computer simulations with empirical and experimental research [170], and with massive (but privacy-respecting) mining of social interaction data [141]. In the ideal case, there would also be an analytical understanding in the end, as it has been recently gained for interactive driver behavior [111].

Acknowledgements The authors are grateful for partial financial support by the ETH Competence Center "Coping with Crises in Complex Socio-Economic Systems" (CCSS) through ETH Research Grant CH1-01 08-2 and by the Future and Emerging Technologies programme FP7-COSI-ICT of the European Commission through the project Visioneer (grant no.: 248438). They would like to thank for feedbacks on the manuscript by Kenett Dror, Tobias Preis and Gabriele Tedeschi as well as for inspiring discussions during a Visioneer workshop in Zurich from January 13 to 15, 2010, involving, besides the authors, Stefano Battiston Guido Caldarelli, Anna Carbone, Giovanni Luca Ciampaglia, Andreas Flache, Imre Kondor, Sergi Lozano, Thomas Maillart, Amin Mazloumian, Tamara Mihaljev, Alexander Mikhailov, Ryan Murphy, Carlos Perez Roca, Stefan Reimann, Aki-Hiro Sato, Christian Schneider, Piotr Swistak, Gabriele Tedeschi, and Jiang Wu. Last but not least, we are grateful to Didier Sornette, Frank Schweitzer and Lars-Erik Cederman for providing some requested references.

References

1. P. Mancosu (ed.) *From Brouwer to Hilbert: The Debate on the Foundations of Mathematics in the 1920s*. (Oxford University Press, New York, 1998)
2. Paul Krugman, "How did economists get it so wrong?", *The New York Times* (September 2, 2009), see http://www.nytimes.com/2009/09/06/magazine/06Economic-t.html
3. D. Colander et al., The Financial Crisis and the Systemic Failure of Academic Economics (Dahlem Report, Univ. of Copenhagen Dept. of Economics Discussion Paper No. 09–03, 2009), see http://papers.ssrn.com/sol3/papers.cfm?abstract_id=1355882
4. See http://www.ipetitions.com/petition/revitalizing_economics/?e and http://emmarogan.wordpress.com/2009/10/23/the-financial-crisis-how-economists-went-astray/
5. DeLisle Worrell, What's wrong with economics. Address of the Governor of the Central Bank of Barbados on June 30, 2010, see http://www.bis.org/review/r100713c.pdf
6. D. Helbing, The FuturIcT knowledge accelerator: Unleashing the power of information for a sustainable future, see http://arxiv.org/abs/1004.4969 and http://www.futurict.eu
7. Bjorn Lomborg, *Global Crises, Global Solutions: Costs and Benefits*. (Cambridge University Press, Cambridge, 2nd ed., 2009)
8. R.H. Nelson, *Economics as Religion: from Samuelson to Chicago and Beyond*. (Pennsylvania State University, 2001)
9. F.A. Von Hayek, *Individualism and Economic Order*. (University of Chicago Press, 1948)
10. F.A. Von Hayek, *The Counter Revolution of Science*. (The Free Press of Glencoe, London, 1955)
11. M. Friedman, The Case of flexible exchange rates, in: *Essays in Positive Economics*. (University of Chicago Press, Chicago, 1953)
12. R. Lucas, Adaptive behavior and economic theory, J. Bus. **59**, S401–S426 (1986)
13. G. Gigerenzer, R. Selten (eds.) *Bounded Rationality. The Adaptive Toolbox*. (MIT Press, Cambridge, MA, 2001)
14. H. Gintis, *The Bounds of Reason: Game Theory and the Unification of the Behavioral Sciences*. (Princeton University Press, Princeton, 2009)
15. F. Caccioli, M. Marsili, On information efficiency and financial stability. Preprint http://arxiv.org/abs/1004.5014
16. H.A. Simon *Models of Man*. (Wiley, New York, 1957)
17. H.A. Simon, *Models of Bounded Rationality*. Vol. 1 (MIT Press, Boston, 1984)
18. G. Gigerenzer, P.M. Todd, and the ABC Research Group, *Simple Heuristics That Make Us Smart*. (Oxford University Press, Oxford, 2000)
19. For a collection of cognitive biases see http://en.wikipedia.org/wiki/List_of_cognitive_biases
20. D. Kahneman, P. Slovic, A. Tversky, *Judgment under Uncertainty: Heuristics and Biases*. (Cambridge University Press, Cambridge, MA, 1982)
21. V.I. Yukalov, D. Sornette, Physics of risk and uncertainty in quantum decision making, *European Physical Journal B* **71**, 533–548 (2009)
22. E. Fehr, K.M. Schmidt, A theory of fairness, competition, and cooperation. Q. J. Econ. **114**(3), 817–868 (1999)

23. J. Henrich, R. Boyd, S. Bowles, C. Camerer, *Foundations of Human Sociality: Economic Experiments and Ethnographic Evidence from Fifteen Small-Scale Societies*. (Oxford University, Oxford, 2004)
24. E. Hoffman, K. McCabe, V.L. Smith, Social distance and other-regarding behavior in dictator games. Am. Econ. Rev. **86**(3), 653–660 (1996)
25. R.O. Murphy, K. Ackermann, Measuring social value orientation. Judgement and decision making **6**(8), 771–781 (2011)
26. Ö.B. Bodvarsson, W.A. Gibson, Economics and restaurant gratuities: Determining tip rates. Am. J. Econ. Soc. **56**(2), 187–203 (1997)
27. J. Elster, *The Cement of Society: A Study of Social Order*. (Cambridge University Press, Cambridge, 1989)
28. C. Horne, *The Rewards of Punishment. A Relational Theory of Norm Enforcement*. (Stanford University Press, Stanford, 2009)
29. D. Helbing, W. Yu, K.-D. Opp, H. Rauhut, The emergence of homogeneous norms in heterogeneous populations. Santa Fe Working Paper 11-01-001 (2011), see http://www.santafe.edu/media/workingpapers/11-01-001.pdf, last accessed on March 6, 2012
30. D. Centola, R. Willer, M. Macy, The emperor's dilemma: A computational model of self-enforcing norms. Am. J. Soc. **110**, 1009–1040 (2005)
31. J.M. Epstein, R.A. Hammond, Non-explanatory equilibria: An extremely simple game with (mostly) unattainable fixed points. Complexity **7**(4), 18–22 (2002)
32. C. Borgs, J. Chayesa, N. Immorlicab, A.T. Kalaia, V. Mirroknia, C Papadimitriou, The myth of the folktheorem, Games and Economic Behavior **70**(1), 34–43 (2010)
33. H.G. Schuster, *Deterministic Chaos*. (Wiley VCH, Weinheim, 2005)
34. For example, three-body planetary motion has deterministic chaotic solutions, although it is a problem in classical mechanics, where the equations of motion optimize a Lagrangian functional.
35. K. Gödel, *On Formally Undecidable Propositions of Principia Mathematica and Related Systems*. (Basic, New York, 1962)
36. A.M. Turing, On computable numbers, with an application to the Entscheidungsproblem. Proc. Lond. Math. Soc. **2.42**, 230–265 (1936)
37. E. Fama, The Behavior of stock market prices. J. Bus. **38** 34–105 (1965)
38. Wikipedia article on Adam Smith, see http://en.wikipedia.org/wiki/Adam_Smith, downloaded on July 14, 2010.
39. A. Smith (1776), *An Inquiry into the Nature and Causes of the Wealth of Nations*. (University of Chicago Press, 1977)
40. A. Smith, The Theory of Moral Sentiments (1759)
41. M.A. Nowak, K. Sigmund, Evolution of indirect reciprocity. Nature **437**, 1291–1298 (2005)
42. M.J. Mauboussin, Revisiting market efficiency: The stock market as a complex adaptive system. J. Appl. Corp. Fin. **14**, 47–55 (2005)
43. J. Stiglitz, There is no invisible hand. The Guardian (December 20, 2002), see http://www.guardian.co.uk/education/2002/dec/20/highereducation.uk1
44. D. Helbing, T. Vicsek, Optimal self-organization. New J. Phys. **1**, 13.1–13.17 (1999)
45. D. Helbing, Traffic and related self-driven many-particle systems. Rev. Mod. Phys. **73**, 1067–1141 (2001)
46. A. Johansson, D. Helbing, H.Z. A-Abideen, S. Al-Bosta, *From crowd dynamics to crowd safety: A video-based analysis. Advances in Complex Systems* **11**(4), 497–527 (2008)
47. W. Michiels, S.-I. Niculescu, *Stability and Stabilization of Time-Delay Systems*. (siam—Society for Industrial and Applied Mathematics, Philadelphia, 2007)
48. D. Helbing, U. Witt, S. Lämmer, T. Brenner, Network-induced oscillatory behavior in material flow networks and irregular business cycles. Phys. Rev. E **70**, 056118 (2004)
49. G. Hardin, The tragedy of the commons. Science **162**, 1243–1248 (1968)
50. E. Fehr, S. Gächter, Altruistic punishment in humans. Nature **415**, 137–140 (2002)
51. T. Preis, H.E. Stanley, Switching phenomena in a system with no switches. J. Stat. Phys. (JSTAT) **138**, 431–446 (2010)

52. G.A. Akerlof, R.J. Shiller, *Animal Spirits: How Human Psychology Drives the Economy, and Why It Matters for Global Capitalism.* (Princeton University Press, 2010)
53. J.D. Sterman, Testing behavioral simulation models by direct experiment. Manag. Sci. **33**(12), 1572–1592 (1987)
54. D. Helbing, A mathematical model for behavioral changes by pair interactions. pp. 330–348 In: G. Haag, U. Mueller, and K.G. Troitzsch (eds.) *Economic Evolution and Demographic Change. Formal Models in Social Sciences.* (Springer, Berlin, 1992)
55. H.N. Agiza, G.I. Bischib, M. Kopel, Multistability in a dynamic Cournot game with three oligopolists. Math. Comput. Simulat. **51**, 63–90 (1999)
56. D. Helbing, A. Szolnoki, M. Perc, G. Szabo, Evolutionary establishment of moral and double moral standards through spatial interactions. PLoS Comput. Biol. **6**(4), e1000758 (2010)
57. A.J. Lotka *Elements of Mathematical Biology.* (Dover, New York, 1956)
58. E. Hopf, Abzweigungen einer periodischen Lösung von einer stationären Lösung eines Differentialgleichungssystems. Math. Naturwiss. Klasse **94**, 1ff (1942)
59. D. Helbing, Dynamic decision behavior and optimal guidance through information services: Models and experiments. pp. 47–95 In: M. Schreckenberg, R. Selten (eds.) *Human Behaviour and Traffic Networks* (Springer, Berlin, 2004)
60. H. Gintis, The dynamics of general equilibrium. Econ. J. **117**, 1280–1309 (2007)
61. C.H. Hommes, Modeling the stylized facts in finance through simple nonlinear adaptive systems. Proc. Natl. Acad. Sc. USA (PNAS) **99**, Suppl. 3, 7221–7228 (2002)
62. The quotes were presented by Jorgen Vitting Andersen in his talk “Predicting moments of crisis in physics and finance” during the workshop “Windows to Complexity” in Münster, Germany, on June 11, 2010
63. Turing A.M., The chemical basis of morphogenesis. Phil. Trans. Roy. Soc. Lond. B **237**, 37–72 (1952)
64. J.D. Murray, *Lectures on Nonlinear Differential Equation-Models in Biology.* (Clanderon Press, Oxford, 1977)
65. W. Weidlich, M. Munz, Settlement formation I: A dynamic theory, Ann. Reg. Sci. **24**, 83–106 (2000); Settlement formation II: Numerical simulation, *Ann. Reg. Sci.* **24**, 177–196 (2000)
66. D. Helbing, T. Platkowski, Self-organization in space and induced by fluctuations. Int. J. Chaos Theor. Appl. **5**(4), 47–62 (2000)
67. J. Mimkes, Stokes integral of economic growth: Calculus and the Solow model. Physica A **389**(8), 1665–1676 (2010)
68. O.J. Blanchard, L.H. Summers, Hysteresis and the European unemployment problem. NBER Macroecon. Annu. **1**, 15–78 (1986)
69. M. Cross, H. Greenside, *Pattern Formation and Dynamics in Nonequilibrium Systems.* (Cambridge University, 2009)
70. D. Sornette, *Critical Phenomena in Natural Sciences: Chaos, Fractals, Selforganization and Disorder.* (Springer, Berlin, 2006)
71. A. Bunde, J. Kropp, H.-J. Schellnhuber (eds.) *The Science of Disasters.* (Springer, Berlin, 2002)
72. W. Horsthemke, R. Lefever, *Noise-Induced Transitions: Theory and Applications in Physics, Chemistry, and Biology.* (Springer, Berlin, 1983)
73. P. Reimann, Brownian motors: noisy transport far from equilibrium. Phys. Rep. **361**, 57–265 (2002)
74. E.C. Zeeman (eds.) *Catastrophe Theory* (Addison-Wesley, London, 1977)
75. H.E. Stanley, *Introduction to Phase Transitions and Critical Phenomena.* (Oxford University, 1987)
76. D. Helbing, S. Lämmer, Managing complexity: An introduction. Pages 1–16 In: D. Helbing (eds.) *Managing Complexity: Insights, Concepts, Applications.* (Springer, Berlin, 2008)
77. D. Helbing, Systemic risks in society and economics. SFI Working Paper 09-12-044, see http://www.santafe.edu/media/workingpapers/09-12-044.pdf
78. J. Lorenz, S. Battiston, F. Schweitzer Systemic risk in a unifying framework for cascading processes on networks. Eur. Phys. J. B **71**(4), 441–460 (2009)

79. P. Bak, *How Nature Works: The Science of Self-Organized Criticality*. (Springer, Berlin, 1999)
80. Knowledge@Wharton, Why economists failed to predict the financial crisis (May 13, 2009), see http://knowledge.wharton.upenn.edu/article.cfm?articleid=2234 or http://www.ftpress.com/articles/article.aspx?p=1350507
81. W. Weidlich, Physics and social science—the approach of synergetics. Phys. Rep. **204**(1), 1–163 (1991)
82. T. Puu, *Nonlinear Economic Dynamics* (Lavoisier, 1991); *Attractors, Bifurcations, & Chaos. Nonlinear Phenomena in Economics* (Springer, Berlin, 2003)
83. H.W. Lorenz, *Nonlinear Dynamical Equations and Chaotic Economy*. (Springer, Berlin, 1993)
84. P. Krugman, *The Self-Organizing Economy*. (Blackwell, 1996)
85. F. Schweitzer (ed.) *Self-Organization of Complex Structures: From Individual to Collective Dynamics* (CRC Press, 1997); *Modeling Complexity in Economic and Social Systems* (World Scientific, 2002)
86. R.H. Day, *Complex Economic Dynamics*. (MIT Press, Vol. 1: 1998; Vol. 2: 1999)
87. W. Weidlich, *Sociodynamics: A Systematic Approach to Mathematical Modelling in the Social Sciences*. (CRC Press, Boca Raton, 2000)
88. J.D. Sterman, *Business Dynamics: Systems Thinking and Modeling for a Complex World*. (McGraw Hill, 2000)
89. W.A. Brock, *Growth Theory, Non-Linear Dynamics and Economic Modelling*. (Edward Elgar, 2001)
90. S.Y. Auyang, *Foundations of Complex-System Theories*. (Cambridge University, 1998)
91. M. Salzano, D. Colander (eds.) *Complexity Hints for Economic Policy*. (Springer, Berlin, 2007)
92. D. Delli Gatti, E. Gaffeo, M. Gallegati, G. Giulioni, A. Palestrini, *Emergent Economics*. (Springer, Berlin, 2008)
93. M. Faggini, T. Lux (eds.) *Coping with the Complexity of Economics*. (Springer, Berlin, 2009)
94. J.B. Rosser, Jr., K.L. Cramer, J. Madison (eds.) *Handbook of Research on Complexity*. (Edward Elgar Publishers, 2009)
95. M. Olson, The rise and decline of nations: economic growth, stagflation, and social rigidities. (Yale University Press, New Haven, 1982)
96. R. Axelrod, *The Evolution of Cooperation*. (Basic Books, New York, 1984)
97. J.C. Harsanyi, R. Selten, *A General Theory of Equilibrium Selection*. (MIT Press, Cambridge, MA, 1988)
98. D. Helbing, S. Lozano, Phase transitions to cooperation in the prisoner's dilemma. Phys. Rev. E **81**(5), 057102 (2010)
99. M.A. Nowak, R.M. May, Evolutionary games and spatial chaos. Nature **359**, 826–829 (1992)
100. F.C. Santos, M.D. Santos, J.M. Pacheco, Social diversity promotes the emergence of cooperation in public goods games. Nature **454**, 213–216 (2008)
101. D. Helbing, W. Yu, The outbreak of cooperation among success-driven individuals under noisy conditions. Proc. Natl. Acad Sci. USA (PNAS) **106**(8), 3680–3685 (2009)
102. E. Ostrom, *Governing the Commons. The Evolution of Institutions for Collective Action*. (Cambridge University, New York, 1990)
103. C.P. Roca, S. Lozano, A. Arenas, A. Sanchez, Topological traps control flow on real networks: The case of coordination failures, PLoS One **5**(12), e15210 (2010)
104. D. Helbing, Modelling supply networks and business cycles as unstable transport phenomena. New J. Phys. **5**, 90.1–90.28 (2003)
105. M. Eigen, The selforganization of matter and the evolution of biological macromolecules. Naturwissenschaften **58**, 465–523 (1971)
106. J. Schumpeter, *The Theory of Economic Development* (Harvard University Press, Cambridge, 1934)
107. Ken Arrow, In: D. Colander, R.P.F. Holt, J. Barkley Rosser (eds.) *The Changing Face of Economics. Conversations with Cutting Edge Economists* (The University of Michigan Press, Ann Arbor, 2004), p. 301.

108. M. Gallegati, A. Kirman, *Beyond the Representative Agent.* (Edward Elgar, Cheltenham, 1999)
109. A. Kirman, *Economics with Heterogeneous Interacting Agents.* (Springer, Berlin, 2001).
110. M. Aoki, *Modeling Aggregate Behavior and Fluctuations in Economics.* (Cambridge University, Cambridge, 2002)
111. D. Helbing, Collection of papers on An Analytical Theory of Traffic Flows in Eur. Phys. J. B, see http://www.soms.ethz.ch/research/traffictheory
112. W.B. Arthur, S.N. Durlauf, D. Lane (eds.) *The Economy as An Evolving Complex System II.* (Santa Fe Institute Series, Westview Press, 1997)
113. L.E. Blume, S.N. Durlauf, (eds.) *The Economy as an Evolving Complex System III.* (Oxford University, Oxford, 2006)
114. U. Witt, Evolutionary economics. pp. 67–73 In: S.N. Durlauf, L.E. Blume (eds.) *The New Palgrave Dictionary of Economics.* (Palgrave Macmillan, 2008)
115. R.M. May, S.A. Levin, G. Sugihara, Ecology for bankers. Nature **451**, 894–895 (2008)
116. E.D. Beinhocker, *Origin of Wealth: Evolution, Complexity, and the Radical Remaking of Economics.* (Harvard Business Press, 2007)
117. R.A. Young, R.L. Giese (eds.), *Introduction to Forest Ecosystem Science and Management.* (Wiley, New York, 2002)
118. H. Meinhardt, *Models of Biological Pattern Formation.* (Academic Press, London, 1982)
119. M. Caramia, P. Dell'Olmo, *Multi-objective Management in Freight Logistics: Increasing Capacity, Service Level and Safety with Optimization Algorithms.* (Springer, New York, 2008)
120. Jean-Philippe Rennard (eds.) *Handbook of Research on Nature-Inspired Computing for Economics and Management.* (IGI Global, 2006)
121. S. Lämmer, D. Helbing, Self-control of traffic lights and vehicle flows in urban road networks. J. Stat. Mech. (JSTAT), P04019 (2008)
122. D. Helbing et al., Biologistics and the struggle for efficiency: Concepts and perspectives. Adv. Comp. Syst. **12**(6), 533–548 (2009)
123. D. Helbing, M. Moussaid, Analytical calculation of critical perturbation amplitudes and critical densities by non-linear stability analysis of a simple traffic flow model. Eur. Phys. J. B **69**(4), 571–581 (2009)
124. K. Windt, T. Philipp, F. Böse, Complexity cube for the characterization of complex production systems. Int. J. Comput. Integrated Manuf. **21**(2), 195–200 (2007)
125. D. Helbing, S. Lämmer, Verfahren zur Koordination konkurrierender Prozesse oder zur Steuerung des Transports von mobilen Einheiten innerhalb eines Netzwerkes (Method for coordination of concurrent processes for control of the transport of mobile units within a network), Patent WO/2006/122528 (2006)
126. I. Kondor, S. Pafka, G. Nagy, Noise sensitivity of portfolio selection under various risk measures. J. Bank. Fin. **31**(5), 1545–1573 (2007)
127. D. Dorner, *The Logic of Failure: Recognizing and Avoiding Error in Complex Situations.* (Basic, New York, 1997)
128. K. Kempf, Complexity and the enterprise: The illusion of control. pp. 57–87 In: D. Helbing (eds.) *Managing Complexity: Insights, Concepts, Applications* (Springer, Berlin, 2008)
129. D. Sornette, Nurturing breakthroughs: lessons from complexity theory. J. Econ. Interaction and Coordination **3**, 165–181 (2008)
130. C.A.E. Goodhart, Monetary Relationships: A View from Threadneedle Street. (Papers in Monetary Economics, Reserve Bank of Australia, 1975); For applications of Le Chatelier's principle to economics see also P. A. Samuelson, *Foundations of Economic Analysis.* (Harvard University, 1947)
131. A.P. Fiske, *Structures of Social Life: The Four Elementary Forms of Human Relations.* (The Free Press, 1993)
132. E.G. Clary et al., Understanding and assessing the motivations of volunteers: a functional approach. J. Pers. Soc. Psychol. **74**(6), 1516–1530 (1998)
133. B.S. Frey, *Happiness: A Revolution in Economics.* (MIT Press, Cambridge, MA and London, UK, 2008)

134. J. Surowiecki, *The Wisdom of Crowds: Why the Many Are Smarter than the Few and How Collective Wisdom Shapes Business, Economies, Societies, and Nations*. (Doubleday, 2004)
135. C. Blum, D. Merkle (eds.), *Swarm Intelligence. Introduction and Applications*. (Springer, Berlin, 2008)
136. B.S. Frey, *Economics as a Science of Human Behaviour: Towards a New Social Science Paradigm*. (Springer, Berlin, 1999)
137. L.M.A. Bettencourt, J. Lobo, D. Helbing, C. Khnert, G.B. West, Growth, innovation, scaling and the pace of life in cities. Proc. Natl. Acad. Sci. USA (PNAS) **104**, 7301–7306 (2007)
138. R.R. McDaniel, Jr. D.J. Driebe (eds.), *Uncertainty and Surprise in Complex Systems*. (Springer, Berlin, 2005)
139. A.-L. Barabasi, Scale-free networks: A decade and beyond. Science **325**, 412–413 (2009)
140. F. Schweitzer et al., Economic networks: The new challenges. Science **325**, 422–425 (2009)
141. D. Lazer et al., Computational social science. Science **323**, 721–723 (2009)
142. E. Bruckner, W. Ebeling, M.A.J. Montano, A. Scharnhorst, Hyperselection and innovation described by a stochastic model of technological evolution; pP. 79–90 In: L.L. Leydesdorff, P. van den Besselaar (eds.) *Evolutionary Economics and Chaos Theory: New Directions in Technology Studies*. (Pinter, London, 1994)
143. R.N. Mantegna, H.E. Stanley, *Introduction to Econophysics: Correlations and Complexity in Finance*. (Cambridge University Press, Cambridge, 2000).
144. D. Challet, M. Marsili, Y.-C. Zhang, *Minority Games: Interacting Agents in Financial Markets*. (Oxford University, Oxford, New York, 2005)
145. J.-P. Bouchaud, Economics needs a scientific revolution. Nature **455**, 1181 (2008)
146. J.D. Farmer, D. Foley, The economy needs agent-based modelling. Nature **460**, 685–686 (2009)
147. M. Buchanan, Meltdown modelling. Nature **460**, 680–682 (2009)
148. E.W. Montroll, W.W. Badger, *Introduction to Quantitative Aspects of Social Phenomena*. (Gordon and Breach, 1974)
149. Y.-C. Zhang, Econophysics Forum, see http://www.unifr.ch/econophysics
150. B.M. Roehner, Fifteen years of econophysics: worries, hopes and prospects (2010), see http://arxiv.org/abs/1004.3229
151. M. Gallegati, S. Keen, T. Lux, P. Omerod, Worrying trends in econophysics. Physica A **370**, 1–6 (2006)
152. B.K. Chakrabarti, A. Chakraborti, A. Chatterjee (eds.) *Econophysics and Sociophysics: Trends and Perspectives*. (Wiley VCH, Weinheim, 2006)
153. Aims and Scope of the Physics of Socio-Economic Systems Division of the German Physical Society, see http://www.dpg-physik.de/dpg/gliederung/fv/soe/aims.html. For past events see http://www.dpg-physik.de/dpg/gliederung/fv/soe/veranstaltungen/vergangene.html
154. D. Helbing, Pluralistic modeling of complex systems. Preprint http://arxiv.org/abs/1007.2818 (2010)
155. L. Tesfatsion, K.L. Judd (eds.), *Handbook of Computational Economics, Vol. 2: Agent-Based Computational Economics*. (North-Holland, Amsterdam, 2006)
156. J.R. Harrison, Z. Lin, G.R. Carroll, K.M. Carley, Simulation modeling in organizational and management research. Acad. Manag. Rev. **32**(4), 1229–1245 (2007)
157. J.P. Davis, K.M. Eisenhardt, C.B. Bingham, Developing theory through simulation methods. Acad. Manag. Rev. **32**(2), 480–499 (2007)
158. T. Schelling, *Micromotives and Macrobehavior*. (Norton, New York, 1978)
159. R. Hegselmann, A. Flache, Understanding complex social dynamics: A plea for cellular automata based modelling. J. Artif. Soc. Soc. Simul. **1**, no. 3, see http://www.soc.surrey.ac.uk/JASSS/1/3/1.html
160. N. Gilbert, S. Bankes, Platforms and methods for agent-based modeling. PNAS **99**(3), 7197–7198 (2002)
161. M.W. Macy, R. Willer, From factors to actors: Computational sociology and agent-based modeling. Annu. Rev. Sociol. **28**, 143–166 (2002)

162. R.K. Sawyer, Artificial societies. Multiagent systems and the micro-macro link in sociological theory. Socio. Meth. Res. **31**(3), 325–363 (2003)
163. J.M. Epstein, *Generative Social Science: Studies in Agent-Based Computational Modeling*. (Princeton University, 2007)
164. J.H. Kagel, A.E. Roth, *The Handbook of Experimental Economics*. (Princeton University, 1997)
165. D. Helbing, *Managing Complexity: Concepts, Insights, Applications*. (Springer, Berlin, 2008)
166. L. Hurwicz, S. Reiter, *Designing Economic Mechanisms*. (Cambridge University, Cambridge, 2006)
167. W.G. Sullivan, E.M. Wicks, C.P. Koelling, *Engineering Economy*. (Prentice Hall, Upper Saddle River, NJ, 2008)
168. D. Helbing, *Quantitative Sociodynamics*. (Kluwer Academic, Dordrecht, 1995)
169. C. Castellano, S. Fortunato, V. Loreto, Statistical physics of social dynamics. Rev. Mod. Phys. **81**, 591–646 (2009)
170. D. Helbing, W. Yu (2010) The future of social experimenting. Proc. Natl. Acad. Sci. USA (PNAS) **107**(12), 5265–5266 (2010); see http://www.soms.ethz.ch/research/socialexperimentingforalongerversion.

Index

D. Helbing (ed.), *Social Self-Organization*, Understanding Complex Systems,
DOI 10.1007/978-3-642-24004-1, © Springer-Verlag Berlin Heidelberg 2012